J. Zupan and J. Gasteiger

Neural Networks in Chemistry and Drug Design

Jure Zupan and Johann Gasteiger

Neural Networks in Chemistry and Drug Design

Second Edition

Weinheim · New York · Chichester · Brisbane · Singapore · Toronto

Professor Jure Zupan
National Institute of Chemistry
P.O. Box 34-30
SL-1001 Ljubljana, Slovenia

Professor Johann Gasteiger
Computer-Chemie-Centrum
Inst. für Organische Chemie
Universität Erlangen-Nürnberg
Nägelsbachstr. 25
D-91052 Erlangen, Germany

The cover shows a three-dimensional molecular model in front of a Kohonen neural network. The weight distribution in the various layers of the network is indicated by color coding.

Library of Congress Card No.: applied for

British Library Cataloguing-in-Publication Data: applied for

Deutsche Bibliothek CIP-Einheitsaufnahme

Zupan, Jure : Neural networks in chemistry for drug and design / Jure Zupan ; Johann Gasteiger.
- 2., unveränd. Aufl., - Weinheim ; New York ; Chichester ; Brisbane ; Singapore ; Toronto : Wiley-VCH, 1999
  1. Aufl. u.d.T.: **Zupan, Jure: Neural networks for chemists**
  ISBN 978-3-527-29779-5 (Softcover)
  ISBN 3-527-29778-2 (Hardcover)

Prepress: Satzrechenzentrum Kühn & Weyh, D-79110 Freiburg
Cover Design: Dipl. Chem. Markus C Hemmer, Computer-Chemie-Centrum, Universität Erlangen-Nürnberg

learning: learning by oneself, unsupervised learning, and learning by a tutor, supervised learning.

These two concepts are probably the most important aspects of learning. Throughout the book, especially in the Part IV "Applications", great emphasis is put on the notion that **both** types of learning, supervised and unsupervised, are needed to accomplish most of the tasks met in applications of neural networks.

The book is composed as a textbook and we therefore first introduce the concepts of neural networks. The ultimate goal of the book is to bring you to a level of understanding that you will be able to apply neural networks to your problems either with a commercial neural network package or with a self-made program.

The first part of the book deals with the concept of neuron, transfer functions, "bias", etc. A comparison is made with the classical linear learning machine. Then, the linking of neurons to layers and the linking of layers of neurons among themselves is described. In the second part, one-layer neural networks are discussed. First, the Hopfield network and the ABAM (Adaptive Bidirectional Associative Memory) are explained and then the Kohonen network is explained in more detail. Kohonen learning is the most important unsupervised, or self-organizing learning scheme offered by neural networks. The two-dimensional map formed by the labels of objects that enter the training procedure as well as the maps which are formed within the individual Kohonen network of layers of weights are the most informative results of the Kohonen learning.

In Part III, multilayer neural networks and learning in those networks is discussed. The counter-propagation of targets and back-propagation of errors learning schemes are introduced. The counter-propagation network consists of two layers: the upper layer performing the Kohonen learning and the output layer performing the adaptation of weights to the targets which are input to the network from the counter-part side of the network, i.e., from the output side. The back-propagation of errors is the most widely used method of neural network learning. Today, at least in chemistry, more than 90 percent of all applications are made by back-propagation of error learning. Therefore, this method is discussed in more detail.

For each learning strategy or neural network architecture we worked out a simple example which in most cases is not from the field of chemistry. In this way, by selecting examples not linked to any specific type or field of application we have been pursuing two goals:

For Breda, Ulrike, Nina,
Julia, Eva, and Michael.

Preface

Almost all fields of human endeavor, science in particular, have seen dramatic changes in recent years. In all domains of our life we are swamped with data, be it on the stock market, medical diagnosis, car crash tests or, to come to the point, in chemistry.

Throughout history, until some decades ago, the main problem scientists were confronted with in their scientific endeavor was how to obtain data. Measurements were time-consuming, not sensitive enough, expensive, required permanent presence of the experimenter, manual recording, etc. In addition, there was the problem of preparation of the adequate material, lack of proper techniques, special purpose equipment and technical support. Scientists had to handle a lot of unpleasant routine work just to obtain a few numbers. Robert A. Millikan meticulously carried out tens of thousands of experiments to determine the charge of the electron. Marie Curie put in tremendous efforts to isolate a minute quantity of radium.

Today, "thanks" to the high amount of instrumentalization of science, the main problem is not how to gather data, but how to get rid of most of them. Unfortunately, only a small amount of data produced by the computerized instruments and stored in computer memory are really relevant to the problem. The sad truth is that the valuable information scientists are seeking can often be extracted much harder from the myriads of computer bits than from a small number of data measured in carefully planned and manually executed experiments.

Because everybody knows that the good old times are not going to return and the deluge of data will not disappear in our scientific work, it is better to be prepared to handle large quantities of data. In industrial laboratories thousands of products are analyzed each day, in production plants thousands of recipes are handled, monitored and slightly changed to meet the control requirements. Organic chemists are performing vast series of chemical reactions. Pharmaceutical research groups are inspecting thousands of potential drugs; biologists are analyzing long strings of amino acids in numerous proteins; spectroscopists are comparing huge databases of spectra or working on multidimensional images produced by sophisticated spectrometers, etc.

It is not hard to find examples where large amounts of data must be handled to extract the vital information. The only way to adapt ourselves to this flood of data is to acquire new knowledge, to rethink the methods we are using, and to adapt our skill for data handling to the new situation.

With this book we want to introduce you into a rediscovered and reshaped old method called "neural networks". No doubt, the name is provocative. But so was the term "artificial intelligence" which decades ago raised the blood pressure of many otherwise very reasonable scientists. Things have settled down and today the term artificial intelligence is, if not fully admired and supported, at least accepted. The situation with artificial neural networks is not quite so settled and peaceful yet.

The first word of the term "neural networks", has a clear link with a "neuron", a nerve cell. This points further towards that part of the human body considered by many to be the one most distinguished in mankind: the brain. In this book we would like to make clear that the emphasis in the term "neural network" is not on the word "neural" but rather on the word "network".

Networking is one of the most important things in any organizational scheme: running a railroad company, a bank, a production plant, or handling scientific data. Networking the flow of data is a familiar concept to the electrical engineer, but is not so familiar to the chemist. In the context of this book, the word network always means an assembly of "little devices" called "neurons" which perform the same set of simple operations all the time. Hence the final output is not primarily the result of these simple operations but rather of the way how these "little devices" are linked together and how they change their internal parameters to adapt their individual outcomes to some external control or competition among other "neurons".

One source of the flexibility for handling data by neural networks stems from their "architecture", i.e., from the number of neurons it is composed of, how these neurons are interconnected, etc. The second reason why neural networks are so adaptable to different problems and applications is the possibility to implement both basic ways of learning: learning by oneself, unsupervised learning, and learning by a tutor, supervised learning.

These two concepts are probably the most important aspects of learning. Throughout the book, especially in the Part IV "Applications", great emphasis is put on the notion that **both** types of learning, supervised and unsupervised, are needed to accomplish most of the tasks met in applications of neural networks.

The book is composed as a textbook and we therefore first introduce the concepts of neural networks. The ultimate goal of the book is to bring you to a level of understanding that you will be able to apply neural networks to your problems either with a commercial neural network package or with a self-made program.

The first part of the book deals with the concept of neuron, transfer functions, "bias", etc. A comparison is made with the classical linear learning machine. Then, the linking of neurons to layers and the linking of layers of neurons among themselves is described. In the second part, one-layer neural networks are discussed. First, the Hopfield network and the ABAM (Adaptive Bidirectional Associative Memory) are explained and then the Kohonen network is explained in more detail. Kohonen learning is the most important unsupervised, or self-organizing learning scheme offered by neural networks. The two-dimensional map formed by the labels of objects that enter the training procedure as well as the maps which are formed within the individual Kohonen network of layers of weights are the most informative results of the Kohonen learning.

In Part III, multilayer neural networks and learning in those networks is discussed. The counter-propagation of targets and back-propagation of errors learning schemes are introduced. The counter-propagation network consists of two layers: the upper layer performing the Kohonen learning and the output layer performing the adaptation of weights to the targets which are input to the network from the counter-part side of the network, i.e., from the output side. The back-propagation of errors is the most widely used method of neural network learning. Today, at least in chemistry, more than 90 percent of all applications are made by back-propagation of error learning. Therefore, this method is discussed in more detail.

For each learning strategy or neural network architecture we worked out a simple example which in most cases is not from the field of chemistry. In this way, by selecting examples not linked to any specific type or field of application we have been pursuing two goals:

first, such a broad choice gives us much more flexibility in selecting appropriate examples, and second, the book might attract a much wider audience if the methods are accompanied by examples that are not tied to only one field of application – chemistry in our case.

Actually, the same strategy for the selection of examples was followed in the "Applications" part. This time the field was of course confined to be chemistry, but we have tried to show you applications from the various domains of chemistry. Today (November 1992) the Chemical Abstract Services lists nearly 500 articles published under the code word "neural networks" and are related to chemistry. There are many interesting applications and the choice of examples must, of course, be an arbitrary one. Other authors might have made a completely different choice. However, there are a few general subfields in chemistry where neural networks are applied more than in others.

The first among such fields is process control and fault detection in processes. The complex relationships between the controlled and the manipulated variables of the process, together with the many data available from the constant monitoring of a process offer an ideal "playground" for testing and applying neural networks. The concepts of "moving window" and of the binary representation of discrete state variables are introduced. These many aspects have made the chapter on process control and fault detection rather large.

The other field where the back-propagation of error learning has seen a fast and early momentum is the prediction of the secondary structure of proteins. The first article by Qian and Sejnowski was already published in 1988. There is no doubt that the introduction of the neural network approach to the determination of protein structure is an achievement by itself. First, because it "transplanted" a method from one scientific discipline (speech reproduction) to a completely different field, showing the power of interdisciplinary knowledge and reasoning. Secondly, because the newly introduced method improves the previously obtained predictions, and thirdly, because it opens new perspectives in posing new questions in the field. For example: What is the most appropriate representation of amino acid sequences? How can the long-range influences be taken into account? What variety of problems can be solved by the method? etc.

The third area in chemistry where the neural network approach seems to be very promising, is structure elucidation by different spectroscopic methods. The problem of establishing reliable correlations between different types of spectra (infrared, mass, ^{1}H NMR, ^{13}C NMR, etc.) and the chemical structure of the corresponding compound has been around so long, that there is no wonder that the neural network approach immediately attracted the attention of spectroscopists and those working on computerized structure elucidation processes. It is interesting to note that spectroscopy was one of the first areas where not only the back-propagation of errors learning, but other learning methods, like Kohonen learning, and neural network architectures like ABAM and Hopfield networks were used.

Other examples in the book come from different areas such as classification of objects into several categories, optimization of recipes and procedures, quantitative structure-activity relationship (QSAR) studies, reactivity of bonds, aromatic electrophilic reactions, and mapping of electrostatic potential into a two-dimensional plane.

We hope that the variety of examples and the flexibilities of the neural network learning methods and their architectures will encourage the user to think about:

- how to select the best neural network learning method or even to simultaneously try several of them,

- what neural network architecture and which parameters to choose to obtain as much information from the data available as possible, and above all:

- how to interpret the results obtained to pinpoint the required information best.

The above points are much more important than to think about how to find a dataset that is suitable for a neural network package from the shelf in order to publish a paper as quickly as possible.

At the end we would like to extend our special thanks to four of our coworkers: Ms. Vera Simon, Ms. Marjana Novic, Mr. Xinzhi Li and Mr. Marjan Tusar who helped us by working out many examples and calculations presented in this book. Of course, the thanks go as well to all other members and coworkers of the Laboratory for Chemometrics at the National Institute of Chemistry in Ljubljana and the Laboratory for Computer Chemistry at the Organisch-Chemisches Institut der Technischen Universität München who helped us by supporting the friendly and stimulating atmosphere in the laboratories while exchanging their ideas and sharing their enthusiasm with us.

XII   *Preface*

Special thanks go to Miss Natalia Berryman who restlessly transferred our sometimes non-executable ideas into her fine drawings. Altogether she managed to complete more than 220 final pictures. This actually means at least twice that number of sketches and provisional ideas.

Thanks are also due to Miss Elisabeth Lohof who read the entire manuscript and made a number of valuable suggestions in English style. In addition, she composed the final version of the text. We also want to thank Dr. Hans Lohninger, Technische Universität Wien, for making a number of valuable comments on style, contents and didactic improvements of the manuscript.

Our families have suffered most from our excessive absence, and it is our sincere hope that the sacrifices they made will be at least partially rewarded by us being with them more in the future.

The roots of this book go back to a meeting in the cafeteria of Beijing airport between a physicist from Slovenia and an organic chemist from Bavaria. This was followed by long scientific discussions in the tiny Tyrolian village of Hochfilzen.

The idea for writing this book developed from a course on neural networks J. Z. gave as visiting professor at the Technische Universität München. The Bundesminister für Forschung und Technologie, Germany is gratefully acknowledged for supporting J. Z.'s three-month/year position for three consecutive years at the Laboratory for Computer Chemistry in Garching. Our thanks go to Dr. Jan Michael Czermak of BMFT and Dr. Volker Schubert of GMD-PTF for arranging for this position of a project professor.

J. G. would like to express his thanks to the Ministry for Science and Technology of Slovenia which enabled him to have this close cooperation and a one month leave in Ljubljana to work on the project.

Jure Zupan   Johann Gasteiger

December 1992

Preface to the Second Edition

It attests to the importance of neural networks in general, and their application to chemical problems in particular, that a second edition of this book becomes necessary. The field has grown: see Table 9-1, that gives the number of publications for the application of neural networks in chemistry. And it has matured: after the initial hype where we saw many problems that could just as well have been tackled with more traditional statistical or pattern recognition methods, the specific advantages of neural networks have become clearer. We also see the use of a more diverse set of neural network methods, not only feed-forward networks trained by the back-propagation algorithm but also other methods, particularly, Kohonen networks.

From the feedback that we have obtained we felt that not much needs to be done to Parts I – III (Chapters 1 – 8). We found a few places that needed minor corrections and some clarification.

However, we have quite heavily extended Part IV „Applications" (Chapters 9 – 21) to account for the surge of new applications. In this endeavor we have largely concentrated on work from our two research groups because much is an extension of work that was emerging in the first edition and has now come to full bloom. Heavy emphasis is placed on examples from drug design because of the strong interest in this area.

A book is somehow outdated when it appears in print. In order to do something against this fate, we have decided to establish a web-site (*http://www2.ccc.uni-erlangen.de/ANN-book/*) that allows us to keep the reader updated on recent developments in the area of neural networks in chemistry. This gives us also the possibility to provide additional material such as giving access to programs and data sets.

We were fortunate that quite a few people have shared our enthusiasm for neural networks. In particular, our coworkers and collaborators have ventured with us into this exciting world of applying neural networks to chemical problems.

XIV   *Preface*

In the past six years since the first edition of this book has appeared both our groups have held many collaborations with various Laboratories in Slovenia, Germany as well as in the rest of Europe.

J. Z. would like to thank his closest coworkers Dr. Marjana Novic, Dr. Marjan Vracko-Grobelsek, and Dr. Marko Perdih for their contributions to various aspects of their common research. It is a pleasure to thank Dr. Darinka Brodnjak-Voncina from the University of Maribor, Dr. Itziar Ruisanchez and Prof. Xavier. F. Rius from University Rovira i Virgili, Tarragona, Prof. Giuseppina G. Gini from Politecnico di Milano, and Dr. Emilio Benfenati from Instituto Mario Negri, Milano, for sharing the common interests in artificial neural networks. For a scientist, the most rewarding moment comes when the research laboratories in the industry are implementing the new methods in their daily work. In this respect the thanks go to Dr. Nineta Majcen and M.Sc. Karmen Rajer-Kanduc from Cinkarna, doo. Celje, to Dr. Livija Tusar and M.Sc. Nevenka Leskovsek from Color, doo. Medvode, and M.Sc. Ales Brglez, Gorenje doo, Velenje.

J. G. wants to thank his coworkers Dr. Bruno Bienfait, Dr. Lingran Chen, Sandra Handschuh, Markus Hemmer, Dr. Xinzhi Li, Oliver Sacher, Dr. Jens Sadowski, Christof Schwab, Dr. Jan Schuur, Dr. Paul Selzer, Dr. Valentin Steinhauer, Andreas Teckentrup, and Dr. Markus Wagener for their work in showing the broad scope of applications that neural networks can find in chemistry. In addition, he appreciated the collaboration with Dr. Soheila Anzali, Darmstadt, Prof. Ulrike Holzgrabe, Bonn and Dr. Jarek Polanski, Katowice, Poland that led to important new contributions in methodology and applications.

Again, the book could directly be produced from the Postscript file that we submitted thanks to the careful editing efforts of Angela Döbler and Oliver Sacher.

Jure Zupan   Johann Gasteiger

January 1999

Contents

Part II   One-Layer Networks

Part III Multilayer Networks

Part IV Applications

22   Prospects of Neural Networks for Chemical Applications   359

Appendices   363

Part I
Basic Concepts

1 Defining the Area

learning objectives

- what a neural network is
- approaches to problem solving
- nomenclature

1.1 Learning from Information

The availability of information is increasingly important in our society. In fact, it has been suggested that we are becoming an information society. Information will become one of the most valuable assets in many of our activities, from business administration to science. However, it can already be observed that we are in danger of being swamped in a profusion of individual data, finding it more and more difficult to obtain the right information for a specific problem.

It is therefore of vital interest to analyze data, to extract knowledge from individual data, and to generalize from single observations to the underlying principles and the structure of the information. We must **learn** from individual observations.

Data analysis is nothing new; it has been performed for many years, mostly by statistical and pattern recognition methods. However, it has been clear for a long time that the human brain analyzes data and information quite differently from such methods, that it processes the flood of data and learns from them along quite different lines. Knowledge acquisition by the human brain is not performed by statistical methods!

Recognition of the inherent limitations of statistical and pattern recognition methods has led to the development of expert systems. In an expert system the knowledge specific to a certain domain of problems is kept apart from the inference mechanisms that draw conclusions from the knowledge (and thus make decisions). However, the mechanisms for acquiring knowledge for expert systems are still far from perfect. The method of a knowledge engineer interviewing experts to build a knowledge base has many drawbacks, including scientific, technical and psychological aspects. In any case, it is not the way the brain acquires knowledge.

Advances in neurophysiology and new experimental techniques such as electroencephalography (EEG), computer-assisted tomography (CAT), magnetic resonance imaging (MRI), positron emission tomography (PET), the superconducting quantum interference device (SQID), and single-photon emission computerized tomography (SPECT) have greatly enhanced our understanding of the anatomy of the human brain and the physical and chemical processes occurring within it. Furthermore, mathematical models and algorithms have been designed to mimic the information processing and knowledge acquisition methods of the human brain. These models are called *neural networks*.

The purpose of this textbook is twofold:

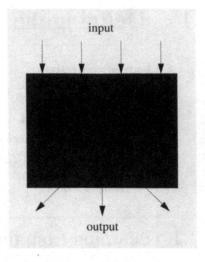

Figure 1-1: The black box.

– First, to develop the **basic principles** and the scope and limitations of the more important neural network models.

– Second, and even more importantly, to understand how **to use** these neural networks for processing information, and thus learn relationships by means of which we can acquire knowledge.

1.2   General Objectives and Concepts

We will first consider a neural network as a black box that can accept a series of input data and produce from these one or more outputs (Figure 1-1).

The input values can be from the stock market and the output, a recommendation to buy or sell particular stocks. Or we can input medical data for a patient and obtain predictions of the kind of disease he/she has; or from a spectrum of a compound we can predict its structure; or input the movements of an object and output the reaction of the robot's arm in an automated laboratory.

For many users of neural networks it will not be necessary to know exactly what happens inside this black box; nevertheless they will be able to apply neural networks to their problems successfully. The purpose of this book is, however, to develop a gradual understanding of the operations inside this black box.

In the following chapter we will learn that there are basic operating units inside the box that are somehow connected (Figure 1-2). The inputs are passed along these connections, the lines of a *network*, and are distributed, transformed, and eventually reunited to produce outputs. The transformation of the data is performed in many basic processing *units*, called *artificial neurons* or simply *neurons*, which perform identical tasks.

Thus, as the name implies, neural networks consist of neurons connected into networks. We will first present the basic concept of a neuron and then look at the ways of connecting them.

1.3   What Neural Networks are Good for

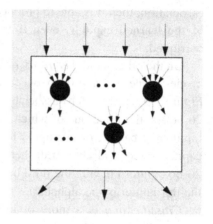

Figure 1-2: Basic units within the box.

The neural network, if considered as a black box, will transform an *m*-variable input into an *n*-variable output. The input or output variables can be:

− real numbers, preferably in the range from 0 to 1, or −1 to +1; if they are outside this range, the input data have to be preprocessed to bring them into it (some methods, however, can handle larger or smaller real values as well);

− binary numbers, i.e. 0 and 1; or

− bipolar numbers, i.e. −1 and +1.

The number of input and output variables is limited only by the available hardware and computation time. The number of output variables is usually smaller than that on the input side, but this is by no means mandatory.

The problems handled by neural networks can be quite varied. On the most general level they can be divided into four basic types:

− association (auto or hetero),

− classification,

− transformation (different representation),

− modeling.

Auto-association means that the system is able to reconstruct the correct pattern if the pattern it learned is incomplete or corrupted. Figure 1-3 illustrates this, with the input consisting of the stack of 26 letters in the top of the figure. If the system is able to do an auto-

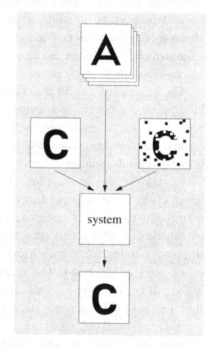

Figure 1-3: Auto-association. The original is reconstructed from an incomplete or corrupted input.

association, then it is able to produce on output a perfect match of any of the learned capitals, even if the input letter was incomplete or corrupted.

On the other hand, *hetero-association* means that the system makes a one-to-one association between members of two sets of patterns. Figure 1-4 shows hetero-association of 26 capital letters with a set of 26 boxes, the filled one of which identifies the output letter. Thus, on input of a perfect or corrupted letter C, the system trained with 26 capital letters and able to do hetero-association will respond with a 26-box pattern having the third box occupied (letter C was defined as the third letter in the alphabet).

Classification is a more or less familiar concept. Its goal is to assign all given objects to appropriate classes (*clusters*) of objects, based on one or more properties that characterize a given class (Figure 1-5). Neural networks are mainly employed in one- or two-level clustering. The advantage of the neural network approach is that only a small proportion of multivariate objects is used for training, and afterwards the network is able to predict the class (cluster) to which an unknown object belongs. Some of these applications are very close to hetero-association applications.

The process of classification can be carried out in a *supervised* or in an *unsupervised manner*. During supervised learning the system is forced to assign each object to a specified class, while during unsupervised learning the clusters are formed naturally without any a priori given information.

Transformation or *mapping* of a multivariate space into another space of the same or lower dimensionality is a frequent application of neural networks. Many researchers consider that the essential process of thinking, learning and reasoning is the mapping of multivariate impulses coming from our sensory organs into a 2-dimensional plane of neurons in the brain. The mapping can be made from lower to higher dimensionality as well. However, this is less frequent; two-dimensional mapping is adequate for an easy, clear representation of multidimensional objects in many applications (Figure 1-6).

Modeling, one of the most frequently used mathematical applications in science, is the search for an analytical function or a procedure (model) that will give a specified *n*-variable output for any *m*-variable input. It is useful in many areas, from process control to expert system design. Standard modeling techniques require the mathematical function to be known in advance. During a "fitting"

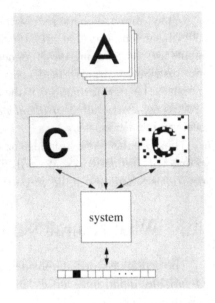

Figure 1-4: Hetero-association. The associated pattern is reconstructed from an ideal or corrupted input.

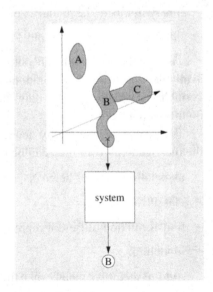

Figure 1-5: Classification of multivariate data.

procedure (Figure 1-7), the parameters of this function are determined on the basis of the best agreement between the experimental (input) and calculated (output) data. The predictions are best if the experimental data covers the variable space evenly and with adequate density over the entire region. The advantage of the neural network model is that it does not require a knowledge of the mathematical function: the nonlinearity of a single unit transformation and a sufficiently large number of variable parameters (weights) ensure enough "freedom" to adapt the neural network to any relation between input and output data.

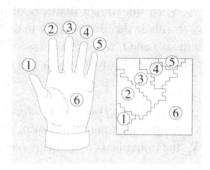

Figure 1-6: Mapping a three-dimensional surface (hand) into a rectangular plane of neurons.

1.4 Notation, Conventions and Abbreviations

The literature on neural networks contains a profusion of notations that makes it difficult for the beginner to compare one method with another. Throughout this textbook we will use a consistent nomenclature and notation.

Names of *scalar* (single-valued) values are printed with small italic letters:

$$a$$

The only **exception** is *Net* which starts with a capital letter so as not to confuse it with the terms "network", or "net".

Names of *vectors* and *matrices* are in bold italics, with initial caps:

$$\boldsymbol{A}$$

The individual values of an *input vector* (***Inp*** or ***X***) are given by lower case *x*, indexed with a subscript *i* and of dimension *m*:

$$inp_i, x_i \qquad\qquad (i = 1, 2, ..., m)$$

The individual values of the *output vector* (***Out*** or ***Y***) of a collection of neurons are indexed with a subscript *j* and are of dimension *n*:

$$out_j, y_j \qquad\qquad (j = 1, 2, ..., n)$$

The *weight matrix* of a layer of neurons, ***W***, thus has individual values w_{ji}, the first index referring to the neuron being considered, the second index specifying the input unit (the preceding neuron that transmits the signal):

$$w_{ji}$$

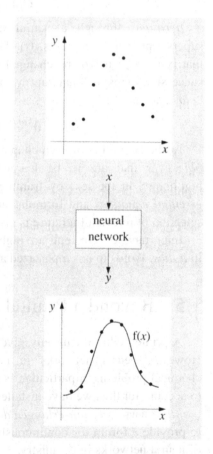

Figure 1-7: Modeling of experimental data with a neural network.

When the weight matrices of different levels are compared with each other, the first weight matrix of level l, $\boldsymbol{W}^l$, has as usual the indices i and j, whereas the one of the next level $\boldsymbol{W}^{l+1}$ has the indices j and k, with k running from 1 to r:

$$w_{kj}$$

If there are several *input objects*, they are identified by a subscript s having a maximum value of p. Thus, the input object is identified by $\boldsymbol{X}_s$, the individual components (signals) by:

$$x_{si}$$

In a multilayer network, the various *layers* are identified by a superscript l. Thus, the output vector of a layer l is $\boldsymbol{Out}^l$, and its individual values are out_j^l.

$$\boldsymbol{Out}^l, \; out_j^l$$

Iterations through a neural network are characterized by a superscript t in parentheses, (t). Thus, the initial value of a weight matrix is $\boldsymbol{W}^{(0)}$; it will be changed in the next iteration to $\boldsymbol{W}^{(1)}$. The successive steps in changing values are indicated by superscripts "*old*" and "*new*":

$$\boldsymbol{W}^{(old)}, \; \boldsymbol{W}^{(new)}$$

At the beginning of each chapter, we present the major topics and *objectives* that are to be learned. Some important remarks are highlighted in the text by framing. At the end of the chapter, the *essential equations* and formulas are collected. After each chapter a selection of relevant literature is given for *further reading*.

Important new concepts are highlighted in the text by writing them in *italics*. Words to be emphasized are printed in **boldface**.

1.5 Beyond a Printed Edition

A printed edition can only give a static view of a scientific field. However, neural networks in general, and their application to chemical problems in particular, is a rapidly expanding area. In order to account for this, we have installed a website at

http://www2.ccc.uni-erlangen.de/ANN-book/

to provide a forum for continuously updating information on the use of neural networks in chemistry. This also gives us the opportunity to provide electronic material such as presentation materials and access to programs and data sets to the readers.

2   Neuron

learning objectives

- neural networks in biology; neurons and synapses

- inputs, weights, outputs, and input-output conversion (transfer) functions

- linear learning machine

- delta-rule

- graphical representation of artificial neurons

2.1   Synapses and Input Signals

Whether artificial or computer simulated, a "neuron" is designed to mimic the function of the neural cells of a living organism. Therefore, we should look at a brief description of the biological neuron. For further details, see textbooks on physiology or neurophysiology (e.g. Color Atlas of Physiology).

The human nerve system consists of about 10^{10} neural cells, also called neurons. Although there are at least five different types of neural cells, it suffices to present only one type. A typical *neuron* of the motor complex consists of a cell body (soma) with a nucleus. The cell body has two types of extensions: the dendrites and the axon. Figure 2-1 shows a drastically simplified picture of such a neuron. The dendrites receive signals and send them to the soma. A neuron has substantially more dendrites than indicated in Figure 2-1, which are also much more branched. Thus, the dendrites have quite a large surface (up to 0.25 mm^2) available to receive signals from other neurons. The axon, which transmits signals to other neurons (or to muscle cells), branches into several "collaterals".

The axons and collaterals end in *synapses*. These synapses make contact with the dendrites or the somata of other neurons. A motor neuron has thousands of synapses; up to 40% of the surface of a neuron is covered with such contact sites.

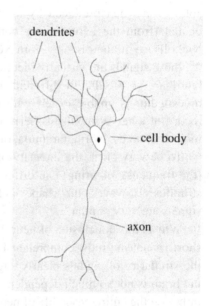

Figure 2-1: Overly simplified scheme of a motor neuron.

The transfer of signals within the dendrites and the axon is electrical, occurring through the transport of ions. However, the signal is transmitted across the synapse by chemical substances. The electric signal in the axon releases a chemical substance, the *neurotransmitter* (for example, acetylcholine), which is stored in vesicles at the presynaptic membrane. This neurotransmitter diffuses across the synaptic gap and through the postsynaptic membrane into the dendrite of the other neuron (Figure 2-2).

In the dendrite, the neurotransmitter generates a new electric signal that is passed through the second neuron. Since the postsynaptic membrane cannot release the neurotransmitter, the synapses can only send the signal in one direction, and therefore function as gates; this is an essential prerequisite for the transmission of information. In addition, other neurons can **modify** the transmission of signals at the synapses.

The signals produced by the neurons, regardless of the species producing them, are very similar and therefore almost indistinguishable, even when produced by a very primitive or a highly sophisticated (from the evolutionary point of view) species. Kuffler and Nicholls say in their book "From Neuron to Brain" (page 4):
"... these signals are virtually identical in all nerve cells of the body ... [and] are so similar in different animals that even a sophisticated investigator is unable to tell with certainty whether a photographic record of a nerve impulse is derived from the nerve fibre of a whale, mouse, monkey, worm, tarantula, or professor."

To be very clear, the intensity of signals produced by the neurons (the frequency of firing) can differ depending on the intensity of the stimulus. However, the shape and overall appearance of different signals are very similar.

Why is this conclusion of neural network research so important? A short (if not the most comprehensive) answer to this question is that the similarity of signals clearly suggests that the real functioning of the brain is not so much dependent on the role of a single neuron, but rather on the entire ensemble of neurons – that is, the way the neurons are interconnected.

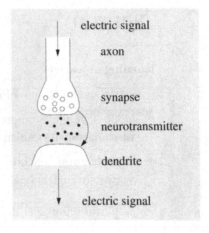

Figure 2-2: Schematic representation of a synapse.

> Therefore, the emphasis in the phrase "neural network" is on "network" rather than on "neural".

The synapses, through which the signals from neighboring neurons enter into one particular neuron, represent barriers which will almost certainly modulate a signal passing through them. The amount of change depends on the so called *synaptic strength*. In artificial neurons, the synaptic strength is called a *weight, w*. This situation is shown schematically in Figure 2-3.

Without going into the physics and chemistry of membranes, we can say that the synaptic strength determines the relative amount of the signal that enters the body of the neuron through the dendrites. Fast changes of the synaptic strengths, even between two consecutive impulses, are regarded as a vital mechanism in the proper and efficient functioning of the brain. The adaptation of synaptic strengths to a particular problem is the essence of learning.

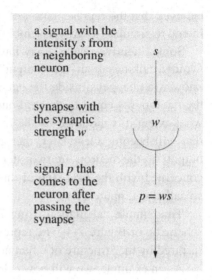

Figure 2-3: The *synaptic strength w* changes the intensity of the incoming signal *s*.

2.2   Weights

Because each neuron has a large number of dendrites/synapses (Figure 2-1), many signals can be received by the neuron simultaneously. The individual signals are labelled s_i and the corresponding synaptic strengths (weights), w_i. Assuming that the weight at each of the neuron's numerous synapses can have a different value at a given moment, we can estimate that the incoming signals can add together into a kind of *collective effect*, or *net input*.

In reality we do not know exactly how the collective effect is formed, nor do we know how large it is with respect to all input signals. Therefore, some very crude simplifications must be made when making a model of a neuron:

– the net input (called *Net*) is a function of all signals s_i that arrive within a given time interval, and of all synaptic strengths (weights, w_i); and

– the function linking these quantities is a simple sum of products of the entering signals s_i and the corresponding weights w_i. Thus, we can write:

$$Net = w_1 s_1 + w_2 s_2 + ... + w_i s_i + ... + w_m s_m \qquad (2.1)$$

This is not the only way to represent the net input of the neuron; some authors have proposed quite elaborate functions. However, it is believed that the representation of *Net* may be relatively simple if we intend to simulate a **large** assembly of neurons.

Some attentive readers may have noticed that until now we have avoided talking about the "output" of the neuron. Since we do not know what happens inside the neuron, we can not say to what extent the net input is equal to the real output of the sending neuron, and to what extent it is modified by the receiving one. However, for reasons that will become clear later, the present model calculates the actual output of the neuron in two steps. At the moment, we are only concerned with the **first step** of this calculation, the evaluation of the so called net input *Net*!

This simple calculation using Equation (2.1) corresponds to the schematic of Figure 2-4. The representation of the artificial neuron is inspired by the structure of a real neuron.

As an example, we will now calculate the net input *Net* of a neuron having only four synapses with weights 0.1, 0.2, –0.3 and –0.02. Because we have designed our artificial neuron to have only four synapses, it can handle signals from only **four** contacting neurons simultaneously (in this example, signals with intensities 0.7, 0.5, –0.1, 1.0 – see Figure 2-5). Because the net input *Net* can not be identified with the real output of the neuron, only the upper half of a circle representing the neuron's body is drawn. The lower half of this circle (dashed line) reminds us that another step is needed to obtain the output from the net input. This will be explained in Section 2.4.

As can be seen from this example, the net input of an **artificial** neuron can have negative or positive values; because there is no reason for limiting either the signs or the magnitudes of the weights, the net inputs can have a very large range of values. This is especially true if neurons with thousands of synaptic connections to surrounding neurons are taken into account.

Rather than a group of signals s_1, s_2, s_3, ..., s_i, ..., s_m received by the given neuron from the surrounding m neurons, it is much more convenient to combine them into a multivariate signal: a multi-dimensional vector X, whose components are the individual signals:

$$\{ s_1, s_2, s_3, ..., s_i, ..., s_m \} = X (x_1, x_2, ..., x_m) \qquad (2.2)$$

Using this notation, the 4-dimensional input vector X for our previous example would be written as:

$$X = (\ 0.7, \ \ 0.5, \ \ -0.1, \ \ 1.0 \ ) \qquad (2.3)$$

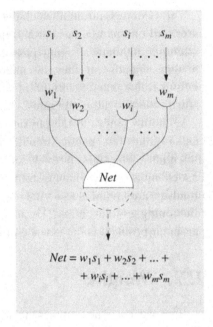

$$Net = w_1 s_1 + w_2 s_2 + ... + \\ + w_i s_i + ... + w_m s_m$$

Figure 2-4: Calculation of the net input *Net* to an artificial neuron having *m* synapses.

When we use the same line of reasoning, all the synaptic strengths in a neuron can be described by using a multidimensional *weight vector* **W**:

$$W = (w_1, w_2, w_3, ..., w_i, ..., w_m) \qquad (2.4)$$

With only one neuron, the 4-dimensional weight vector of Figure 2-5 will look like this:

$$W = (0.1, \; 0.2, \; -0.3, \; -0.02) \qquad (2.5)$$

It is evident that each neuron should have at least as many weights as attached neurons. In a biological neuron, a synapse with a certain synaptic strength is immediately formed when an axon and a dendrite link together; in a computer simulated one, the programmer has to accommodate the corresponding number of weights.

2.3   Linear Learning Machine

The so-called *linear learning machine*, a topic found in standard textbooks on pattern recognition methods, introduces many valuable concepts and techniques that may be used in more complicated ways later.

The linear learning machine, employing a linear transformation (represented by Equation (2.1)) on a multivariate signal *X* using the weight vector *W* to obtain a one-variable (univariate) signal, was very popular in the sixties in many of the sciences including chemistry (see Nilsson: Linear Learning Machines).

The learning machine procedure is mainly used for deciding whether a given multivariate input signal *X* belongs to a certain category (see Jurs and Isenhour, *Analytical Chemistry*, August 1971; cf. Figure 2-10).

Such a multidimensional vector *X* can represent many things, such as:

- an audio signal,

- an optical image,

- a spectrum of any kind,

- a many-component chemical analysis,

- the status of a technological process at a given time,

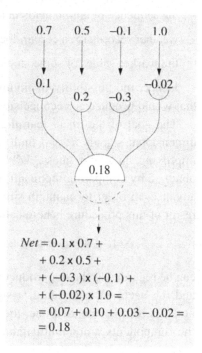

Figure 2-5: Calculation of the net input of a neuron with four synapses.

- the sequence of amino acids in a protein,
- weather records for a certain location,
- the market values of some stocks.

Hence, the possibility of finding an appropriate weight vector W that would produce correct decisions is very attractive.

The goal of the linear learning machine was to input a set of m-dimensional signals X, and find the appropriate vector W by slowly improving an initial guess, $W^{(0)}$, through a number of corrections obtained by comparing the result with the correct decisions known in advance. In order to maintain similarity with the neural network, the result of this procedure is here called *Net*. Equation (2.6):

$$Net = w_1 s_1 + w_2 s_2 + ... + w_i s_i + ... + w_m s_m \qquad (2.6)$$

can be regarded as the dot product of two vectors: the weight vector W and the vector X containing m signals s_i. Because we prefer vector notation to this extended one, the individual signals s_i are labeled as the components x_i of a multisignal input vector X:

$$Net = w_1 x_1 + w_2 x_2 + ... + w_i x_i + ... + w_m x_m = WX \qquad (2.7)$$

The input vector is linearly proportional to the corrected result, *Net*; hence, the corrective procedure (2.7) is called a **linear** learning machine.

Now, the "input vector X" is a set of signals: a multivariate object described by several parameters. In case there are more objects for input, they are distinguished by an index s, e.g. X_s. Therefore, the actual univariate signals coming to the individual synapses are components of these multivariate objects having two indices x_{si}, the first of which labels the multivariate object and the second of which labels the synapse to which this individual signal is linked.

From now on the components x_{si} will be called *signals*, and the multivariate inputs will be called *objects* or *vectors* X_s. The multivariate space, where the objects are represented as radius vectors leading to points X_s, is called the *measurement space*.

In Equations (2.6) and (2.7), the terms *Net*, W and X are so named to remind us of the corresponding features in the artificial neuron.

Net, the scalar product of a weight vector W and a multivariate vector X_s representing an arbitrary object in measurement space, is a very convenient quantity for making decisions. Its sign can indicate to

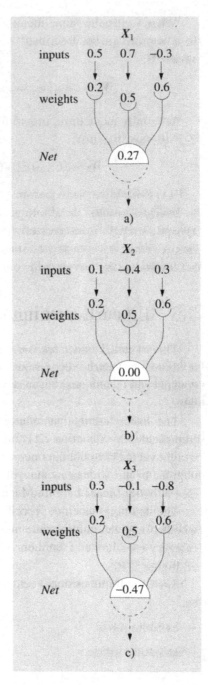

Figure 2-6: Dot products *Net* between the weight vector $W = (0.2, 0.5, 0.6)$ and three arbitrarily chosen input vectors X_1, X_2 and X_3.

which of the two categories selected in advance, C_1 or C_2, the object X_s belongs.

Figure 2-6 shows how a vector W (0.2, 0.5, 0.6) can be used as a decision vector to classify three different 3-dimensional objects $X_1 = (0.5, 0.7, -0.3)$, $X_2 = (0.1, -0.4, 0.3)$, $X_3 = (0.3, -0.1, -0.8)$.

Separating objects into classes based on whether *Net* is positive or negative is not the only possible division; *Net* can be easily divided into three or more intervals to define three or more categories. For example, a division into the intervals $-\infty$ to $-a$, $-a$ to $+a$, and $+a$ to $+\infty$ can be used to decide to which of the three categories object X_s belongs. Most of the decisions are binary, i.e. decisions between two categories, because all complex decisions can be composed of a series of binary ones.

Let the weight vector W be selected so that all objects X_s belonging to category C_1 give the scalar product *Net* = WX_s positive, and all objects X_s from C_2 give *Net* = WX_s negative; then W can be called a *perfect decision vector*. In principle, a perfect decision vector can be found if the objects are linearly separable. However, practical ways of obtaining a good (let alone perfect!) decision vector are hard to come by, if the data are not linearly separable.

Usually W is obtained from an initial (presumably bad) guess $W^{(0)}$ by some kind of learning procedure. $W^{(0)}$ is improved iteratively; an iterative procedure in this context means that if $W^{(t+1)}$ is obtained from $W^{(t)}$, then $W^{(t+1)}$ should be a slightly better decision function than $W^{(t)}$ (Figure 2-7). In the literature there are many different learning procedures in addition to the linear learning machine. Some of these from the field of neural networks will be explained later on in this book.

Figure 2-8 shows three decision vectors W_1, W_2, W_3 used in a hierarchical manner to decide in which quadrant of the *xy*-plane an input point is located.

If we could design a method for producing reliable decision vectors W, decision schemes of any complexity and size could be built from them, making binary decisions (the so-called piecewise linear classifier). The tree of Figure 2-8 is composed of three decisions: whether the point lies to the left or right of the ordinate axis, and (for each of these cases) where it lies with respect to the abscissa. In this case, the objects are points in the *xy*-plane, and the weight vectors are given in parentheses at the decision nodes of the tree. In order to make

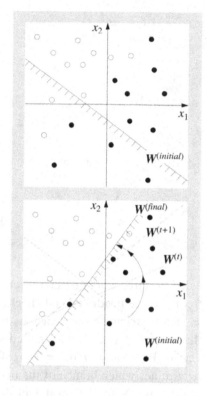

Figure 2-7: Changing the decision vector W towards a better position: $W^{(t+1)}$ is a better decision vector than $W^{(t)}$ because $W^{(t+1)}$ classifies only two objects falsely, compared with five by $W^{(t)}$. The perfect decision vector W should separate **all** objects of category C_1 (full circles) from those of category C_2 (empty circles).

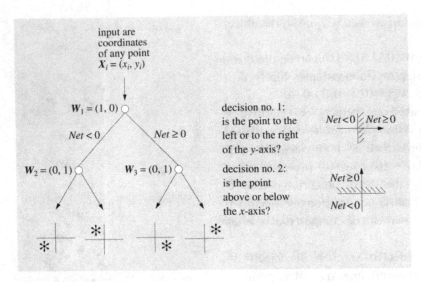

Figure 2-8: A small decision hierarchy for deciding in which quadrant a given point lies.

a decision tree work properly, each binary decision (at any point in the tree) must be **produced** and **checked** separately.

Try to calculate the decisions for some points. For example: take the point $P = (-3, 4)$ and make its dot product *Net* with the weight vector W_1: $Net = PW_1 = (-3, 4)(1, 0) = -3$. The result of *Net* determines which branch of the tree in Figure 2-8 has to be followed; in this case, Net < 0 gives the left-hand branch. Then a decision has to be made on the second level. Here, repeat the procedure with W_2 to get the final result.

As mentioned above, we can take the decision function (whose result, remember, is *Net*), to be the dot product between the representation of the object X and the weight vector W. The dot product between two vectors can be regarded as a linear function:

$$y = ax + b \qquad (2.8)$$

where y is equal to *Net*, a and x are the magnitudes of the vectors W and X, and b is a scalar constant that can be regarded as the offset of this straight line.

For compatibility with the notation of neural networks, we label this offset parameter as ϑ:

$$Net = WX + \vartheta = w_1x_1 + w_2x_2 + w_3x_3 + \ldots + w_mx_m + \vartheta =$$

$$= \sum_{i=1}^{m} w_i x_i + \vartheta \qquad (2.9)$$

Clearly, both vectors **W** and **X** must have the same number of components.

Observe the similarity of Equations (2.9) and (2.6); they differ only in the constant ϑ, which actually generalizes the linear form. Because we will meet this constant while discussing the neuron's transfer function (see Section 2.4), let us explain this in more detail.

Because ϑ is an arbitrary scalar constant, it can be written as a product of two other constants, say w_{m+1} and 1. If the two scalar constants are treated as the $(m+1)$-th components of the two vectors **W** and X, Equation (2.9) can be rewritten as a scalar product of these two new $(m+1)$-dimensional vectors. X is identical to the old representation of the object augmented by one dimension, the value of which is always equal to 1 (the so-called augmented feature vector), while w_{m+1} is equal to ϑ.

$$Net = w_1x_1 + w_2x_2 + w_3x_3 + ... + w_mx_m + \vartheta =$$
$$= w_1x_1 + w_2x_2 + w_3x_3 + ... + w_mx_m + w_{m+1} \cdot 1 = \qquad (2.10)$$
$$= WX$$

> From now on, remember that both W and X are $(m+1)$-dimensional.

This new component w_{m+1} has far-reaching significance in neural network learning and adaptation (see Section 2.5 describing the concept of "bias").

As before, the sign of the calculated scalar product *Net* defines the category to which the object X belongs. If the dot product **WX** for a certain vector X has the wrong sign with respect to the category of the X's, we calculate an increment for **W** as the difference between some new, corrected weight vector $W^{(new)}$ and the old, uncorrected vector $W^{(old)}$:

$$\Delta W = W^{(new)} - W^{(old)} \qquad (2.11)$$

Since the correction of weights will be done in iterative steps (0, 1, 2, ..., t, $t+1$, ...), $W^{(new)}$ and $W^{(old)}$ actually refer to $W^{(t+1)}$ and $W^{(t)}$, respectively.

In order to achieve this correction we will use the so called *delta-rule*, which states that, in order to improve the decision vector, the correction ΔW should be proportional to a certain parameter δ, (which is proportional to the error) **and** to the input X for which the wrong answer was obtained. After correction, the new weight vector should classify the vector X **correctly** or at least with a smaller error than before.

$$\Delta W \sim \delta X$$

or:

$$\delta \sim \Delta W / X \qquad (2.12)$$

or:

$$\delta = \eta \, (\Delta W / X)$$

where δ is the correction constant we are looking for, η is a constant of proportionality and X is the input object **wrongly** classified by $W^{(old)}$. Our goal is to find a parameter δ with which the new weight vector $W^{(new)}$ will classify X correctly.

If the dot product $Net = W^{(old)}X$ has a wrong sign, the dot product $Net = W^{(new)}X$ must have the opposite sign:

$$W^{(new)} X = -W^{(old)} X \qquad (2.13)$$

If (2.11) is substituted into (2.12):

$$\delta = \eta \left(W^{(new)} - W^{(old)} \right) / X \qquad (2.14)$$

and if the right side of (2.14) is multiplied by X/X, we obtain:

$$\delta = \eta \left(W^{(new)} - W^{(old)} \right) X / (X \cdot X) \qquad (2.15)$$

In the denominator of (2.15) we obtained the dot product of the vector X with itself:

$$X \cdot X = \sum_{i=1}^{m+1} x_i^2 \qquad (2.15a)$$

The expression (2.15a) is called the **norm** of the vector X and is written as $\|X\|^2$. So the next expression is:

$$\delta = \eta \left(W^{(new)} - W^{(old)} \right) X / \|X\|^2 \qquad (2.16)$$

This can be written as:

$$\delta = \eta \left(W^{(new)} X - W^{(old)} X \right) / \|X\|^2 \qquad (2.17)$$

from which, by substitution of (2.13), the final expression for the correction δ is obtained:

$$\delta = \eta\left(-W^{(old)}X - W^{(old)}X\right)/\|X\|^2$$
$$= -2\eta\,W^{(old)}X/\|X\|^2 \qquad (2.18)$$

Using Equation (2.18), the correction $\Delta W = W^{(new)} - W^{(old)}$ can now be easily obtained:

$$\Delta W = W^{(new)} - W^{(old)} = -\left(2\eta\,W^{(old)}X/\|X\|^2\right)X \qquad (2.19)$$

or in a more extended form:

$$W^{(new)} = W^{(old)} - \left(2\eta\,W^{(old)}X/\|X\|^2\right)X \qquad (2.20)$$

The *delta-rule correction* of $W^{(old)}$, (2.19) and (2.20), **guarantees** that $W^{(new)}$ will classify the object X correctly. However, the delta-rule does not say anything about the other objects that were also classified using $W^{(old)}$; some or all of them might now be classified falsely.

If the proportionality constant η is set to 1, Equation (2.20) gives the corrected decision vector $W^{(new)}$ as the mirror image of $W^{(old)}$ (Figure 2-9). Under certain circumstances, the corrections produced by mirroring are small, but often the mirror image can change $W^{(old)}$ considerably. Such large changes are not desirable, especially at the end of iterative learning, because large changes in the decision vector W mean that a number of previously correctly classified objects might become classified falsely. Therefore, an adaptable correction can sometimes be more appropriate.

An adaptable correction is obtained when, in Equations (2.18) to (2.20), the **constant** η is replaced by a **variable** η, smaller than 1:

$$W^{(new)} = W^{(old)} - \left(2\eta\,W^{(old)}X/\|X\|^2\right)X$$

or: $\qquad\qquad\qquad\qquad\qquad\qquad\qquad\qquad (2.21)$

$$\Delta W = -\left(2\eta\,W^{(old)}X/\|X\|^2\right)X$$

Writing the correction $-2W^{(old)}X/\|X\|^2$ as δ and remembering that X is an input object, we obtain the standard equation of the delta-rule in its most widely known form:

$$\Delta W = \eta\,\delta\,X \qquad (2.22)$$

Because the object X is the input it can also be called *Inp*:

$$\Delta W = \eta\,\delta\,\textbf{\textit{Inp}} \qquad (2.22a)$$

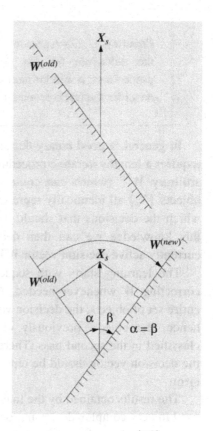

Figure 2-9: The vector $W^{(new)}$, which is the mirror image of $W^{(old)}$, classifies the object X correctly into another category.

> Equation (2.22) represents the most general form of the delta-rule for the correction of self-learning procedures; it will be intensively used in Chapter 8 to describe the back-propagation of errors algorithm.

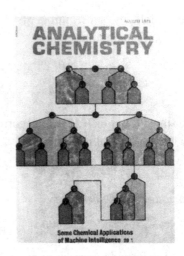

Figure 2-10: Cover design of the August 1971 issue of Analytical Chemistry, showing a decision tree with 26 decisions for predicting a molecular formula from the mass spectrum.

In general, a good binary decision vector W for any sub-decision requires a lengthy iterative procedure. Such a procedure starts with an arbitrary $W^{(0)}$ (which can contain random numbers) and a set of objects $\{X_s\}$ all identically represented as m-dimensional objects for which the decisions that should be made are already known. Using this knowledge we can then decide whether the correction of a currently active decision vector W is necessary.

The learning starts with sequentially testing all objects X and correcting W whenever needed. After finishing one pass over the entire set of objects, the decision vector W has changed considerably; hence, the objects previously classified correctly might be misclassified in the second pass. Therefore, the testing and correcting of the decision vector should be repeated as long as there is at least one error.

The results obtained by the linear learning machine method can be used in very complex decision processes. Jurs and Isenhour have used this method to predict molecular formulas from mass spectra, using a decision tree composed of 26 binary decisions. The scheme of the tree was even used as the cover design for the August 1971 issue of the journal *Analytical Chemistry* (Figure 2-10).

In spite of some success, the complex decision schemes obtained by linear learning machines are generally not very satisfactory, especially for large real-world problems. Therefore, after some initial enthusiasm, serious criticisms of this method were published (see Minsky and Papert); today, linear learning machines are mainly used with very restricted sets of data, and for education.

2.4   Transfer Functions in Neurons

The present model of a neuron consists of two distinct steps in obtaining output from the incoming signals. The first step, evaluation of the net input *Net*, was explained in the previous paragraph. We will now have a closer look at the second step, in which a nonlinear transformation of the net input signal *Net* takes place.

We might come up with a model involving only a direct transformation of the input signals to the output; However, such a function would involve many input signals and many weights, and would at the same time be nonlinear. Therefore, instead of one complex transformation, which would be hard to justify and computationally difficult, a two-step procedure was introduced. In addition, these two steps seem more plausible in light of what we know about the functioning of biological neurons.

It would not be very convincing to represent the output of an artificial neuron as the weighted sum of the input signals, since the resulting signal *Net* can be a) very large and b) negative. The latter seems particularly unrealistic. After all, the neuron either "fires" or it does not; even though the firing frequency differs from stimulus to stimulus, the idea of a negative firing frequency is just not reasonable.

We are forced, then, to make the net input to the neuron (Equation 2.10) undergo an additional, nonlinear transformation:

$$out = \mathrm{f}(Net) \hspace{3cm} (2.23)$$

called a *transfer function*.

What can we do to the artificial neuron's output signal *out* in order to make it more realistic? Because of the physical limitations on the size and frequency of brain signals, these conditions are quite simple: first, the final output signal of a neuron should be non-negative, whatever its magnitude; and second, it should be continuous and confined to a specified interval, say between zero and one. (In many publications, such a transfer function is called a **squashing** function because it squashes the output into a small interval.)

There are several functions satisfying the above conditions, but we will take a closer look at only three of them.

Hard-limiter: The first transfer function we will look at is, as Figure 2-11 shows, very simple indeed. It is called a *hard-limiter*, hl for short. It can have only two values: zero or one. There is one important point for this function, called the threshold value, ϑ. The value output by the hard-limiter depends on where the threshold value is set; thus, the threshold parameter ϑ decides whether the neuron will fire or not.

If the input value *Net* is $\geq \theta$, the output *out* will be one; otherwise it will be zero. Mathematically, this function (the *binary hard-limiter*) can be written:

$$out = \mathrm{hl}(Net, \vartheta) = \begin{cases} 1 & \text{if } Net \geq \vartheta \\ 0 & \text{if } Net < \vartheta \end{cases} \qquad (2.24)$$

or as an expression that lends itself to easy programming:

$$out = \mathrm{hl}(Net, \vartheta) = 0.5 \; \mathrm{sign}(Net - \vartheta) + 0.5 \qquad (2.24a)$$

where the function sign($\cdot$) has only two values, +1 and −1, for positive and negative arguments, respectively. If the threshold value ϑ = 0, then the hard-limiter (2.24) assigns 0 to all negative values of *Net* and +1 to all positive ones. Since the hard-limiter is very convenient for giving straight answers (yes or no), it is often used for final outputs where definite answers are required.

The hard-limiter giving 0 and 1 as output is not very suitable for many applications. Therefore, a slightly modified form of (2.24) giving +1 and −1 instead of 0 and 1 may be used for output. This *bipolar hard-limiter* can be written:

$$out = \mathrm{hl}(Net, \vartheta) = \begin{cases} 1 & \text{if } Net \geq \vartheta \\ -1 & \text{if } Net < \vartheta \end{cases} \qquad (2.25)$$

or:

$$out = \mathrm{hl}(Net, \vartheta) = \mathrm{sign}(Net - \vartheta)$$

The picture of the bipolar hard-limiter (Figure 2-12) is very similar to the binary hard-limiter (Figure 2-11).

Threshold logic: The second form of the transfer function we will discuss is the so-called *threshold logic*, tl, shown in Figure 2-13. In some respects, it is similar to the hard-limiter but has, in addition, a *swap interval*, within which *out* is linearly proportional to *Net*. The width of this interval is determined by a parameter α; the interval starts at ϑ and has a width of $1/\alpha$.

The calculation of the threshold logic function is very simple because the functions max(*a*, *b*) and min(*a*, *b*), which determine the maximum or minimum of two (or more) values between the two parameters *a* and *b* are easy to calculate.

Regardless of the value of the parameter *a*, max(0, *a*) is always **larger** than or equal to zero and min(1, *a*) is always less than or equal

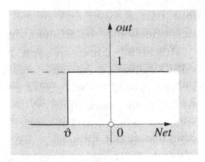

Figure 2-11: A binary hard limiter hl(*Net*, ϑ). If ϑ is set to less than zero the limiter jumps to a value of 1 on the negative side of the argument *Net*, if ϑ is larger than zero the transition to 1 occurs on the positive side.

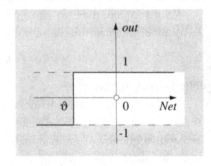

Figure 2-12: A bipolar hard-limiter.

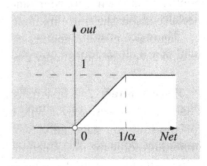

Figure 2-13: Threshold logic.

to 1. Setting the value of *a* equal to *Net* and combining both functions, the following expression is obtained:

$$y = \max (0, \min (1, Net))\qquad(2.26)$$

y(Net) is called a *threshold logic* (Figure 2-13).

By substituting the more general expression $\alpha(Net - \vartheta)$ for *Net*, the threshold logic is obtained in a form that can be used as a transfer function in neurons (Figure 2-14):

$$out = \mathrm{tl}\,(Net, \alpha, \vartheta\,) =$$
$$= \max \{0, \min [1, \alpha\,(Net - \vartheta)]\}\qquad(2.27)$$

α is called the *reciprocal width* of the swap interval.

Table 2-1 shows how the threshold logic behaves for two different sets of values of α and ϑ. The shaded area is the interval of values of *Net* in which the tl(*Net*, α, ϑ) function response is linear. It can be seen that the size of the linear response interval is inversely proportional to the parameter ($\alpha\vartheta$); if positive, it shifts this interval towards negative *Net* values, and *vice versa*.

This means that the linear transfer interval of the tl function starts at ϑ. If one does not need the actual values of ϑ, Expression (2.27) can be rewritten in a slightly modified form by the substitution:

$$\alpha\vartheta = \vartheta'\qquad(2.28)$$

This actually does not change anything; it just puts the variable expression into the more common form:

$$\mathrm{tl}\,(Net, \alpha, \vartheta') = \max [0, \min (1, \alpha Net - \vartheta')]\qquad(2.29)$$

Both the hard-limiter and the threshold logic are very convenient for computer programs. The threshold logic, for example can be used where a linear output is desired over the entire output signal range. However, due to the fact that they may contain singularities, they are of little theoretical use.

Sigmoidal function: The most widely used transfer function in various neural network applications is the so called *sigmoidal function*, sf, shown in Figure 2-15. It is somewhat time-consuming for numerical calculation, especially if tens of thousands of neurons are involved, but nevertheless it is used so often that it is worthwhile to take a closer look at it:

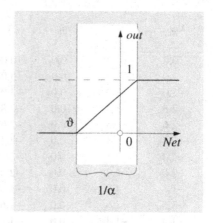

Figure 2-14: The threshold logic as a transfer function; the width of the linear response as a function of input is proportional to 1/α. The beginning of the swap interval is located at ϑ.

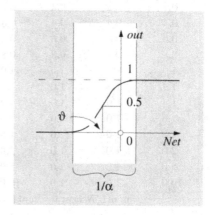

Figure 2-15: Sigmoidal transfer function.

Net	v	out	Net	$\alpha(Net-\vartheta)$ $\alpha=0.333,$ $\vartheta=-1$	v	out	Net	$\alpha Net-\vartheta'$ $\alpha=0.333,$ $\vartheta'=-1$	v	out
−10.0	−10.0	0.0	−10.0	−3.00	−3.00	0.000	−10.0	−2.333	−2.333	0.000
−5.0	−5.0	0.0	−5.0	−1.33	−1.33	0.000	−5.0	−0.667	−0.667	0.000
−3.0	−3.0	0.0	−3.0	−0.667	−0.667	0.000	**−3.0**	0.000	0.000	**0.000**
−2.8	−2.8	0.0	−2.8	−0.600	−0.600	0.000	−2.8	0.067	0.067	**0.067**
−2.6	−2.6	0.0	−2.6	−0.533	−0.533	0.000	−2.6	0.133	0.133	**0.133**
−2.4	−2.4	0.0	−2.4	−0.467	−0.467	0.000	−2.4	0.200	0.200	**0.200**
−2.2	−2.2	0.0	−2.2	−0.400	−0.400	0.000	−2.2	0.267	0.267	**0.267**
−2.0	−2.0	0.0	−2.0	−0.333	−0.333	0.000	−2.0	0.333	0.333	**0.333**
−1.8	−1.8	0.0	−1.8	−0.267	−0.267	0.000	−1.8	0.400	0.400	**0.400**
−1.6	−1.6	0.0	−1.6	−0.200	−0.200	0.000	−1.6	0.467	0.467	**0.467**
−1.4	−1.4	0.0	−1.4	−0.133	−0.133	0.000	−1.4	0.533	0.533	**0.533**
−1.2	−1.2	0.0	−1.2	−0.067	−0.067	0.000	−1.2	0.600	0.600	**0.600**
−1.0	−1.0	0.0	**−1.0**	0.000	0.000	**0.000**	−1.0	0.667	0.667	**0.667**
−0.8	−0.8	0.0	−0.8	0.067	0.067	**0.067**	−0.8	0.733	0.733	**0.733**
−0.6	−0.6	0.0	−0.6	0.133	0.133	**0.133**	−0.6	0.800	0.800	**0.800**
−0.4	−0.4	0.0	−0.4	0.200	0.200	**0.200**	−0.4	0.867	0.867	**0.867**
−0.2	−0.2	0.0	−0.2	0.267	0.267	**0.267**	−0.2	0.933	0.933	**0.933**
0.0	0.0	**0.0**	0.0	0.333	0.333	**0.333**	**0.0**	1.000	1.000	**1.000**
0.2	0.2	**0.2**	0.2	0.400	0.400	**0.400**	0.2	1.067	1.000	1.000
0.4	0.4	**0.4**	0.4	0.467	0.467	**0.467**	0.4	1.133	1.000	1.000
0.6	0.6	**0.6**	0.6	0.533	0.533	**0.533**	0.6	1.200	1.000	1.000
0.8	0.8	**0.8**	0.8	0.600	0.600	**0.600**	0.8	1.267	1.000	1.000
1.0	1.0	**1.0**	1.0	0.667	0.667	**0.677**	1.0	1.333	1.000	1.000
1.2	1.0	1.0	1.2	0.733	0.733	**0.733**	1.2	1.400	1.000	1.000
1.4	1.0	1.0	1.4	0.800	0.800	**0.800**	1.4	1.467	1.000	1.000
1.6	1.0	1.0	1.6	0.867	0.867	**0.867**	1.6	1.533	1.000	1.000
1.8	1.0	1.0	1.8	0.933	0.933	**0.933**	1.8	1.600	1.000	1.000
2.0	1.0	1.0	**2.0**	1.000	1.000	**1.000**	2.0	1.667	1.000	1.000
5.0	1.0	1.0	5.0	2.000	1.000	1.000	5.0	2.667	1.000	1.000
10.0	1.0	1.0	10.0	3.667	1.000	1.000	10.0	3.333	1.000	1.000

$v = \min(1, Net)$ or $v = \min[1, \alpha(Net - \vartheta)]$ *and*   cf. Equation (2.27)

$out = \max\{0, \min[1, \alpha(Net - \vartheta)]\}$

$out = \max[0, \min(1, \alpha Net - \vartheta')]$ cf. Equation (2.29)

Table 2-1:   The threshold logic function *tl*: effects of the parameters α and ϑ on the neuron's output *out* for different net input values *Net*. The left side of Table 2-1 (columns 1 through 3) shows the „normalised" *tl* function with $\alpha=1$ and $\vartheta=0$, therefore the swap of length 1 begins at *Net*=0 and becomes 1 at *Net*=1. To make the swap interval start at ϑ, use Equation (2.27): columns 4 to 7. Equation (2.29) is used in columns 8 to 11: the swap interval starts at ϑ'/α, i.e.: at $-1/0.333 = -3$ in our case. From the 7th and 11th columns it can be seen that the swap interval is equal to $1/\alpha$. For $\alpha=0.333$, the swap is 3 units long extending from −1 to +2 or from −3 to 0, depending on ϑ or ϑ', respectively.

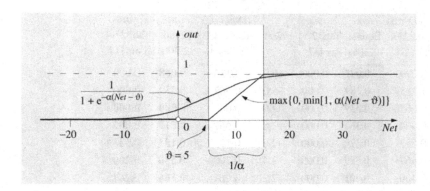

Figure 2-16: Comparison of the threshold logic and sigmoidal transfer functions if both have the same parameters ($\alpha = 0.1$, $\vartheta = 5$). Note that in the case of *tl* the swap starts at *Net*=ϑ, while at exactly the same point the *sf* has its inflection.

$$sf\,(Net,\, \alpha,\, \vartheta) \;=\; 1\,/\,\{\,1 + \exp\,[\,-\alpha\,(Net - \vartheta)\,]\,\} \quad \text{(a)}$$

or: (2.30)

$$sf\,(Net,\, \alpha,\, \vartheta') \;=\; 1\,/\,\{\,1 + \exp\,[\,-(\alpha Net - \vartheta')\,]\,\} \quad \text{(b)}$$

Note the similarity between the pairs of Equations (2.27) / (2.30a), and Equations (2.29) / (2.30b).

Table 2-2 is calculated using the sigmoidal function (2.30a) and (2.30b) with the same input as Table 2-1. Because the same parameters α, ϑ and ϑ' equal to 0.333, –1, and –1, respectively, were used, the arguments $\alpha(Net - \vartheta)$ and $\alpha Net - \vartheta'$ are the same in both Tables. Thus, the two transfer functions can be easily compared.

Obviously, with the **same pair of parameters** and the **same** *Net* values, the two equations have quite similar behavior; however, they give the response in slightly different regions. Figure 2-16 compares the behavior of the threshold logic and the sigmoidal function with exactly the same parameters.

Beginners often believe that the transfer function (either Equation (2.29) or (2.30)) is adjusted optimally if all net input signals fall into the quasilinear region shown in Figure 2-13 and in Table 2-2; this interpretation is completely false.

> The fundamental assumption of modern neural network theory is that the transfer signals are not linearly dependent on the net input.

Net α = 1.0, ϑ = 0.0	*out* Equat. (2.30a)	*out* Tab. 2-1 col. 3	*Net*	α(Net − ϑ) α=0.333, ϑ = −1	*out* Equat. (2.30a)	*out* Tab. 2-1 col. 7	*Net*	αNet − ϑ' α=0.333, ϑ' = −1	*out* Equat. (2.30b)	*out* Tab. 2-1 col. 11
−10.0	0.000	0.0	−10.0	−3.000	0.047	0.000	−10.0	−2.333	0.088	0.000
−5.0	0.001	0.0	−5.0	−1.333	0.209	0.000	−5.0	−0.667	0.339	0.000
−3.0	0.047	0.0	−3.0	−0.667	0.339	0.000	−3.0	0.000	0.500	0.000
−2.8	0.057	0.0	−2.8	−0.600	0.354	0.000	−2.8	0.067	0.517	0.067
−2.6	0.069	0.0	−2.6	−0.533	0.370	0.000	−2.6	0.133	0.133	0.133
−2.4	0.083	0.0	−2.4	−0.467	0.385	0.000	−2.4	0.200	0.550	0.200
−2.2	0.100	0.0	−2.2	−0.400	0.401	0.000	−2.2	0.267	0.566	0.267
−2.0	0.119	0.0	−2.0	−0.333	0.418	0.000	−2.0	0.333	0.562	0.333
−1.8	0.142	0.0	−1.8	−0.267	0.434	0.000	−1.8	0.400	0.599	0.400
−1.6	0.168	0.0	−1.6	−0.200	0.450	0.000	−1.6	0.467	0.615	0.467
−1.4	0.198	0.0	−1.4	−0.133	0.467	0.000	−1.4	0.533	0.630	0.533
−1.2	0.231	0.0	−1.2	−0.067	0.483	0.000	−1.2	0.600	0.646	0.600
−1.0	0.269	0.0	−1.0	0.000	0.500	0.000	−1.0	0.667	0.661	0.667
−0.8	0.310	0.0	−0.8	0.067	0.517	0.067	−0.8	0.733	0.675	0.733
−0.6	0.354	0.0	−0.6	0.133	0.533	0.133	−0.6	0.800	0.690	0.800
−0.4	0.401	0.0	−0.4	0.200	0.550	0.200	−0.4	0.867	0.704	0.867
−0.2	0.450	0.0	−0.2	0.267	0.566	0.267	−0.2	0.933	0.718	0.933
0.0	0.500	0.0	0.0	0.333	0.562	0.333	0.0	1.000	0.731	1.000
0.2	0.550	0.2	0.2	0.400	0.599	0.400	0.2	1.067	0.744	1.000
0.4	0.599	0.4	0.4	0.467	0.615	0.467	0.4	1.133	0.756	1.000
0.6	0.646	0.6	0.6	0.533	0.630	0.533	0.6	1.200	0.768	1.000
0.8	0.690	0.8	0.8	0.600	0.646	0.600	0.8	1.267	0.780	1.000
1.0	0.731	1.0	1.0	0.667	0.661	0.677	1.0	1.333	0.791	1.000
1.2	0.768	1.0	1.2	0.733	0.675	0.733	1.2	1.400	0.802	1.000
1.4	0.802	1.0	1.4	0.800	0.690	0.800	1.4	1.467	0.813	1.000
1.6	0.832	1.0	1.6	0.867	0.704	0.867	1.6	1.533	0.822	1.000
1.8	0.858	1.0	1.8	0.933	0.718	0.933	1.8	1.600	0.832	1.000
2.0	0.881	1.0	2.0	1.000	0.731	1.000	2.0	1.667	0.841	1.000
5.0	0.993	1.0	5.0	2.000	0.881	1.000	5.0	2.667	0.935	1.000
10.0	1.000	1.0	10.0	3.667	0.975	1.000	10.0	4.333	0.987	1.000

$$sf\,(Net,\,\alpha,\,\vartheta) = 1 / \{1+\exp[-\alpha(Net - \vartheta)]\} \qquad \text{cf. Equation (2.30a)}$$
$$sf\,(Net,\,\alpha,\,\vartheta') = 1 / \{1+\exp[-(\alpha Net - \vartheta')]\} \qquad \text{cf. Equation (2.30b)}$$

Table 2-2:   Comparison of the sigmoidal function output, *sf*, with the threshold logic function, *tl*: effects of the parameters α and ϑ on the neuron's output *out* for different net input values, *Net*. The swap interval, i.e., the transition between output of zero and one, is much broader for the sigmoidal function compared to that of the threshold logic. It can be clearly seen that the inflection of the *sf*, *out* = 0.5, occurs always at the same position as the beginning of the swap of the corresponding threshold logic function *tl*.

Of course, some neurons will show a linear relation between *Net* and *out*; however, it is the nonlinearity of the transfer function that makes the neural network so flexible for adjusting to different learning situations. A linear relation between the net input and its output may be desirable for some special cases, but nonlinearity should usually prevail in a neural network.

In Figure 2-17, the input is assumed to be confined to the range from *A* to *B*. In the topmost case the sigmoidal transfer function does not produce an output at all (an output of zero means that the neuron does not fire). In the second, the sigmoidal transfer function produces a maximum output signal no matter what the net input is. In the third, the response is almost linearly proportional to the input. In the fourth case, signals between *B* and *B'* all produce maximum output, while inputs between *A* and *B'* make full use of the transition interval of the sigmoidal transfer function. And in the last example, all inputs between *A* and *A'* produce no output; only signals between *A'* and *B* give a nonzero output.

Even those neurons whose transfer functions preclude firing at all during the entire training period (Figure 2-17, top) can be of interest in some applications; it is thought that their function is merely to wait for some event not anticipated in the training phase that might stimulate a response.

Instead of the mentioned *Net* input (Equation (2.1)) and one of the above transfer functions, more complex functions can easily be found. However, we are looking for the simplest adequate description of a neuron.

We previously mentioned the difficulties involving the derivatives of the hard-limiter and the threshold transfer functions because they contain singularities. Now, let's have a look at the derivative of the sigmoidal function (2.30), which will be used later (Figure 2-18). For clarity we will write Equation (2.30) as:

$$\mathrm{sf}(x) = 1/[1 + \exp(-x)] \qquad (2.31)$$

The derivative is obtained according to the rule for quotients. In case your calculus is a bit rusty, we will carry out the derivation in detail:

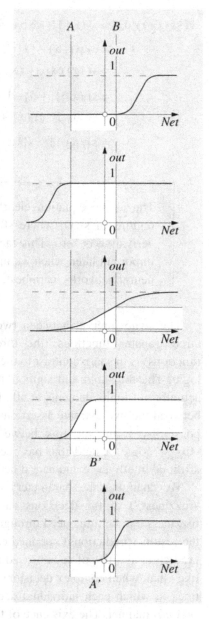

Figure 2-17: Different adaptations of neural sigmoidal functions for the same range of signals; the inputs *Net* are all confined to the interval (*A*, *B*).

$$d\,(sf\,(x)\,)\,/\,dx \;=\; -1\,/\,[\,1 + \exp\,(-x)\,]^{2}\,\{d\,[\,1 + \exp\,(-x)\,]\,/\,dx\} \;=\;$$
$$=\; \exp\,(-x)\,/\,[\,1 + \exp\,(-x)\,]^{2} \;=\;$$
$$=\; sf\,(x)\,\exp\,(-x)\,/\,[\,1 + \exp\,(-x)\,] \;=\;$$
$$=\; sf\,(x)\,\{\,[\,-1 + 1 + \exp\,(-x)\,]\,/\,[\,1 + \exp\,(-x)\,]\,\} \;=\;$$
$$=\; sf\,(x)\,[\,-sf\,(x) + 1\,] \;=\;$$
$$=\; sf\,(x)\,[\,1 - sf\,(x)\,] \qquad\qquad (2.32)$$

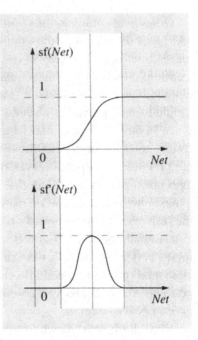

Figure 2-18: The derivative of the sigmoidal transfer function.

> The above equation clearly shows that in the flat regions of sf(x) (where sf(x) = 0 or sf(x) = 1), the derivative is zero. This fact will turn out to be very important later, when we investigate when and where neural networks learn best.

The swap interval between two states, exhibited by two of the three transfer functions (the threshold logic and the sigmoidal functions) is in sharp contrast to the hard-limiter's *yes/no-* (*true/false-*) logic. The threshold and sigmoidal functions, (2.29) and (2.30), are actually not logic elements at all. They may hold all states or values between the two extreme assertions *yes* and *no*, or *true* and *false*; this possibility forms the link between artificial neural networks and "fuzzy" logic, a field that has recently gained a lot of attention in artificial intelligence and neural network research.

By changing the parameters α and ϑ, the swap interval (the "fuzziness") of the decisions can be influenced. If the extent of fuzziness can be influenced through the choice of learning procedure, the results (predictions) obtained can be quantitatively evaluated; the prediction abilities of such procedures would be much more "human-like" than when complex decisions are based on hierarchical decision trees in which each individual decision has to be made in a logical (*yes/no*) manner. The existence of the swap interval and "fuzzy" logic in the transfer function is one of the greatest assets of this model.

2.5   Bias

The addition of an extra parameter, called bias, to the decision function increases its adaptability to the decision problem it is designed to solve. We will illustrate the importance of the bias through the linear learning machine. However, the conclusions are valid for many other neural network models, in particular, for the back-propagation algorithm (see Chapter 8). The example at the top of Figure 2-19 shows that the family of straight line functions $y = \alpha x$ cannot separate the class of points A, B and C from the class represented by point D. On the other hand, it is easy to separate the class (A, B, C) from the class (D) by introducing a constant ϑ, the bias, to the straight line decision function $y = \alpha x + \vartheta$, as shown in the lower part of Figure 2-19.

We have seen that in order to describe an artificial neuron we must know two types of parameters: the set of weights and the parameters of the transfer function. There are as many weights as there are signals entering the neuron. Generally, they are initialized as small random numbers. The actual interval within which these random weights are selected roughly depends on the number of weights in the neuron; a fair choice in applications where the normalization of weights is not strictly required is to set the interval so that the squared sum of the weights in one neuron is about 0.5.

Now the problem is how to treat the parameters of the transfer function. If we leave aside the swap interval of the transfer function for a moment, the crucial point in all three transfer functions is the threshold ϑ, i.e. the point where the neuron starts to react.

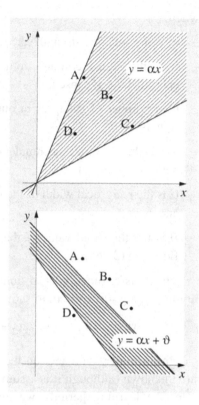

Figure 2-19: Introducing the bias, ϑ, to the decision function separating D from the points A, B and C.

> In this paragraph we will show that all parameters determining the artificial neuron (weights, interval, and threshold) can all be formally treated in exactly the same way.

Let us review the two main equations that describe the functioning of an artificial neuron:

$$Net = w_1 x_1 + w_2 x_2 + \ldots + w_i x_i + \ldots + w_m x_m = \mathbf{WX} \qquad (2.7)$$

$$\mathrm{sf}(Net, \alpha, \vartheta) = 1 / \{1 + \exp[-\alpha(Net - \vartheta)]\} \qquad (2.30a)$$

where:

- w_i is the synaptic strength (the weight) of the synapse i,

- x_i is the signal coming from a connected neuron entering our neuron at the synapse i,

- m is number of synapses in our neuron, i.e., the number of input signals,

- *Net* is the net input (accumulation of all inputs) formed within our neuron,

- α is the reciprocal width of the swap interval (see Equation (2.27)), and

- ϑ is the threshold value of the sigmoidal transfer function sf (see Equation (2.24a)).

The transfer function is so simple that we need to consider only its argument, $\text{arg} = \alpha Net - \vartheta$, to show our conjecture:

$$\text{arg} = \alpha w_1 x_1 + \alpha w_2 x_2 + \dots + \alpha w_m x_m - \alpha\vartheta \qquad (2.33)$$

Because all parameters on the right hand side of Equation (2.33) are unknown (although it is assumed that the signals x_i will be known after the learning period), we may substitute the products of two unknown values αw_i by the single value w_i', and $-\alpha\vartheta$ by ϑ'. Equation (2.33) now becomes:

$$\text{arg} = w_1' x_1 + w_2' x_2 + \dots + w_m' x_m + \vartheta' \qquad (2.34)$$

Now, if we regard the scalar value ϑ' as a product of ϑ and a component x_{m+1} which is **always equal to 1**, we get the following summation:

$$\text{arg} = w_1' x_1 + w_2' x_2 + \dots + w_m' x_m + \vartheta' x_{m+1} \qquad (2.35)$$

If we label ϑ' as w_{m+1}', we actually have created a product of w_{m+1}' and a signal x_{m+1} (always equal to 1); since this is completely analogous to the other products in the series, we can extend the summation by one more element:

$$\text{arg} = w_1' x_1 + w_2' x_2 + \dots + w_m' x_m + w_{m+1}' x_{m+1} = $$
$$= \sum_{i=1}^{m+1} w_i' x_i \qquad (2.36)$$

Now, by inserting the evaluated arg (2.36) back into the sigmoidal transfer function (2.30), and dropping the unnecessary "prime" on the w's, we obtain:

$$\text{sf}(Net, \alpha, \vartheta) = 1 / \left\{ 1 + \exp\left(-\sum_{i=1}^{m+1} w_i x_i \right) \right\} \qquad (2.37)$$

It is evident that the real output produced by the neuron, as described by the sigmoidal function sf(Net, α, ϑ), depends only on the $(m+1)$-dimensional weight vector W and the $(m+1)$-dimensional signal X:

$$W = (w_1, w_2, ..., w_m, w_{m+1})$$

and $\qquad\qquad (2.38)$

$$X = (x_1, x_2, ..., x_m, 1)$$

The extra weight which should be present in artificial neurons of this design (and which always receives an input value of 1) is the bias.

> We can regard the threshold value ϑ and the constant α simply as an additional "synaptic strength" called bias to which a signal of value 1 is always transmitted.

We will now explore the influence of the bias on a modification of the example considered in Figure 2-8. Using W_1 (the decision vector at the root of the decision tree in Figure 2-8) it is possible to decide whether a point $X(x_1, x_2)$ is above or below the abscissa, according to the sign of the dot product $W_1 X$:

if $\quad W_1 X \geq 0 \qquad$ X is above or on the abscissa

if $\quad W_1 X < 0 \qquad$ X is below the abscissa

At this point we will explain how the decision vector W_1 is obtained, using a slightly more complicated example. Let us assume that we have four points $X_1 = (3, 2)$, $X_2 = (-1, -1)$, $X_3 = (1, -3)$ and $X_4 = (4, 1)$, of which the first two belong to the category C_1 and the other two to C_2. Plotting these four points into the xy-plane (Figure 2-20),

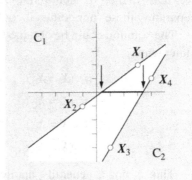

Figure 2-20: Four points in the xy-plane belonging to two different categories, X_1 and X_2 belong to C_1, while X_3 and X_4 belong to C_2. The separating line must be somewhere between the two lines indicated.

we realize that it should not be difficult to find a decision line separating these four points; it could be any line within the white area.

The solution could be obtained by writing down four trivial logical statements:

$$\text{if } (X = X_1), \quad \text{then } (X \text{ is in } C_1)$$
$$\text{if } (X = X_2), \quad \text{then } (X \text{ is in } C_1)$$
$$\text{if } (X = X_3), \quad \text{then } (X \text{ is in } C_2)$$
$$\text{if } (X = X_4), \quad \text{then } (X \text{ is in } C_2)$$

This is not a generally applicable procedure: it may not tell us anything about other points in the plane. In other words, the above procedure is **not a model** that can be used for predictions.

Using analytical geometry, we can write an equation for a straight line separating these four points. This line **is** a model, because it can be used as a decision line for **any** point in the plane. However, we do not favor the way it is obtained because, in a general case, complex handling of multidimensional planes and lines is required.

Instead, we would like to obtain the result simply by learning from a series of points (learning-by-example), and have it apply to any point, not just these four.

That is, we would like to obtain a decision vector W which will give a positive dot product WX_i for points X_i belonging to C_1, and negative for points belonging to C_2. No matter how hard we try, we cannot obtain a solution with a 2-dimensional W. However, Equation (2.38) shows that flexibility of adaptation can be obtained by augmenting the weight vector W by one additional weight w_{m+1}. An appropriate weight vector should therefore have three dimensions.

The only remaining question is: what weight vector should we start with? The simplest guess is the vector (1, 1, 1).

Hence, we start with:

$$X_1 = (3, \ 2, \ 1), \quad X_2 = (-1, -1, 1), \ \text{category 1, dot product} \geq 0$$
$$X_3 = (1, -3, 1), \quad X_4 = (\ 4, \ 1, 1), \ \text{category 2, dot product} < 0$$

and the starting weight vector $W^{(0)}$:

$$W^{(0)} = (1, 1, 1)$$

For correcting W we will use the delta-rule as introduced by Equation (2.12), $\Delta W \sim \delta X$, through Equation (2.20)

$$\Delta W = W^{(new)} - W^{(old)} = -\left(2\eta \, W^{(old)} X / \|X\|^2 \right) X \quad (2.20)$$

which is clearly identical to (2.12). Because the proportional coefficient δ is written as:

$$\delta = -\left(2\eta\, W^{(old)} X / \|X\|^2 \right) \qquad (2.39)$$

for each new vector X, ΔW can be calculated using equation (2.20) as shown in Table 2-3.

It is evident that the first product, WX_1, will yield a positive result, so no correction is necessary. The next product WX_2, which gives a **negative** dot product, requires a correction of W because X_2 belongs to category C_1 (for simplicity we set η equal to 1). After this correction has been made, the new W is multiplied by the vector X_3; since the wrong answer is produced by the multiplication, the correction is made again. This training continues until the correct prediction is achieved for all four points; a record of this is given in Table 2-3.

i	category of X_i	X_i	W	WX_i	sign predicted of WX_i	category	result	$\|X_i\|^2$ (2.15a)	δ	δX_i	new $W = W - \delta X_i$
1	C_1	(3, 2, 1) *	(1.00, 1.00, 1.00) =	6.00	+	C_1	OK				
2	C_1	(−1,−1, 1) *	(1.00, 1.00, 1.00) =	−1.00	−	C_2	wrong	3.00	−0.67	(0.67, 0.67, −0.67)	(0.33, 0.33, 1.67)
3	C_2	(1, −3, 1) *	(0.33, 0.33, 1.67) =	1.00	+	C_1	wrong	11.00	0.18	(0.18, −0.55, 0.18)	(0.15, −0.88, 1.49)
4	C_2	(4, 1, 1) *	(0.15, 0.88, 1.49) =	2.97	+	C_1	wrong	18.00	0.33	(1.33, 0.33, 0.33)	(−1.17, 0.55, 1.16)
						end of the first cycle: 3 errors					
1	C_1	(3, 2, 1) *	(−1.17, 0.55, 1.16) =	−1.25	−	C_2	wrong	14.00	−0.18	(−0.54, −0.36, −0.18)	(−0.63, 0.91, 1.33)
2	C_1	(−1,−1, 1) *	(−0.63, 0.91, 1.33) =	1.06	+	C_1	OK				
3	C_2	(1, −3, 1) *	(−0.63, 0.91, 1.33) =	−2.02	−	C_2	OK				
4	C_2	(4, 1, 1) *	(−0.63, 0.91, 1.33) =	−0.29	−	C_2	OK				
						end of the second cycle: 1 error					
1	C_1	(3, 2, 1) *	(−0.63, 0.91, 1.33) =	1.25	+	C_1	OK				
2	C_1	(−1,−1, 1) *	(−0.63, 0.91, 1.33) =	1.06	+	C_1	OK				
3	C_2	(1, −3, 1) *	(−0.63, 0.91, 1.33) =	−2.02	−	C_2	OK				
4	C_2	(4, 1, 1) *	(−0.63, 0.91, 1.33) =	−0.29	−	C_2	OK				
						end of the third cycle: 0 error					
		final W	(−0.63, 0.91, 1.33)								

Table 2-3:   Training events in the process of adaptation of the weight vector to classify four objects correctly.

If the training had started with a different weight vector and/or a different sequence of training objects, the resulting weight vector would be slightly different, although it would correctly predict all four points.

As mentioned before, by adding the bias we have moved our problem from the two-dimensional space, where the solution should be a line, to a three-dimensional space, where the solution is a plane. The solution is defined by a vector $W^{(final)} = (-0.63, 0.91, 1.33)$ that is perpendicular to the decision plane (see Table 2-3).

The sequence of changes of the weight vector W shown in Table 2-3 is called learning and the entire procedure, a *linear learning machine*. It is tedious to carry out even this very simple example by hand, let alone cases with hundreds of points and tens of decisions. Therefore, the reader is encouraged to do a little programming.

> A solution may be obtained in all linearly separable cases by augmenting the decision vector W with an additional component, the bias, and the vectors representing objects with an additional component equal to 1.

With different initial guesses of the weight vector W, or different sequence orders of the four objects for learning, we will get different results. However, all of the final weight vectors W will be equally satisfactory for making a decision for any point. Not only X_1, X_2, X_3, X_4, but **all** the points above the line (X_1X_2) or below the line (X_3X_4) (see Figure 2-20) will be classified in the categories C_1 and C_2, respectively.

2.6   Graphical Representation of Artificial Neurons

Although there exist a number of suggestions for representing artificial neurons, none of them seems to be completely satisfactory. Figure 2-21 shows some historically ordered examples of how artificial neurons have been presented in the literature. Representing the neurons by rectangles (Kohonen) has the advantage that they show the connections of neurons in one layer, which makes it easy to understand how the neurons obtain the same multidimensional signal simultaneously. Additionally, it makes the vector representation of the

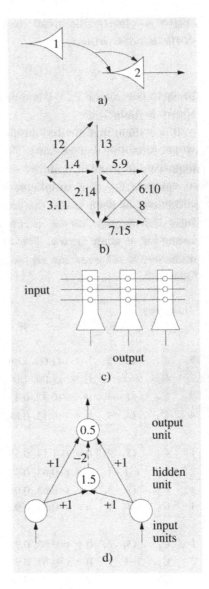

Figure 2-21: Representation of neurons by: a) McCulloch and Pitts, b) Hebb, c) Kohonen, d) Rumelhart.

neuron's weight vector W, composed of one row or one column of weights, more visual.

In this book, a neuron will be represented by a circle divided by a horizontal line into two halves representing the summation and transfer functions. Figure 2-22 shows how the presentations of a neuron was developed in this chapter.

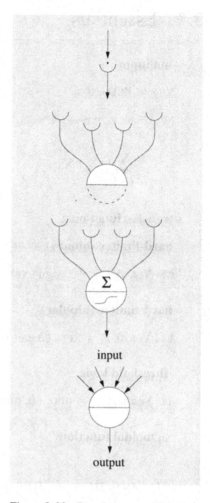

Figure 2-22: Development of the representation of a neuron in this chapter.

2.7   Essentials

- **net input:**

$$Net = \mathbf{W}\mathbf{X} + \vartheta =$$

$$= w_1 x_1 + w_2 x_2 + w_3 x_3 + \dots + w_m x_m + \vartheta =$$

$$= \sum_{i=1}^{m} w_i x_i + \vartheta \tag{2.9}$$

- **transfer functions:**

  hard-limiter (binary)

$$\mathrm{hl}\,(Net, \vartheta\,) = 0.5\ \mathrm{sign}\,(Net - \vartheta) + 0.5 \tag{2.24}$$

  hard-limiter (bipolar)

$$\mathrm{hl}\,(Net, \vartheta\,) = \mathrm{sign}\,(Net - \vartheta) \tag{2.25}$$

  threshold logic

$$\mathrm{tl}\,(Net, \alpha, \vartheta\,) = \max\,\{0, \min\,[\,1, \alpha\,(Net - \vartheta)\,]\,\} \tag{2.27}$$

  sigmoidal function

$$sf\,(Net, \alpha, \vartheta) = 1 / \left\{ 1 + \exp\left(-\sum_{i=1}^{m+1} w x_i \right) \right\} \tag{2.37}$$

- **learning:**

  delta-rule

$$\Delta \mathbf{W} = \eta\ \delta\ \mathbf{X} \tag{2.22}$$

  delta-rule for LLM

$$\mathbf{W}^{(new)} = \mathbf{W}^{(old)} - \left(2\eta\,\mathbf{W}^{(old)}\,\mathbf{X} / \|\mathbf{X}\|^2 \right)\mathbf{X} \tag{2.20}$$

2.8   References and Suggested Readings

2-1.   S. Silbernagl and A. Despopoulos, *Color Atlas of Physiology*, Thieme, Stuttgart, FRG, 1991; *dtv-Atlas der Physiologie*, Thieme, Stuttgart, FRG, 1991.

2-2.   S. W. Huffler and J. G. Nicholls, *From Neuron to Brain – A Cellular Approach to the Function of the Nervous System*, Sinauer Associates, Sunderland, MA, USA, 1986.

2-3.   R. Forsyth and R. Rada, *Machine Learning: Applications in Expert Systems and Information Retrieval*, Ellis Horwood Ltd., Chichester, UK, 1986, Chapters 1 and 2.

2-4.   N. J. Nilsson, *Learning Machines: Foundations of Trainable Pattern Classifying Systems*, McGraw-Hill, New York, USA, 1965, Chapter 3.

2-5.   K. Varmuza, *Pattern Recognition in Chemistry*, Springer Verlag, Berlin, FRG, 1980.

2-6.   S. Watanabe, *Pattern Recognition: Human and Mechanical*, Wiley, New York, USA, 1985.

2-7.   J. Zupan, *Algorithms for Chemists*, John Wiley, Chichester, UK, 1989, Chapter 8.

2-8.   M. Minsky and S. Papert, *Perceptrons: An Introduction to Computational Geometry*, MIT Press, Cambridge, USA, 1969.

2-9.   P. C. Jurs and T. L. Isenhour, "Some Chemical Applications of Machine Intelligence", *Anal. Chem.* **43** (1971) 20A – 36A.

2-10.   D. O. Hebb, *The Organization of Behavior*, Wiley, New York, USA, 1949.

2-11.   W. S. McCulloch and W. Pitts, "A Logical Calculus of the Ideas Immanent in Nervous Activity", *Bull. Math. Biophys.* **5** (1943) 115 – 133.

2-12.   T. Kohonen, *Self-Organization and Associative Memory*, Third Edition, Springer-Verlag, Berlin, FRG, 1989.

2-13.   D. E. Rumelhart, G. E. Hinton and R. J. Williams, "Learning Internal Representations by Error Propagation", in *Parallel Distributed Processing: Explorations in the Microstructures of Cognition*, Eds.: D. E. Rumelhart, J. L. McClelland, Vol. **1**, MIT Press, Cambridge, MA, USA, 1986, pp. 318 – 362.

2-14.   J. Gasteiger and J. Zupan, "Neuronale Netze in der Chemie", *Angew. Chem.* **105** (1993) 510 – 536; "Neural Networks in Chemistry", *Angew. Chem. Int. Ed. Engl.* **32** (1993) 503 – 527.

3   Linking Neurons into Networks

learning objectives

- why networking is emphasized
- parallel processing
- organization of neurons into layers
- more about inputs and outputs
- architectures
- graphical representation of neural networks
- matrix notation of neural networks

3.1   General

In order to achieve realistic results, our model must conform to certain facts. For example, we know (see Smith-Churchland, Reference 3-7) that a neuron can fire again after approximately one millisecond, 10^{-3}s. Since the reaction time of most vertebrates is around one tenth of a second (10^{-1}s), we conclude that whatever happens in the brain to provoke a reaction must occur in less than 100 firing times. This is called the "hundred steps paradoxon".

The most striking implication of this is that the brain possesses a signal processing algorithm so powerful that it can handle the most difficult tasks we can imagine in only 100 steps. Because even the most powerful computers with their nanosecond (10^{-9}s) clock rates do not come close to such performance, we must attribute the brain's magnificent performance to something unique about its structure and functioning.

Since a single neuron approach, no matter how many weights such a neuron has, cannot find solutions to complicated real-world applications, this "unique something" must involve the way neurons are interconnected; we have come to think of the brain as a massively parallel processor. Hence, as we stated at the beginning of the book

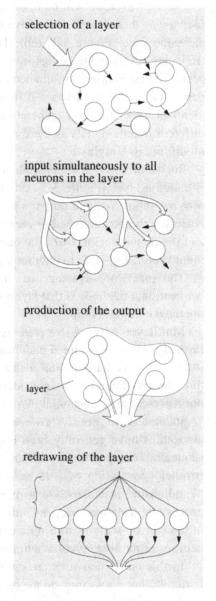

selection of a layer

input simultaneously to all neurons in the layer

production of the output

layer

redrawing of the layer

Figure 3-1: The steps in the formation of a neural network layer.

the emphasis in the phrase "neural networks" is clearly on the word "networks". In the rest of the book we will explore networks of neurons and their properties.

Our first exercise will be to organize a large number of neurons in such a way that **all** of them receive the same input X for processing at the same time (Figure 3-1, top). The production of a net input Net (Equation (2.7)) and the transformed output out (Equation (2.23)) then occurs in all neurons simultaneously (Figure 3-1, second and third part). As each neuron has a different set of weights, the otherwise identical procedure for generating output will produce as many different output signals as there are neurons. (A network "learns" by modifying its weights.)

Such a group of neurons producing a set of outputs simultaneously is called a layer (Figure 3-1 bottom). As each neuron j produces its own net input Net_j and output signal out_j, these individual signals of one layer can be combined to vectors, the net input signal vector, Net, and the output vector, Out. The output vector, Out, can be used as an input vector, Inp or X, to another layer of neurons.

The practical advantage of the neural network approach over conventional methods is that layers can be implemented on a parallel-operated computer chip.

Multilayer networks operate sequentially, i.e. the neurons in a layer do not receive the signals until the neurons from the previous ("upper") layer have produced them. Usually, no more than two or three layers of neurons are considered, and so the sequential link does not represent any substantial loss of time.

It should be noted, however, that until now neural network algorithms have generally been implemented on von-Neumann (i.e., sequential) computers; in such an implementation, the "simultaneous parallel" processing of a layer is actually performed **sequentially**. "Parallelism" is understood to mean that the neurons in one layer process information independently of each other. The output signals of one layer will be transmitted to the next layer only when **all** neurons of the first layer have finished their processing.

In this book, networks are drawn so that they "run" from top to bottom; other books may do the opposite (see Section 2.6).

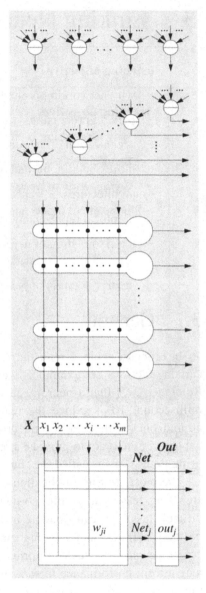

Figure 3-2: Four steps in the "evolution" from the biological to the matrix representation of neural networks.

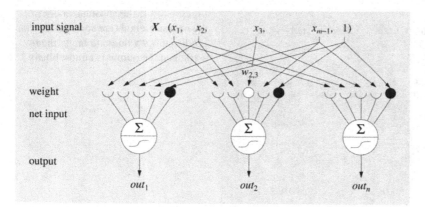

Figure 3-3: One-layer neural network.

3.2   One Layer

In our current model, a layer is a group of neurons **all** of which have the same number, m, of weights (synapses) and **all** receive the **same** m-dimensional input signal **simultaneously**. (What we have been calling a "layer" of neurons is often regarded as a "string", i.e., a linear arrangement of neurons.)

Neurons or layers of neurons are usually drawn to resemble biological neurons as well as possible: with circles acting as neural cells, and a number of interconnecting lines representing dendrites and axons. The synapses are placed somewhere along these lines. However, programmers and mathematicians prefer to think of the neural layer as a matrix of weights. Figure 3-2 shows a plausible way of obtaining a matrix notation from the "biological" one.

In the matrix of weights W, the rows represent the neurons. Each row j can be labeled as a vector W_j representing a neuron j, consisting of m weights w_{ji}, $W_j = (w_{j1}, w_{j2}, ..., w_{jm})$. All weights in the same column i, w_{ji} ($j = 1, 2, ..., n$), simultaneously obtain the same signal x_i. At a given moment, the entire input vector $X = (x_1, x_2, ..., x_m)$ (which may come from an external source (sensor or instrument), or from another group of neurons) is input into the network, i.e. to the matrix W. Since all weights w_{ji} in the entire matrix are simultaneously exposed to the corresponding input signals all products $w_{ji} x_i$ are made at the same time.

Figure 3-3 shows such a one-layer network composed of three neurons, each having the same number (five) of randomly generated weights. Each neuron in the layer obtains the same set of m signals $(x_1, x_2, x_3, ..., x_{m-1}, 1)$; here, $m = 5$. The weight w_{ii} is on the i-th

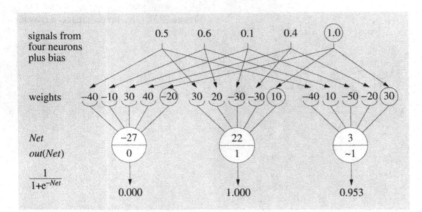

Figure 3-4: If the absolute values of the neuron weights are so large that the produced *Net* values are larger than +10, then the output is almost binary.

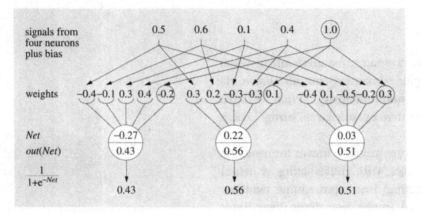

Figure 3-5: An example of a layer consisting of three neurons, each having five weights. The weight called *bias* is marked as a full circle; the value of the input signal to it is always 1.

position of the *j*-th neuron; for example, w_{23} is the 3rd weight of the 2nd neuron – see the circle above the second output neuron. The rightmost weight on each neuron (full circles) is the bias.

To make the picture unambiguous, a small example is given in Figure 3-4. A layer of three neurons is shown receiving signals from four neurons. For such an arrangement there must be five weights in each neuron. Remember, the input signal is actually 4-variate; the fifth input, which is always equal to 1, should be directed to the weight labeled *bias*. In order to check your understanding of the subject, you are encouraged to calculate the output values.

This example shows that the network responds continuously if the weights are small; it produces a continuum of values of the net input vector, **Net**, between ±10 (see Table 3-1 and Table 2-2, first example, column 3). On the other hand, if the absolute values of the weights are much larger than 10, the network will act as a binary device, as we will now show.

Net	sf(Net)	
−10	0.0000	
−9	0.0001	
−8	0.0003	
−7	0.0011	
−6	0.0025	
−5	0.0067	
−4	0.0180	interval for
−3	0.0474	giving
−2	0.1192	values
−1	0.3679	between 0
0	0.500	and 1, when
1	0.6321	the cut-off is
2	0.8808	set to 1%
3	0.9526	(0.01 or
4	0.9820	0.99)
5	0.9937	
6	0.9975	
7	0.9989	
8	0.9997	
9	0.9999	
10	1.0000	

Table 3-1: Outputs of the sigmoidal transfer function (2.30) for different *Net* values

Let us submit the same set of four inputs, that is, the 5-variate input vector X = (0.5, 0.6, 0.1, 0.4, 1.0) shown in Figure 3-4, to the same neural network having weights larger than 10 or smaller than −10 (the weights of the previous example, −0.5 to +0.4, multiplied by 100).

Figure 3-5 shows that now the outputs of the neurons are either very close to 0 or very close to 1, quite in contrast to the results shown in Figure 3-4.

To get a feeling of how large the values of *Net* should be to obtain nonbinary output, some values of *sf(Net)* for −10 ≤ *Net* ≤ 10 are given in Table 3-1. These data illustrate why one usually selects very small random numbers for the initial weights, usually in the range of 0.1 or even smaller. A rule of thumb is to set the *m* initial values of the weights w_{ji} in each neuron *j* so that

$$\sum_{i=1}^{m} \left| w_{ji} \right| = 1 \qquad (3.1)$$

> The example corresponding to Figure 3-5 is important because it shows that in certain circumstances it is extremely difficult (and will take a large number of iterations) to change the output of the network with **small** corrections of weights, since the outputs will **discontinuously** flip between 0 and 1.

3.3   Input

Until now we did not bother much about the mechanism by which the signals actually enter the network. The fact that each signal x_i of the input vector X should come to all neurons in the first layer means that somehow x_i should be "distributed" over as many weights as there are neurons in the layer. Graphically this is shown in Figure 3-6.

In order to make the flow of input data graphically consistent with the flow of data within and between the layers of neurons, the crossing points in the top row of Figure 3-6a where each input signal x_i is forked towards the weights should be considered "neurons". These are the small circles in Figure 3-6b; on the output side, they behave as full-fledged neurons, able to send many signals of the same value to their attached neurons, but on the other side, each has only one input signal, x_i.

Nor do these "input neurons" change the input signals x_i at all, which means that they have neither weights nor any kind of transfer function. The "input neurons" only serve as distributors of signals and do not play any active role in modifying them.

In order to stress this difference between the non-active input "neurons" and the active ones, the former will be drawn throughout this book as in Figure 3-6c, as squares (and not as circles as many other authors do). In addition, we will refer to them as input *units* and not as input neurons.

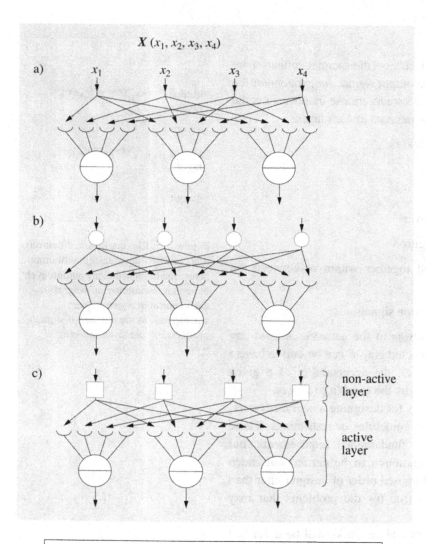

a)

$X (x_1, x_2, x_3, x_4)$

x_1 x_2 x_3 x_4

b)

c)

non-active layer

active layer

Figure 3-6:
a) Distribution of signals to the neurons in the first active layer.
(b) The points where the input signals are forked towards the neurons in the active layer are marked with small circles.
(c) Representing a non-active layer of neurons with squares.

Be very careful about labeling the non-active input, active layer, and output layers of neurons. Be sure you know exactly which layer of neurons the author intends when he or she applies the terms "inputs" or "outputs" to a particular layer. Many times in the same equation the inputs and outputs are taken from different layers.

In **counting** the number of layers to classify the architecture of a network, we do not include the input layer.

3.4   Architectures

The basic operation of a neuron is always the same: it collects a net input, Net_j, and transforms it into the output signal, out_j, via one of the transfer functions; the only thing we have to choose in advance is the number of layers, and the number of neurons in each layer.

All the topological data about the network:

– the number of inputs and outputs

– the number of layers

– the number of neurons in each layer

– the number of weights in each neuron

– the way the weights are linked together within or between the layer(s)

– which neurons receive the correction signals

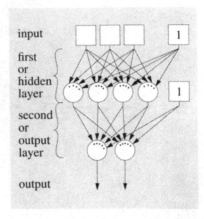

Figure 3-7: Linking layers of neurons together: two-layer design with input layer, one hidden and one output layer. Note that the number of weights in each neuron in a given layer corresponds to the number of signals produced by the layer above it.

together form the *architecture* or *design* of the network. Almost any network that is described as a "connected graph" can be said to have a neural network *architecture*; however, the acceptability of a given network architecture is judged solely by the result it produces.

In spite of such wide-open options for designing a neural network, there are some commonly accepted guidelines or restrictions. These are not imposed because of some fundamental requirements, but simply because theoretical investigations can be carried out much more easily if the network has an imposed order of design rather than random connections. The same is true for the problems that may appear at the programming stage.

The most common features of neural networks will be described briefly in the next paragraphs.

All neurons in one layer should obtain the same number of inputs, including an additional input connected to the bias. The number of weights in each neuron is fixed by the number of signals produced in the layer above it (Figure 3-7). (The network in this figure is referred to as a (3 x 4 x 2) network; the bias is not connected in this case.)

3.5   Hidden Layer; Output Layer

The layers below the passive input layer are usually referred to as the *hidden* layers, because they are not directly connected to the outside world as the input units and the output neurons are. The layer of neurons that yields the final signal(s) is called the *output* layer. For now, we will simplify the figure of a neuron to a plain circle; you can think of weights as being distributed over the upper half arc of the circle.

In more complex neural network designs some of the signals might be input to neurons in more than one layer, some of which may lie much deeper in the network. The adjacent hidden layer might even be skipped altogether for some signals. Figure 3-8 shows both these possibilities in a multilayer neural network design: all neurons in the second layer get an additional signal from the input, and one neuron in the third layer gets a signal directly from the input (these links are drawn with thicker lines). Without them, this would be an ordinary three-layer (2 x 3 x 2 x 2) neural network.

In some cases, such "far-going" signals can be linked to only a few neurons on a particular deeper layer or only to a single one. This might cause problems in the correction algorithms because the weights are usually corrected layer by layer.

On linear computers, the nonstandard links cause a considerable slowdown of the computation because either the branching conditions have to be checked at each step, or some additional pointers to the proper weights have to be introduced.

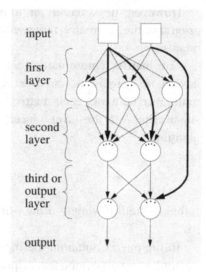

Figure 3-8: Three-layer neural network design with some signals bypassing the layer immediately below and linking to a deeper layer.

3.6 Graphical Representation of Neural Networks

Until now neural networks, consisting of layers of interconnected neurons have been shown as lines of circles (layers), linked together with arrows going from circles of one layer to the next. The arrows represent the direction of flow of the signals; at the places where the arrows touch the circles, the weights (synapses) are applied to them. The circles represent the bodies of the neurons, where all incoming signals are summed and the result then (nonlinearly) transformed into one output signal, which is transmitted to all neurons in the next layer.

Although it contains some simplifications of how neurons really work, this picture is quite adequate for describing artificial neural networks (e.g. Figure 3-7). It works, for example, whether the signal transformation actually takes place within the neuron's body or within the axon.

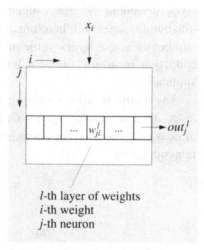

Figure 3-9: Matrix model of a network: the position of the weights within the neuron and the neurons within the layer.

However, it is based on a model of biological neurons; for programming, a matrix representation would be more explicit and precise.

The matrix representation considers a layer of *n* neurons, each having *m* weights, as an (*n* x *m*)-variate weight matrix **W**. For a multilayer network, each matrix (and thus its elements w_{ji}) obtain a superscript *l* (*l* for layer) specifying the index of the layer. The notation:

$$w_{ji}^{l} \qquad (3.2)$$

refers to the *i*-th weight of the *j*-th neuron in the *l*-th layer (Figure 3-9).

In the matrix notation, the input layer is actually a unit vector (a vector having all components equal to 1), and we do not need to draw or show it at all (if for some reason we do, then the layer index has to be 0). The weight matrix of the input layer, W^{0} that "transmits" *m* signals is a vector containing the value 1, *n* times:

$$W^{0} = (1, 1, 1, ..., 1) \qquad (3.3)$$

in agreement with the fact that the weight matrix for the first active layer of weights W^{1} has a superscript of 1. It is evident that the last (the output) layer will therefore always have a layer index equal to the number of active layers in the network. This notation avoids a lot of confusion regarding the actual number of layers used in the given application.

The matrix notation shows clearly that the input signals to the *l*-th layer X^{l} and output signals Out^{l} from this layer are *m*- and *n*-dimensional vectors, respectively. As mentioned above we must remember that:

$$X^{l} \quad = \quad Out^{l-1}$$
or: $\qquad\qquad\qquad\qquad (3.4)$
$$Out^{l} \quad = \quad X^{l+1}$$

To apply an *m*-variate signal input to a one-layer neural network consisting of *n* neurons each having *m* weights, we multiply an *m*-variate vector $X(x_{1}, x_{2}, ..., x_{m-1}, 1)$ with the (*n* x *m*)-variate weight matrix **W**. The result is an *n*-variate net input vector **Net** (Net_{1}, Net_{2}, ..., Net_{n}).

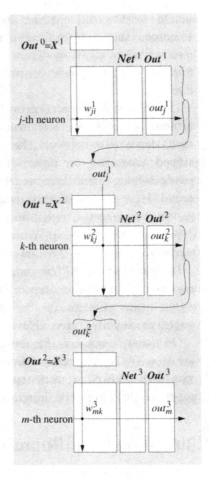

Figure 3-10: Matrix representation of a multilayer neural network. The number of weights in each layer is determined by the number of outputs from the layer above. However, the number of neurons in the layer is determined by the user (usually by trial and error).

$$\boldsymbol{Net} = (Net_1, Net_2, ..., Net_j, ..., Net_n) =$$

$$= \begin{bmatrix} w_{11} & w_{12} & \cdots & w_{1m} \\ w_{21} & w_{22} & \cdots & w_{2m} \\ w_{31} & w_{32} & \cdots & w_{3m} \\ \vdots & \vdots & w_{ji} & \vdots \\ & & & \vdots \\ w_{n1} & w_{n2} & \cdots & w_{nm} \end{bmatrix} \begin{bmatrix} x_1 \\ x_2 \\ x_3 \\ \vdots \\ x_i \\ \vdots \\ x_{m-2} \\ x_{m-1} \\ 1 \end{bmatrix}$$

Using extended notation, we can show how each component Net_j is calculated for layer l:

$$Net_j^l = \sum_{i=1}^{m} w_{ji}^l x_i^l \tag{3.5}$$

$$j = 1, 2, ..., n$$

The index j spans the n neurons, while i spans the m weights in the j-th neuron. The number of weights in the neuron is one more than the number of input variables, x_i; the remaining one input variable is the bias, which is always equal to 1.

The matrix equation (3.6) is a concise description of all net inputs in a one-layer network using the weight matrix $\boldsymbol{W}$:

$$\boldsymbol{Net}^l = \boldsymbol{W}^l \boldsymbol{X}^l \tag{3.6}$$

In a multilayer network, the weight matrices representing the layers are distinguished by the superscript l. For such a calculation the l-th layer weight matrix elements w_{ji}^l are used together with the input to the l-th layer x_i^l. Because the input to the l-th layer is usually the output of the $(l-1)$-st layer, Equation (3.5) can be written as:

$$\boldsymbol{Net}^l = \boldsymbol{W}^l \boldsymbol{X}^l = \boldsymbol{W}^l \boldsymbol{Out}^{l-1}$$

or:

$$Net_j^l = \sum_{i=1}^{m} w_{ji}^l out_i^{l-1} \tag{3.7}$$

$$j = 1, 2, ..., n$$

***Out**[^l]* is obtained from ***Net**[^l]* by one of the transfer functions ((2.24), (2.25), (2.29), or (2.30)). Let us apply the sigmoidal function (2.30) here as an example:

$$\boldsymbol{Out}^l = \text{sf}(\boldsymbol{Net}^l) \qquad (3.8)$$

From Equations (3.4) to (3.7), it can be concluded that the input layer $\boldsymbol{W}^0$ (which has only one row of weights, all of them equal to 1 and has no transfer function) will produce an output $\boldsymbol{Out}^0$ that is exactly equal to the external input. Thus, the m-variate input vector $\boldsymbol{X}$ that is input to the network can be labeled as output also:

$$\boldsymbol{X}^1 = \boldsymbol{Out}^0 \qquad (3.9)$$

Some authors draw networks so that information flows from left to right (Figure 3-11), and some, from bottom to top (Figure 3-12); but Kohonen favors (as do we, cf. Figure 3-7) the "top-down" design, which means that any input is above the neuron and its output is below. This corresponds to our everyday concept of "flow", whether in a liquid, a signal, or information. However, it must be said that most authors today prefer the "bottom-up" design (see Rumelhart's example, Figure 2-21d). This presumably originated with the convention in information theory where the flow of signals starts at the bottom. (If you lay the drawing in front of you on your desk, it runs away from you.)

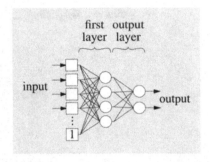

Figure 3-11: Neural network showing the flow of signals from left to right.

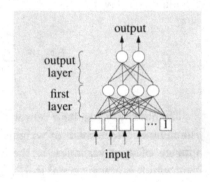

Figure 3-12: Neural network showing the flow of information from bottom to top.

3.7   Essentials

- **selection of weights**

$$\sum_{i=1}^{m} \left| w_{ji} \right| = 1 \tag{3.1}$$

- **indices of elements of the weight matrix**

$$w_{ji}^{l} \qquad \begin{array}{l} l\text{-th layer} \\ j\text{-th neuron} \\ i\text{-th weight} \end{array} \tag{3.2}$$

- **evaluation of the net input**

$$Net_{j}^{l} = \sum_{i=1}^{m} w_{ji}^{l} x_{i}^{l} \tag{3.5}$$

$$j = 1, 2, ..., n$$

$$\boldsymbol{Out}^{l} = \mathrm{sf}(\boldsymbol{Net}^{l}) \tag{3.8}$$

$$Net_{j}^{l} = \sum_{i=1}^{m} w_{ji}^{l} out_{i}^{l-1} \tag{3.7}$$

$$j = 1, 2, ..., n$$

- **evaluation of the output**

$$\boldsymbol{Out}^{l} = \mathrm{sf}(\boldsymbol{Net}^{l}) \tag{3.8}$$

- **labeling inputs and outputs**

$$\boldsymbol{X}^{l} \quad = \quad \boldsymbol{Out}^{l-1} \quad \text{input to the } l\text{-th layer}$$

or:

$$\boldsymbol{Out}^{l} \quad = \quad \boldsymbol{X}^{l+1} \quad \text{output from the } l\text{-th layer} \tag{3.4}$$

($\boldsymbol{Out}^{0}$ is identical to the network input)

3.8   References and Suggested Readings

3-1.   J. A. Anderson and E. Rosenfeld, Eds., *Neurocomputing: Foundations of Research*, MIT Press, Cambridge, MA, USA, 1988, pp. 159.

3-2.   M. Minsky and S. Papert, *Perceptrons: An Introduction to Computational Geometry*, MIT Press, Cambridge, MA, USA, 1969.

3-3.   L. B. Elliot, "Neural Networks – Conference Update and Overview", *IEEE Expert*, Winter 1987, 12 – 13.

3-4.   R. P. Lippmann, "An Introduction to Computing with Neural Nets", *IEEE ASSP Magazine*, April 1987, 4 – 22.

3-5.   J. Zupan and J. Gasteiger, "Neural Networks: A New Method for Solving Chemical Problems or Just a Passing Phase?", *Anal. Chim. Acta* **248** (1991) 1 – 30.

3-6.   J. Dayhoff, *Neural Network Architectures, An Introduction*, Van Nostrand Reinhold, New York, USA, 1990.

3-7.   P. Smith-Churchland, *Neurophysiology: Towards a Unified Science of the Mind-Brain*, MIT Press, Cambridge, MA, USA, 1986, Chapter 2.

Part II
One-Layer Networks

4   Hopfield Network

learning objectives

- Hopfield networks, though extremely simple, are capable of auto-association

- how neural nets are designed (architecture)

- simple data representations: binary and bipolar

- how a simple network is trained (stabilized) by iteration

- how a network trained to recognize images can recognize them even when they are corrupted

- the number of neurons needed for a given dataset

4.1   General

In 1982, the American physicist J. J. Hopfield brought neural network research back from the anathema to which it had been pushed in the seventies and early eighties. His paper (Reference 4-1), as a milestone on the way into the new era of neural network research, introduced nonlinear transfer functions for the evaluation of the final output from neurons.

The Hopfield neural network performs one of the most interesting tasks the brain is able to do: auto-association (Figure 1-3), by means of which a stored image (or any other information representable by a multivariable vector or matrix) is regenerated from partial or corrupted data. In other words, you perform auto-association whenever you recognize a friend after seeing, say, only his/her eyes.

Such procedures are clearly desirable not only in science but in areas as diverse as art, law, and economics. In art and science, it offers the possibility of reconstructing original images from blurred copies (NASA uses "image enhancement" techniques, for example, to increase the number of pixels in astronomical photographs); it could serve a similar purpose, say, in deciphering the Dead Sea Scrolls.

A "better" auto-association procedure is one which can produce a given degree of reconstruction from a "worse" original.

Besides auto-association, the Hopfield net can solve optimization problems (such as the famous "traveling salesman problem" – see Reference 4-3). However, we will not go into that here.

4.2   Architecture

The Hopfield neural net is a one-layer neural network. It consists of as many neurons as there are input signals, each neuron having, of course, the same number of weights as there are input signals. This means that the Hopfield network for m-variate signals (group of m individual signals) is a quadratic (square) (m x m)-variate matrix of weights.

Figure 4-1 shows us a design for a Hopfield net that can learn the auto-association of (4 x 4)-pixel images. The image is input as a 16-variable array, and so the Hopfield net has 16 neurons, each having 16 weights.

The original Hopfield neural network treats the binary signals as bipolar ones (having values of +1 or −1 only). It actually does not matter which notation is used for storing the images; however, the image should be represented in bipolar notation. In order to avoid confusion, and because computers store black and white pixels as bits (binary digits, 0 or 1), a simple transformation for each binary signal x_i can be introduced before entering the Hopfield network:

$$X^{bipolar} = 2X^{binary} - 1 \tag{4.1}$$

$$(x_1, x_2, ...,x_m)^{bipolar} = (2x_1 - 1, 2x_2 - 1, ...,2x_m - 1)^{binary}$$

The transformation $2x_j - 1$ in Equation (4.1) is a programmer's trick to enlarge the interval of x_j values by a factor of 2, and to shift the entire interval one unit lower on the axis: from $0 - 1$ to $(-1) - (+1)$.

4.3   Transfer Function

Because the transfer function in the Hopfield net is a bipolar version of the hard-limiter, hl, (Equation (2.24)) the input signals must be in bipolar notation:

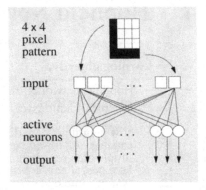

Figure 4-1: Hopfield net for learning associations of (4 x 4)-pixel images. The network has one layer with 16 neurons.

$$\text{hl}(Net_j) = \text{sign}(Net_j) = \text{sign}\left(\sum_{i=1}^{m} w_{ji} x_i^{bipolar}\right) \qquad (4.2)$$

where the function sign(u) means the algebraic sign of the argument u. Each term in the summation Net_j, the net input of the j-th neuron, is the product of a weight w_{ji} of the j-th neuron and the input signal $x_i^{bipolar}$ fed into it. The superscript bipolar is to remind us that the input signal must lie between +1 and −1. From now on, it will be omitted.

To use the binary representation, Equation (4.2) should be slightly modified:

$$\text{hl}(Net_j) = \text{sign}\left[\sum_{i=1}^{m} w_{ji}\left(2x_i^{binary} - 1\right)\right] \qquad (4.3)$$

4.4 Weight Matrix

How is the association learned in a Hopfield network? Very simply. Assume that each image is stored in the computer as an m-variable vector X of bipolar values (pixels):

$$X = (x_1, x_2, ..., x_i, ..., x_m) \qquad (4.4)$$

obtained by substituting white and black pixels by −1 and +1, respectively, and by taking the 4-pixel lines in sequence from the top row to the bottom row of the (4 x 4) image.

Now, first, we have to calculate the weights so that they will be appropriate for the image patterns that are to be learned by the Hopfield network. Say there are p images; the weights w_{ji} are calculated as follows:

$$w_{ji} = \sum_{s=1}^{p} x_{sj} x_{si} \quad \text{if} \quad j \neq i$$

and

$$w_{ji} = 0 \qquad \text{if} \quad j = i$$

$$(4.5)$$

Analyzing the value of w_{ji}, we see that it increases by 1 if the j-th and i-th pixel in a given pattern s are equal (i.e. both are white or both black), and it decreases by 1 if they are different. The more identical pairs of pixels j and i exist in the p patterns, the larger is the weight; the maximum absolute value that each w_{ji} can reach is p, the number of patterns in the set.

Hence, a **new** pattern is learned (that is, one cycle of weight modification takes place) simply by adding or subtracting 1 from each w_{ji}, depending on whether the j-th and the i-th pixel are equal or different. (It doesn't get much easier than that!)

Let us calculate a simple example. Figure 4-2 gives us four 16-variable patterns, actually (4 x 4)-pixel images, showing four arbitrary patterns. We would like these four patterns to be learned by the Hopfield net (Figure 4-3) by association.

The reader is encouraged to go through this example, or similar ones, by himself. To this purpose, the program HOPF is provided on the website of this book

http://www2.ccc.uni-erlangen.de/ANN-book/

On this site, also the datafiles for the patterns used in the example discussed here can be found. See Appendix for further details.

Equation (4.5) "stores" all of the patterns in the weight matrix **W**, in the sense that **each** element of **W** is actually influenced by **all** patterns X_s (x_{s1}, x_{s2}, ..., x_{si}, ..., x_{sm}). (Since the weight matrix **W** "contains" all patterns, then presumably the patterns can be "retrieved" from it!)

A new pattern $X^{(new)}$ can be added to the matrix by simply adding the product of the corresponding components of $X^{(new)}$ to each element of the weight matrix **W**:

$$\text{Updated } W \leftarrow w_{ji}^{(new)} = w_{ji}^{(old)} + x_j^{(new)} x_i^{(new)}$$

It is evident from Equation (4.5) that the matrix of weights is quadratic (because both i and j run from 1 to m), and symmetric over the main diagonal, which is set to zero. You should carry out the calculation of the weight matrix (or at least try to evaluate a few weights), and then compare your results with the complete weight matrix shown below. See also Figure 4-3; note that the first index, j, of w_{ji} is associated with the output and the second, i, refers to the input.

As an example, $w_{1,4}$ is calculated to be

$$w_{1,4} = (-1)(+1) + (-1)(-1) + (+1)(+1) + (+1)(-1) = 0$$

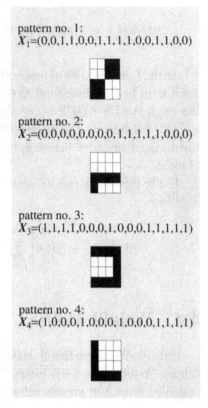

pattern no. 1:
$X_1 = (0,0,1,1,0,0,1,1,1,1,0,0,1,1,0,0)$

pattern no. 2:
$X_2 = (0,0,0,0,0,0,0,0,1,1,1,1,1,1,0,0,0)$

pattern no. 3:
$X_3 = (1,1,1,1,0,0,0,1,0,0,0,1,1,1,1,1)$

pattern no. 4:
$X_4 = (1,0,0,0,1,0,0,0,1,0,0,0,1,1,1,1)$

Figure 4-2: (4 x 4)-pixel images to be memorized by a Hopfield net can be represented as 16-variable vectors, given in binary notation. (Before entering the Hopfield net they are converted into bipolar representation using the transformation of (4.1).)

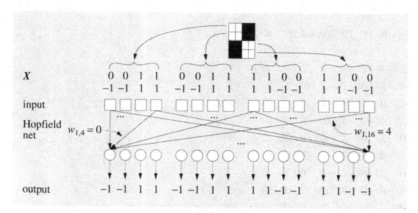

Figure 4-3: The Hopfield net shown in conventional notation. Each neuron has 16 weights ("synapses") monitoring the 16 input signals.

What are some of the things W can tell us about the learned images? For one thing, the value -4 for $w_{10,1}$ indicates that in all images, the first and tenth pixels are different from each other. Similarly, the value of $w_{1,16} = 4$ indicates that all four $(1,16)$-pairs of pixels have the same color. Figure 4-2 shows that the first and last pixel in patterns X_1 and X_2 are white, while they are both black in X_3 and X_4.

The (16×16) matrix W, below, represents 16 neurons, each having 16 weights. Because it is symmetric across the main diagonal, it does not matter whether we imagine the organization of the neurons to be by column or by row; that is, each row or column of the matrix is a set of 16 weights which belong to one neuron.

Once the patterns to be stored in a Hopfield network are known, "learning" consists simply of calculating the weight matrix W according to Equation (4.5). A very interesting thing about the particular W shown above is that each of the four patterns X_1, X_2, X_3 or X_4 (Figure 4-2) from which W was generated, are recognized exactly when put back into the net; that is, each one produces a 16-variable signal identical to the input. This, of course, is what we hoped for, but it is surprising that it was achieved all at once. With larger Hopfield networks (that can learn hundreds of different images), exact recognition of the "training" patterns comes only after a number of iterations. We will now examine the process of iteration during learning.

$$
\begin{array}{cccccccccccccccc}
1 & 2 & 3 & 4 & 5 & 6 & 7 & 8 & 9 & 10 & 11 & 12 & 13 & 14 & 15 & 16
\end{array}
$$

$$
\boldsymbol{W} =
\left(
\begin{array}{cccccccccccccccc}
0 & 2 & 0 & 0 & 2 & 0 & -2 & 0 & -2 & -4 & -2 & 0 & 0 & 2 & 4 & 4 \\
2 & 0 & 2 & 2 & 0 & 2 & 0 & 2 & -4 & -2 & 0 & 2 & -2 & 0 & 2 & 2 \\
0 & 2 & 0 & 4 & -2 & 0 & 2 & 4 & -2 & 0 & -2 & 0 & 0 & 2 & 0 & 0 \\
0 & 2 & 4 & 0 & -2 & 0 & 2 & 4 & -2 & 0 & -2 & 0 & 0 & 2 & 0 & 0 \\
2 & 0 & -2 & -2 & 0 & 2 & 0 & -2 & 0 & -2 & 0 & -2 & -2 & 0 & 2 & 2 \\
0 & 2 & 0 & 0 & 2 & 0 & 2 & 0 & -2 & 0 & 2 & 0 & -4 & -2 & 0 & 0 \\
-2 & 0 & 2 & 2 & 0 & 2 & 0 & 2 & 0 & 2 & 0 & -2 & -2 & 0 & -2 & -2 \\
0 & 2 & 4 & 4 & -2 & 0 & 2 & 0 & -2 & 0 & -2 & 0 & 0 & 2 & 0 & 0 \\
-2 & -4 & -2 & -2 & 0 & -2 & 0 & -2 & 0 & 2 & 0 & -2 & 2 & 0 & -2 & -2 \\
-4 & -2 & 0 & 0 & -2 & 0 & 2 & 0 & 2 & 0 & 2 & 0 & 0 & -2 & -4 & -4 \\
-2 & 0 & -2 & -2 & 0 & 2 & 0 & -2 & 0 & 2 & 0 & 2 & -2 & -4 & -2 & -2 \\
0 & 2 & 0 & 0 & -2 & 0 & -2 & 0 & -2 & 0 & 2 & 0 & 0 & -2 & 0 & 0 \\
0 & -2 & 0 & 0 & -2 & -4 & -2 & 0 & 2 & 0 & -2 & 0 & 0 & 2 & 0 & 0 \\
2 & 0 & 2 & 2 & 0 & -2 & 0 & 2 & 0 & -2 & -4 & -2 & 2 & 0 & 2 & 2 \\
4 & 2 & 0 & 0 & 2 & 0 & -2 & 0 & -2 & -4 & -2 & 0 & 0 & 2 & 0 & 4 \\
4 & 2 & 0 & 0 & 2 & 0 & -2 & 0 & -2 & -4 & -2 & 0 & 0 & 2 & 4 & 0
\end{array}
\right)
\begin{array}{c}
1 \\ 2 \\ 3 \\ 4 \\ 5 \\ 6 \\ 7 \\ 8 \\ 9 \\ 10 \\ 11 \\ 12 \\ 13 \\ 14 \\ 15 \\ 16
\end{array}
$$

4.5   Iteration

When, during training, the network produces an output that is not equal to the input (which is what usually happens), this output is input again. The input in the next iteration, $X^{(t+1)}$, is taken from the output of the present one, $\boldsymbol{Out}^{(t)}$:

$$
X^{(t+1)} = \boldsymbol{Out}^{(t)}
$$

Generally, a pattern more similar to the input will emerge. This procedure is repeated (without changing the weights) until identical outputs are produced in two consecutive steps (Figure 4-4).

$$
\boldsymbol{Out}^{(t)} = \boldsymbol{Out}^{(t-1)} = X^{(t)}
$$

NOTE: While the outputs are converging towards **something**, there is no guarantee that this "something" is the signal which was input to the net at the beginning (the original). As a matter of fact, the iterative procedure can have many different outcomes:

1) the final output is identical to the **initial** input;

2) the final output is identical to some **other** pattern stored in the net;

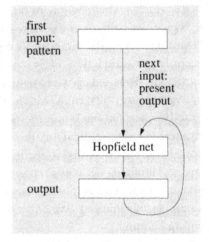

first input: pattern

next input: present output

Hopfield net

output

Figure 4-4: Iteration. Cycling of a multivariate signal from the net's output to the input, then through the net again until an output identical to the input is obtained.

3) the final output is not equal to any of the patterns used in the evaluation of weights;

4) the final output is as described in 1), 2) or 3), but with the pixels color-inverted (black for white, white for black) – that is, a **negative** image. This can happen, if the input signal has **more than** 50% wrong pixels from one of the originals.

Or, if convergence is not achieved:

5) the final output may **oscillate** between two or more patterns (possibly inverted) which are not equal to any of the original patterns.

If the Hopfield network produces the **original** input signal, we refer to it as a *stable* or *stabilized* network.

Note that these cycles do not change the weights!

In order to appreciate how difficult it is to get a "balanced" set of inputs to form a stabilized Hopfield net, you should program your own small nets and test them with different images. The main trick is to select images with approximately the same number of black and white pixels, as evenly distributed over the entire area as possible.

Clearly, only outcome 1) is of much practical use. But, you may ask, what good is a device for reproducing the same image that was input to it? The reproduction of all originals is only an initial test, determining whether the net is stabilized.

The idea of the Hopfield net is that the stabilizing effect of weights will force the net to produce the original image even from a corrupt, incomplete, dim, or blurred image. How far such an association of an incomplete input with the original one will go is hard to predict; in some cases, even an image with more than 30% of the pixels corrupted or missing will produce the original.

The necessary condition for a successful association is that the network be stable from the beginning.

Let us explore this feature of the Hopfield net using our recent example. Instead of inputting the original four patterns, let's corrupt the originals by 2, 5 and 13 randomly selected pixels and put it into the Hopfield network (in an image with 16 pixels, these represent errors of 13%, 31% and 81%: the proportions of the 16 pixels whose colors are opposite of what they should be). The numbers of iterations required

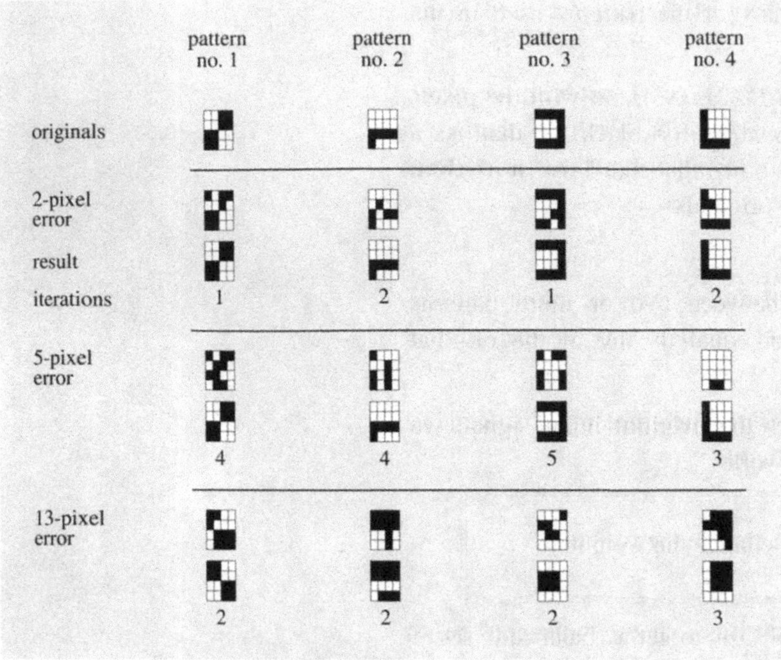

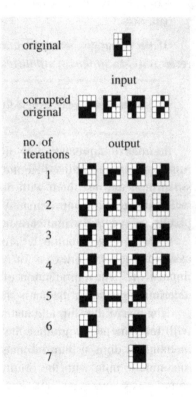

Figure 4-5: Outputs obtained by the Hopfield network of Figure 4-3 from inputs with 2, 5, and 13 errors, compared to the original patterns (top row). Numbers of iterations needed for each stable output are given below each output.

to get a stable output in each of these cases are given in Figure 4-5. However, two images with the same percentage of corrupted pixels may not require the same number of iterations, nor will they necessarily converge towards the same pattern. The larger the degree of corruption, the more unpredictable the outcome.

Figure 4-6 shows unpredictable output for the middle case (31% errors). You might think the stability of the output would get worse as the proportion of errors gets larger, but this is not the case.

If more than 50% of the pixels are corrupted, the retrieved image is likely to become the negative of the original (a 95% corrupt picture is actually the negative of the original corrupted by 5%). Therefore, it is not hard to understand the bottom row of Figure 4-5, in which an 81% (13 pixels) corrupted image produces a negative of the original.

4.6   Capacity of the Hopfield Network

The Hopfield net may seem very attractive for some applications, but there are severe restrictions on the number of images (patterns) that such a network is able to learn. In general, the net should contain about seven neurons for each pattern to be learned (and recognized):

Figure 4-6: Some possible outputs (last pattern in each column) if the original pattern has 5 corrupted pixels. The outcome is unpredictable: it may be the original (first column), some other pattern (second column), the negative of an actual pattern (third column), or it may oscillate between two patterns (fourth column). Note that the number of iterations required differs from case to case.

$$N^{neurons} = 7N^{images} \qquad (4.6)$$

(In our preceding example, the number of neurons was only four times the number of images. This will be explained at the end of this section; for details, see Reference 4-12.)

Equation (4.6) has two consequences. First, it shows that the minimum resolution of the images depends on how many there are. In terms of the previous example, Equation (4.6) requires that each of one hundred 2-dimensional images, which would require 700 neurons, be represented on a grid of at least 700 pixels, or, for a square image, a grid of about (27 x 27) pixels.

Second, Equation (4.6) determines the size of the weight matrix. Since a Hopfield network generates a quadratic matrix, learning (recognizing, associating) 100 images requires a matrix containing $700^2 = 490,000$ weights.

This brings up an obvious problem: since each number in a computer is represented by four bytes (floating points) or two bytes (short integers), a Hopfield net for recalling 100 images (on a (27 x 27) grid) will require one or two Mbyte of memory space for the weight matrix. However, there is an even worse aspect of this problem. Recalling any image requires as many as 2 million multiplications and 2 million additions (performed sequentially, on most computers) for **each** iteration cycle! And the space and time requirements grow quadratically as the size of the problem grows!

Equation (4.6) is valid for a random sample only; as a consequence, the number of patterns to be stored in a Hopfield net need not depend only on the number of neurons. By a careful arrangement or selection of patterns, we can increase the capacity of the Hopfield network above 0.14 images/neuron, up to maximum of about 0.25. Our example, storing four images on a grid of 16 pixels, is about the maximum we can achieve on such a grid.

4.7   Essentials

- the number of input data is equal to the number of output data

- there is no refinement of weights in the Hopfield network; the weights are calculated from the patterns

- the stored patterns are retrieved from the network by a "circular" flow of signals; the output becomes the next input

- the network is able to "retrieve" the uncorrupted patterns even if the inputs are corrupted

- the number of patterns that can be stored simultaneously in the Hopfield network is relatively small

- **binary to bipolar vector conversion**

$$(x_1, x_2, ..., x_m)^{bipolar} = (2x_1 - 1, 2x_2 - 1, ..., 2x_m - 1)^{binary}$$

(4.1)

- **Hopfield network:**

  weights

$$w_{ji} = \sum_{s=1}^{p} x_{sj} x_{si} \quad \text{if} \quad j \neq i$$

and

$$w_{ji} = 0 \qquad \text{if} \quad j = i$$

(4.5)

  output

$$\text{hl}(Net_j) = \text{sign}(Net_j) = \text{sign}\left(\sum_{i=1}^{m} w_{ji} x_i^{bipolar}\right) \quad (4.2)$$

- **requirement or capacity**

$$N^{neurons} = 7N^{images}$$

(4.6)

4.8   References and Suggested Readings

4-1.   J. J. Hopfield, "Neural Networks and Physical Systems With Emergent Collective Computational Abilities", *Proc. Natl. Acad. Sci. USA* **79** (1982) 2554 – 2558.

4-2.   J. J. Hopfield, "Neurons With Graded Response Have Collective Computational Abilities", *Proc. Natl. Acad. Sci. USA* **81** (1984) 3088 – 3092.

4-3.   J. J. Hopfield and D. W. Tank, "Neural Computation of Decisions in Optimization Problems", *Biol. Cybern.* **52** (1985) 141 – 152.

4-4.   J. J. Hopfield and D. W. Tank, "Computing with Neural Circuits: A Model", *Science* **233** (1986) 625 – 633.

4-5.   J. J. Hopfield and D. W. Tank, "Simple Neural Optimization Networks", *IEEE Trans CS*, **CAS-33**, 533 – 541.

4-6.   R. P. Lippmann, "An Introduction to Computing with Neural Nets", *IEEE ASSP Magazine*, April 1987, 4 – 22.

4-7.   R. J. McEliece, E. C. Posner, E. R. Rodemich and S. S. Venkatesh, "The Capacity of Hopfield Associative Memory", *IEEE TIT*, **IT-33** (1987) 461 – 482.

4-8.   J. J. Hopfield, D. I. Feinstein and R. G. Palmer, "Unlearning has a Stabilizing Effect in Collective Memories", *Nature* **304** (1988) 158 – 159.

4-9.   J. A. Anderson and E. Rosenfeld, Eds., *Neurocomputing: Foundations of Research*, MIT Press, Cambridge, MA, USA, 1988.

4-10. M. Tusar and J. Zupan, "Neural Networks", in *Software Development in Chemistry 4*, Ed.: J. Gasteiger, Springer Verlag, Berlin, FRG, 1990, pp. 363 – 376.

4-11. H. Ritter, T. Martinetz and K. Schulten, *Neuronale Netze, Eine Einführung in die Neuroinformatik selbstorganisierender Netzwerke*, Addison-Wesley, Bonn, FRG, 1990.

4-12. D. J. Amit, H. Gutfreund and M. Sompolinski, "Storing Infinite Number of Patterns in a Spin-glass Model of Neural Networks", *Phys. Rev. Lett.* **55** (1985) 1530 – 1533.

5   Adaptive Bidirectional Associative Memory (ABAM)

learning objectives
- the differences between two types of learning methods: supervised and unsupervised
- how signals (training data) may be sent forward **and** backward through a network
- how a network "learns" by changing the weights of its neurons
- how matrices are used to represent and modify weights
- how a trained network can actually recognize separated parts of a stored pattern

5.1   Unsupervised and Supervised Learning

When learning is *unsupervised*, the system is provided with a group of facts (patterns) and then left to itself to settle down (or not!) to a stable state in some number of iterations (Figure 5-1). Inherent in any unsupervised learning system is an optimization (or decision) criterion that is used for the evaluation of the result at the end of each cycle. This, however, is a very general one, such as minimization of energy or distance, maximization of profit, etc.

Thus, learning is basically an optimization procedure. An example of this kind of learning is *hierarchical clustering*, in which we have a set of objects or patterns $\{X_s\}$ and two criteria, the distance between two objects and the distance between two groups of objects. We want a system which will, in response **only** to these conditions, organize the objects into groups and the groups into a hierarchy. The result should be a surprise, or at least not be influenced by our expectations.

Suppose, however, that we have some objects (patterns, say) whose behavior (responses) in a given system are known. The two types of data (the representation of the objects and their responses in

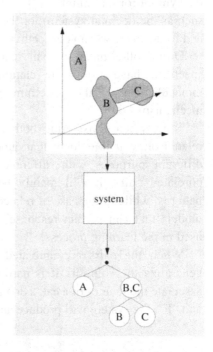

Figure 5-1: Unsupervised learning.

the system) form pairs of what we might call inputs and targets. The goal of *supervised methods* is to find a *model* – a general procedure – that will correctly associate the inputs with the targets (Figure 5-2).

In a sense, the targets do not enter the learning procedure; they only serve as a criterion for how well the system has been trained. According to this estimate, the decision is made whether the network (specifically, the weights) need to be corrected (see Section 2.3). Of course, the change or correction of each weight depends on the size of the error produced, which in turn depends on the target; but mainly the correction is proportional to the input.

At the moment we are not concerned with how to change the weights; let's concentrate on the idea that the weights are forced to change to give a specific answer defined by the user.

When applying any supervised learning method, we have to distinguish two cases that differ from each other in the way the target is related to the input; specifically, whether some intrinsic relationship occurs between them, or they are just arbitrarily associated. Examples of arbitrary associations are the number of eyes most of us have, and the symbol for that number ("10" in the binary system, "2" in others, such as the decimal system); or the sound you make when you smile and force air between your teeth, and the letter "c" (cf. Figure 5-3).

On the other hand, the pair "chemical structure" and its "infrared spectrum" have an intrinsic relationship, because the structure of the molecule causes the spectrum whether or not we know the mechanism.

Now, when considering both types of pairs, the unrelated and related ones, it is evident that supervised learning will be used for different purposes with different types of pairs. In the case of unrelated pairs, it will mainly be used for identifying corrupted patterns, while in the case of related pairs it will be used for building models that can predict responses for different patterns from those used in the learning process.

When the pairs are unrelated, it is obviously not possible to generalize the solution: it is hard to imagine that after learning to associate the images of a cat, a dog and a fish with the letters "C", "D" and "F", the system will produce an "H" for the image of a horse.

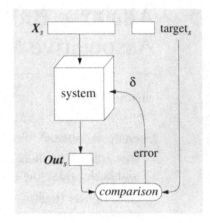

Figure 5-2: Supervised learning.

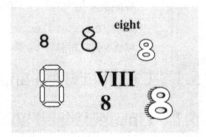

Figure 5-3: Arbitrary relations between objects: a number, and the symbols used for it.

But this conclusion requires a word of caution. At one level, "cow" and "milk" are unrelated – they don't even have any letters in common – but at another level, they are. That is, there might be no generalization on the level at which the system was trained, while one does exist on some higher (or, if you prefer, lower) level for which the system was not intended to make predictions. If this turns out to be true the ABAM net is worth considering. Below, we will give some examples to demonstrate the point.

5.2   General

The word "associative" in the name *adaptive bidirectional associative memory* (ABAM) indicates that it is a system that can learn to associate patterns. In a way, ABAM resembles the Hopfield network: both are one-layer neural networks, and both were developed to manipulate binary (or bipolar) vectors and have therefore an identical evaluation of the initial weights. After this, however, the similarity breaks down.

The most important difference between the two networks lies not in their architecture, but in the scope of the problems they can tackle and how the networks (or weights) are adapted to these problems. The entire Hopfield network (the number of neurons, and the number of weights and their values), is determined once and for all after the patterns or objects to be learned have been chosen. The size of the image, i.e. the number of pixels in each pattern, determines the number of neurons and the number of weights (see Section 4.2), while the number and the form of the selected patterns determine the values of all the weights (see Equation (4.5)).

This does not work with the ABAM network. First, it is not necessarily square, but can be rectangular if convenient. This is because there can be fewer neurons in the output layer than there are in the non-active input layer (Figure 5-4). This can save a lot of computer memory, as well as computation time.

Second, although the initial setup of the weights is made in the same way as for the Hopfield net (compare Equations (5.2) and (4.5)),

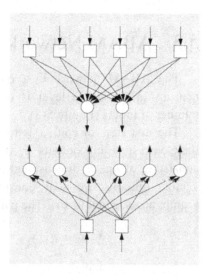

Figure 5-4: The two directions of signal propagation in an ABAM network (cf. also Section 5.4).

in the ABAM network they are used only in the first iteration step. After this, they are changed (adapted).

> This is what the "adaptive" in the name ABAM stands for.

How and why are the weights changed?

Let us turn to the second question first. The weights in the ABAM network are changed in order to give an exact and **predetermined** response that is not the same as the input pattern. This means that the ABAM network can learn to associate **unrelated pairs** of patterns. Thus, the ABAM network is able to associate patterns with their written or spoken descriptions, and, as in the case of the Hopfield network, the trained ABAM network will still be able to reproduce the associated pattern even from corrupted input. Such pairs of patterns, the inputs and the targets, are essential for supervised learning.

5.3   ABAM Network

The ABAM network is a one-layer network. The number of neurons in the active layer is often considerably smaller than the number of inputs (Figure 5-5).

The first step, of course, is to generate the input vectors X_s with their corresponding outputs Y_s (targets). Let there be p pairs, each consisting of one m-dimensional input and one n-dimensional target vector. Initially, we will consider these vectors to be bipolar (components, equal to ±1). The pairs (X_s, Y_s) can be written:

$$X_s = (x_{s1}, x_{s2}, ..., x_{si}, ..., x_{sm})$$

and   (5.1)

$$Y_s = (y_{s1}, y_{s2}, ..., y_{sj}, ..., y_{sn})$$

The initial weights, $W^{(0)}$, are calculated from the set of p pairs of input and target vectors, $\{X, Y\}$:

$$w_{ji}^{(0)} = \sum_{s=1}^{p} x_{si} y_{sj}$$ (5.2)

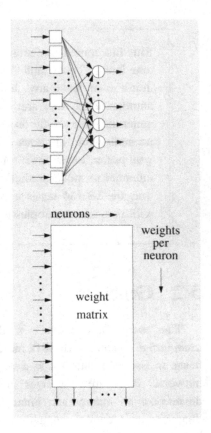

Figure 5-5: The architecture of the ABAM network: it is very similar to the Hopfield network, but usually has many fewer neurons on the output side than on the input.

The symbol { } denotes a group of objects, in our case a group of pairs consisting of input and target vectors X and Y. Be sure to distinguish this notation from that for a single pair of vectors (X_s, Y_s).

5.4 Learning Procedure

The basis of learning in the ABAM network is the fact that an $(m \times n)$ matrix can be multiplied from two directions: in the standard way by an m-dimensional vector, or in transposed form (reflected across the main diagonal) by an n-dimensional vector. In the language of neural networks, the side of the weight matrix W having the same number of rows (or columns) as the input vector is called the *input* side, and the other, the *output* side.

This leads to the very nice property that an "input" object X can produce the "output" $Y^{(1)}$; or a "target" Y if inputted on the **output** side can produce an "output" vector $X^{(1)}$ on the, formally speaking, **input** side of the weight matrix. (The quote around "input", etc., reflect the fact that input and output are not fixed with respect to the matrix any more. See Figure 5-6)

Thus, **any** pair of objects (X, Y) of different dimensions may be paired up with another pair having the same dimensions $(X^{(1)}, Y^{(1)})$. Because both multiplications can be regarded as sending signals (X or Y) through the neural network in opposite directions, it is called *bidirectional*.

Combining the procedure that generates the weight matrix W from a **set** of pairs of objects $\{X, Y\}$ (Equation (5.2)) with the bidirectional procedure for generating pairs of vectors, we obtain an iterative learning scheme:

The iterative procedure (5.4) stops when the weight matrix $W^{(t)}$ produces a set of pairs $\{X^{(t)}, Y^{(t)}\}$ identical to the initial set $\{X, Y\}$.

It must be emphasized that there is **no guarantee** that such a weight matrix can be found for an arbitrary set of pairs. As with the Hopfield network, extreme care should be used when selecting the appropriate representations of input patterns and targets.

We will now describe the iterative procedure in detail. Each iteration step t is composed of three parts:

– the generation of a new weight matrix $W^{(t)}$ from the set of pairs $\{X^{(t)}, Y^{(t)}\}$

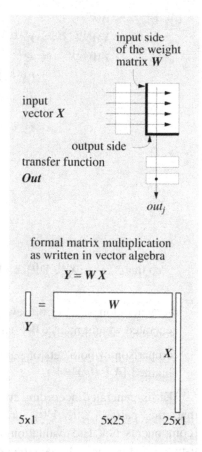

Figure 5-6:
a) Multiplication of a weight matrix by the input vector X in order to obtain the output vector Y.
(b) Multiplication of the transposed matrix W^{T} by the target vector Y in order to obtain a vector X on the "input" side of the matrix W.

$$\{X, Y\} \longrightarrow W^{(0)}$$
$$XW^{(0)} \longrightarrow Y^{(1)}$$
$$YW^{(0)\mathrm{T}} \longrightarrow X^{(1)}$$
$$\{X^{(1)}, Y^{(1)}\} \longrightarrow W^{(1)}$$
$$XW^{(1)} \longrightarrow Y^{(2)}$$
$$YW^{(1)\mathrm{T}} \longrightarrow X^{(2)}$$
$$\{X^{(2)}, Y^{(2)}\} \longrightarrow W^{(2)}$$
$$XW^{(2)} \longrightarrow Y^{(3)}$$
$$YW^{(2)\mathrm{T}} \longrightarrow X^{(3)}$$
$$\vdots \qquad\qquad \vdots$$
$$XW^{(t)} \longrightarrow Y^{(t+1)}$$
$$YW^{(t)\mathrm{T}} \longrightarrow X^{(t+1)}$$

so that $\qquad \{X^{(t)}, Y^{(t)}\} \equiv \{X, Y\}$ $\qquad\qquad$ (5.4)

- the generation of a set of new pairs $\{X^{(t+1)}, Y^{(t+1)}\}$ using the just generated weight matrix $W^{(t)}$ and the **original** set of pairs $\{X, Y\}$

- comparison of both sets of pairs: the original $\{X, Y\}$, with the last obtained $\{X^{(t+1)}, Y^{(t+1)}\}$

$W^{(t)}$ is generated according to Equation (5.2). It should be noted that the pairs $\{X^{(t)}, Y^{(t)}\}$ are represented as bipolar vectors (components ± 1). The evaluation of each output vector is made in the usual way: first, we calculate the net input *Net* (Equation (3.5)):

$$Net_j = \sum_{i=1}^{m} w_{ji} x_i \qquad\qquad (3.5)$$

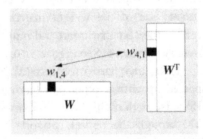

Figure 5-7: Transposed matrix has rows and columns interchanged.

and then apply the nonlinear transformation (see Equation (3.8)). In the case of a bipolar ABAM network, the nonlinear transformation is, clearly, the bipolar hard-limiter (Equation (2.25)) with the threshold value ϑ equal to zero:

$$\mathrm{hl}\,(Net_j) = \mathrm{sign}\,(Net_j) \qquad\qquad (5.5)$$

These equations are used either if we evaluate a new $Y^{(1)}$ from the input side with one of the input vectors X and the weight matrix $W^{(t)}$, or if we evaluate $X^{(1)}$ from the output side with one of the targets Y and the transposed matrix $W^{(t)\mathrm{T}}$.

5.5 An Example

As an example of the ABAM learning procedure, we will take a series of five images on a (5 x 5)-pixel grid. Because ABAM is a supervised learning technique, each of the five patterns to be learned requires a specific associated target.

The reader can go through this example by himself by downloading the program ABAM from the website for this book

http://www2.ccc.uni-erlangen.de/ANN-book/

This site also provides the datafiles for the patterns used in this example. See Appendix for further details.

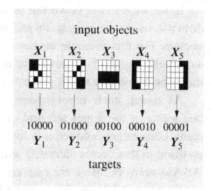

Figure 5-8: Inputs and targets used to illustrate the ABAM neural network. The input vectors $X_1, ..., X_5$ and the targets $Y_1, ..., Y_5$ are written in binary notation for simplicity.

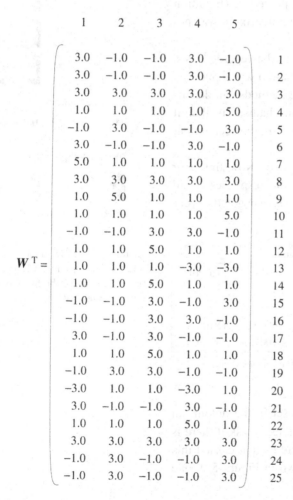

$$
W^{\mathrm{T}} = \begin{pmatrix}
3.0 & -1.0 & -1.0 & 3.0 & -1.0 \\
3.0 & -1.0 & -1.0 & 3.0 & -1.0 \\
3.0 & 3.0 & 3.0 & 3.0 & 3.0 \\
1.0 & 1.0 & 1.0 & 1.0 & 5.0 \\
-1.0 & 3.0 & -1.0 & -1.0 & 3.0 \\
3.0 & -1.0 & -1.0 & 3.0 & -1.0 \\
5.0 & 1.0 & 1.0 & 1.0 & 1.0 \\
3.0 & 3.0 & 3.0 & 3.0 & 3.0 \\
1.0 & 5.0 & 1.0 & 1.0 & 1.0 \\
1.0 & 1.0 & 1.0 & 1.0 & 5.0 \\
-1.0 & -1.0 & 3.0 & 3.0 & -1.0 \\
1.0 & 1.0 & 5.0 & 1.0 & 1.0 \\
1.0 & 1.0 & 1.0 & -3.0 & -3.0 \\
1.0 & 1.0 & 5.0 & 1.0 & 1.0 \\
-1.0 & -1.0 & 3.0 & -1.0 & 3.0 \\
-1.0 & -1.0 & 3.0 & 3.0 & -1.0 \\
3.0 & -1.0 & 3.0 & -1.0 & -1.0 \\
1.0 & 1.0 & 5.0 & 1.0 & 1.0 \\
-1.0 & 3.0 & 3.0 & -1.0 & -1.0 \\
-3.0 & 1.0 & 1.0 & -3.0 & 1.0 \\
3.0 & -1.0 & -1.0 & 3.0 & -1.0 \\
1.0 & 1.0 & 1.0 & 5.0 & 1.0 \\
3.0 & 3.0 & 3.0 & 3.0 & 3.0 \\
-1.0 & 3.0 & -1.0 & -1.0 & 3.0 \\
-1.0 & 3.0 & -1.0 & -1.0 & 3.0
\end{pmatrix}
\begin{matrix}
1 \\ 2 \\ 3 \\ 4 \\ 5 \\ 6 \\ 7 \\ 8 \\ 9 \\ 10 \\ 11 \\ 12 \\ 13 \\ 14 \\ 15 \\ 16 \\ 17 \\ 18 \\ 19 \\ 20 \\ 21 \\ 22 \\ 23 \\ 24 \\ 25
\end{matrix}
$$

Table 5-1: The (5 x 25) weight matrix **W** obtained from the patterns shown in Figure 5-8, using Equation (5.2) is shown as (25 x 5) W^{T}.

If we want to cut down the dimensions of the neural network, the dimensions of the targets should be smaller than those of the input vectors. For this example we have decided to use five-element vectors as targets. Figure 5-8 shows the inputs (X_1, X_2, X_3, X_4 and X_5) and their corresponding target patterns (Y_1, Y_2, Y_3, Y_4 and Y_5).

As usual, each input image is coded as a 25-element vector, beginning at the upper left and proceeding row by row to the lower right. Each target is coded as a five-element vector of 1's and 0's; each position in this vector identifies a different target. (Before entering the ABAM network, these are converted into bipolar coding.)

Equation (5.2) produces the (25 x 5)-element matrix shown in Table 5-1. Fortunately, in this case the learning procedure (Equation (5.4)) required no iteration, which means that the matrix W was not modified.

In applying the network, the output layer can be regarded as a switch with five positions, depending on which pattern was input.

Okay, but so what? Many other conventional methods can do this just as well. What is interesting about this mechanism is that it recognizes the input patterns (i.e., answers with the correct output signals) even if **corrupted** patterns are input.

Changing one pixel out of 25 represents a 4% error. Now, this error can occur at 25 different places; if we check the responses to all possible 1-pixel errors on each of the 5 images, we get the very interesting results given in Table 5-2.

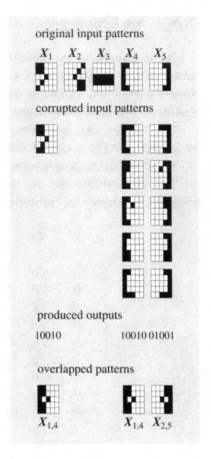

original input patterns

X_1 X_2 X_3 X_4 X_5

corrupted input patterns

produced outputs

10010   10010 01001

overlapped patterns

$X_{1,4}$ $X_{1,4}$ $X_{2,5}$

Figure 5-9: The eleven corrupted patterns (from among all 125 one-pixel errors) that caused the firing of two neurons instead of one.

pattern	target	actual output			
1	10000	10000	(24 times)	10010	(once)
2	01000	01000	(25 times)		
3	00100	00100	(25 times)		
4	00010	00010	(20 times)	10010	(5 times)
5	00001	00001	(20 times)	01001	(5 times)

Table 5-2:   Responses to all possible 1-pixel errors in the images shown in Figure 5-8.

Although in all 125 cases the correct bit is set in the output, an additional bit is set in eleven of them. So, we are led to ask if there is anything special about the eleven corrupted patterns that caused this additional output neuron to fire. These are shown in Figure 5-9: one has an error in X_1, five in X_4 and five in X_5.

(In the following, we will use the symbol $X_{i,j}$:

$$X_i \text{ OR } X_j = X_{i,j} \qquad (5.6)$$

to designate the overlap, or logical OR, of the patterns X_i and X_j.)

Look at the corrupted X_1, which activates the first and fourth bits in the target; it is very similar to $X_{1,4}$, which is the overlap of the two patterns X_1 and X_4 shown at the bottom of Figure 5-9; the difference between $X_{1,4}$ and the corrupted X_1 is only two pixels.

Interestingly, all five corrupted X_4's activate these same two bits in the output; the difference between $X_{1,4}$ and the five corrupted X_4's is either two or four pixels. And finally the five corrupted X_5's shown in the rightmost column of Figure 5-9 all activate the second and fifth bits in the output. The corresponding pattern overlap is shown at the bottom right of Figure 5-9; note its similarity to the five corrupted inputs. The reader can verify the outputs of the images corrupted by the 1-pixel errors by the use of program ABAM as described in Paragraph 5.6.

So, even when the ABAM network makes errors, they are reasonable ones. Encouraged by this, we can set up an additional experiment to show the power of the method. Ten different overlapping pairs (Figure 5-10) can be made from five different input patterns; let us construct the bipolar 25-element representations of these overlapped input patterns, input them to the matrix W and watch the 5-element output.

The combination $X_{1,2}$ is shown explicitly below:

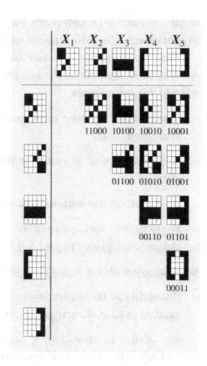

Figure 5-10: The ten possible combinations of two images from five different patterns. The corresponding output vector is shown below each overlapped pattern.

$$
\begin{aligned}
X_{1,2}^{binary} &= (1, 1, 0, 0, 1, & X_{1,2}^{bipolar} &= (\ 1,\ \ 1, -1, -1,\ \ 1, \\
&\quad\ \ 1, 1, 0, 1, 0, & &\quad\ \ \ 1,\ \ 1, -1,\ \ 1, -1, \\
&\quad\ \ 0, 0, 1, 0, 0, & &\quad -1, -1,\ \ 1, -1, -1, \\
&\quad\ \ 0, 1, 0, 1, 1, & &\quad -1,\ \ 1, -1,\ \ 1,\ \ 1, \\
&\quad\ \ 1, 0, 0, 1, 1) & &\quad\ \ 1, -1, -1,\ \ 1,\ \ 1)
\end{aligned}
$$

$$(5.7)$$

In all ten cases the resulting output is exactly what we expected: each input pattern triggered the two neurons associated with its component images. In only one of the ten cases was a third, unrelated neuron fired; this is the combination of the third and fifth object, i.e., $X_{3,5}$. The additional bit which was triggered is bit No. 2. A three-bit output corresponds to the overlap (logical OR) of three of the basic inputs; Figure 5-11 compares these with the actual input pattern that triggers them; the difference is only one pixel.

This example, which can be exercised by the use of program ABAM as described in Paragraph 5.6, shows four capabilities of the ABAM neural network:

− it associates the five input patterns with the five targets, even for corrupted inputs;

− it produces the original input vectors from the corresponding outputs;

− it recognizes the components of two ORed patterns; and

− it produces the negative (complemented) target when negative patterns are input (Figure 5-12).

By analogy with the biological brain, these can be interpreted as:

− the ability to recognize learned images by activating the appropriate neuron even if the trigger image is distorted or incomplete;

− the ability to associate a complex pattern with a single neuron (which is fired when that pattern is recognized);

− the ability to identify the parts of a complex pattern, and

− understanding the concept of a "negative" image.

5.6   Significance of the Example

Besides these demonstrated features of the simple ABAM network, there is an even more significant lesson to be drawn from this experiment. We have seen, on a small scale, what we hope to obtain from neural networks on a large scale. The effect we are talking about can be linked to the warning (given in the last paragraph of Section 5.1) that "hidden" relationships may exist among seemingly unrelated patterns.

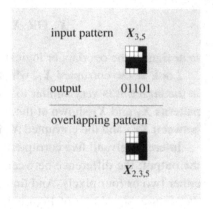

Figure 5-11: The comparison of the two-image input pattern $X_{3,5}$ that fires three bits, 01101, with the corresponding three-image pattern.

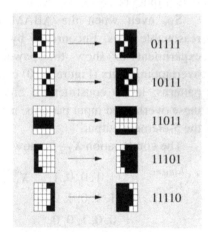

Figure 5-12: The set of five patterns (left) and their negatives (right). Negative (complemented) inputs produce the complements of the corresponding outputs.

We have taught the system to associate several pairs of obviously unrelated patterns, and nothing more. But the system has learned two **general** concepts that we never taught it: analyzing a sum into its parts, and complementing a result to match a complemented input.

The result of this experiment and the generalizations we have drawn from it, result from the (very careful) choice of patterns; there is little chance that an arbitrary selection of patterns will produce an ABAM net intelligent enough to resolve two ORed patterns. However, it is worth mentioning that ABAM learning is at least sometimes capable of generalization.

In spite of all such precautions, we must admit that this small neural network (an "artificial brain", if you like) consisting of only five neurons with 25 weights each (125 synapses altogether) has done a remarkably good job.

There are, of course, many questions about the performance of such nets when applied to larger problems (more patterns or more pairs of patterns), and we must **always** keep in mind that a net's response is always a result of the entire assembly of patterns. A neural network is in this sense, **holographic**.

Intuition tells us that similar patterns should give similar responses. In most cases this will be true. However, as mentioned before, the way we think about "similarity" between the samples does not necessarily correspond to the similarity found by the network, leading to unexpected results.

Let us go back to Figure 5-9. We might expect that if a given error in pattern X_1 causes the firing of the two neurons 1 and 4, an analogous error in X_2 should also cause the firing of two neurons. Figure 5-13 shows the error in X_2 analogous to the error in X_1. However, this corrupted signal did not produce the two signal output of firing neurons 2 and 5, as might have been expected. Why didn't this happen?

The reason for this is that we've confused "similarity" (a vacuous, undefined term) with "symmetry".

The patterns X_1 and X_2, are related by symmetry (specifically, an inversion center, denoted C_i; that is, taking every point in X_i and moving it to the opposite side of the center point: produces X_2); these are different figures, and can not be superimposed. X_4 and X_5 are related in the same way. The OR of X_1 and X_2, when subjected to this operation, produces a figure which is congruent to itself. Thus, we say that the ORed figure contains an inversion center, while the individual figures, X_1, and X_2, do not.

However, the symmetry relations between input and output patterns are products of the entire ABAM matrix, which, in turn, results from all input patterns. In our example, since X_1 and X_2 together possess C_i, they generate a C_i in the network (i.e., they make its **responses** symmetrical); X_4 and X_5 do the same. But X_3 is unique in that it neither possesses the C_i symmetry, nor can combine with another member of the set to produce it; hence, X_3 prevents the net from having this symmetry.

This is also why the overlapping patterns $X_{1,3}$ and $X_{2,3}$ bear no similarity at all in spite of the fact that they were produced by overlapping the object X_3 with two very "similar" (actually "symmetrical") objects X_1 and X_2. On the other hand, the patterns $X_{1,4}$ and $X_{2,5}$ appear similar to the eye because they are related by an inversion center (or plane of symmetry).

Thus, the source of similar effects with corrupted patterns depends on the actual symmetry of the original patterns and on the symmetry of the composites (Figure 5-14).

In summary, our expectation that X_2 would behave like X_1 was based on their symmetrical relationship; we did not take into account the lack of symmetry (due to X_3's contribution to the weights) of the network itself.

Never trust your impressions about the relation between the objects and their apparent symmetry; find someone who understands symmetry (spectroscopist or a crystallographer is your best bet).

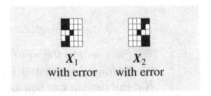

Figure 5-13: The images X_1 and X_2 each with a one-pixel error.

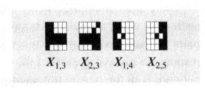

Figure 5-14: The overlapping patterns $X_{1,3}$ and $X_{2,3}$ are not symmetric; $X_{1,4}$ and $X_{2,5}$ are symmetrically related.

5.7   Essentials

- hetero-association is the main goal of the ABAM network

- the number of output data is usually smaller than the number of input data

- learning, i.e. evaluation of weights, is supervised and achieved by bidirectional iteration, in which the input signals flow towards the output side, forming output data, while the output data are returned in the opposite direction towards the input side forming new inputs

- corrupted input patterns can retrieve the corrected associated patterns from the network

- in certain circumstances the ABAM network can generalize what it has learned

- **weights**

$$w_{ji}^{(0)} = \sum_{s=1}^{p} x_{si} y_{sj} \qquad (5.2)$$

- **iterative learning**

  start:

$$\{X, Y\} \longrightarrow W^{(0)}$$
$$XW^{(0)} \longrightarrow Y^{(1)}$$
$$YW^{(0)\mathrm{T}} \longrightarrow X^{(1)}$$
$$\{X^{(1)}, Y^{(1)}\} \longrightarrow W^{(1)}$$
$$XW^{(1)} \longrightarrow Y^{(2)}$$
$$YW^{(1)\mathrm{T}} \longrightarrow X^{(2)}$$
$$\vdots \qquad \vdots$$
$$XW^{(t)} \longrightarrow Y^{(t+1)}$$
$$YW^{(t)\mathrm{T}} \longrightarrow X^{(t+1)}$$

so that $\qquad \{X^{(t)}, Y^{(t)}\} \equiv \{X, Y\} \qquad (5.4)$

5.8   References and Suggested Readings

5-1. B. Kosko, "Adaptive Bidirectional Associative Memories", *Appl. Optics* **26** (1987) 4947 – 4960.

5-2. B. Kosko, "Constructing an Associative Memory", *Byte*, September 1987, 137 – 144.

5-3. B. Kosko, "Bidirectional Associative Memories", *IEEE Trans. Syst., Man and Cyber.* **18** (1988) 49 – 60.

5-4. B. Kosko, *Neural Networks and Fuzzy Systems*, Prentice-Hall, Englewood Cliffs, NJ, USA, 1992.

5-5. M. Otto and U. Hörchner, "Application of Fuzzy Neural Networks to Spectrum Identification", in *Software Development in Chemistry 4*, Ed.: J. Gasteiger, Springer-Verlag, Berlin, FRG, 1990, pp. 377 – 384.

6 Kohonen Network

learning objectives

- concept of "topology"

- "mapping" a dataset from a space of high dimension to a space of lower dimension

- how a neural network (the Kohonen net) can do mapping

- how "unsupervised" learning is used when you don't have a set of "correct" answers

- how topology is maintained in mapping, as in representing the relationships between continents on a globe, on a planar map

6.1 General

When we think of "data", we ordinarily think of values, magnitudes, signs, etc.; this is an **algebraic** view of a dataset. In addition, there is an **information science** view, which focuses on the relationships among data items. These relationships may exist entirely within the given dataset, or may involve data in other datasets as well.

Complex information is always incomplete, to some extent. In fact, we may **choose** to reduce the dataset, as when digital images are compressed to reduce storage requirements. In dealing with missing data, we must not forget their possible relationships. When we focus on the relationships among data, rather than their algebraic attributes, we say that we are dealing with the *topology* of the information. Figure 6-1 contrasts the concept of topology with the concept of numerical values.

Efficiency is obviously crucial in handling large amounts of information; for a given level of hardware competence, efficiency is generally achieved by compressing the data. *Compression* may be thought of as a process of *mapping* a multidimensional input into an output space of significantly smaller dimension (Figure 6-2).

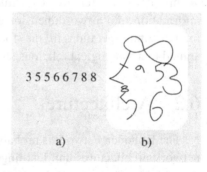

Figure 6-1: A completely new level of information can be revealed by taking into account the topology between the numerals (b) instead of considering their numerical values (a) only.

Obviously, we want a maximum of compression and a minimum of information loss; this is one of the basic problems in information and computer science in general, and in artificial intelligence and neural network research in particular. There are, of course, many questions involved in this problem. For example: how can we make the trade-off between reduction and preservation when the information has not yet been processed? Can we map information onto a two-dimensional array of neurons? How can such a mapping be performed or learned? How can the retained knowledge be retrieved from the mapped information?

Teuvo Kohonen has introduced the very interesting concept of *self-organized topological feature maps*, which are maps that preserve the topology of a multidimensional representation within the new one- or two-dimensional array of neurons (Reference 6-1). We will discuss Kohonen's approach to neural networks, which attempts to preserve the topology of the input information while mapping it into the neural array.

The concept of topology (or better still, the concept of "preservation of topology") has become the essential feature of the Kohonen approach in neural network research. As he has pointed out in his book (Reference 6-2) the mapping of multidimensional information into a two-dimensional plane of neurons "seems to be a fundamental operation in the formation of abstractions too!" He argued that topological relations should be preserved in this mapping.

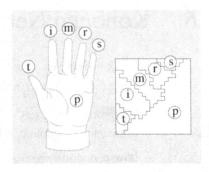

Figure 6-2: The topology of the five fingers (t, i, m, r, s) and the palm (p) (above) is preserved or "mapped" onto a square plane of (13 x 13) neurons (below) using the Kohonen algorithm, which will be explained later.

6.2   Architecture

The Kohonen network is probably the closest of all artificial neural network architectures and learning schemes to the biological neuron network. As a rule, the Kohonen type of network is based on a single layer of neurons arranged in a one-dimensional array or in a two-dimensional plane having a well defined topology (Figure 6-3 and Figure 6-4).

A defined topology means that each neuron has a defined number of neurons as nearest neighbors, second-nearest neighbors, etc. To be in accordance with the previous cases the neurons for the Kohonen layer are visualised by circles (Figure 6-3). However, from the didactic as well as from the mathematical point of view the visualisation of neurons in the Kohonen network is more natural in the form of columns (Figure 6-4, above). The advantage of a scheme with

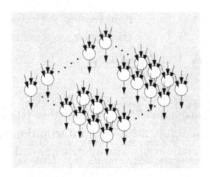

Figure 6-3: Two-dimensional layout of the Kohonen neural network.

column-like neurons is its clear presentation of the weights in individual neurons and how the weights handling the same input variable are connected together in the network. It can easily be seen that weights affected by each variable are lying on a single and well-defined level of weights. Each set of weights affected either by the first, the second, or by the third input variable are forming a separate level of weights. The levels of weights are superimposed onto each other in a one-to-one-correspondence, hence the weights of each neuron are obtained by looking at the weights in all levels that are exactly aligned in a vertical column. There are as many weight levels in each Kohonen network as there are input variables describing the objects for which the network is designed. Because the input vector consists of three variables, there are also three levels of weights in the case shown on Figure 6-4.

The *neighborhood* of a neuron is usually arranged either in squares or in hexagons, which means that each neuron has either four or six nearest neighbors. The concept of "nearest neighbors" needs some elaboration, especially for those who have studied crystals or coordination chemistry. For example, the square neighborhood is often regarded as having eight and not four nearest neighbors (Figure 6-5). Certainly, the corner points in the rectangular grid are further away from the central point compared to the actual first neighbors; but we are interested in the topology, that is, the **connections** and not the actual **distances**. In the Kohonen conception of neural networks, the signal similarity is related to the spatial (topological) relations among the neurons in the network.

The Kohonen concept tries to map the input so that similar signals excite neurons that are very close together (in terms of spatial distance); this similarity-to-distance relationship should be generalized to include the entire range of similarity relations between different signals as well. Kohonen learning represents an attempt to fit the signal space onto the neural network by a kind of "smoothing" or "reshaping" procedure.

> The aim of Kohonen learning is to map similar signals to similar neuron positions.

A practical point: we need to ensure that each neuron in the net has the same number of first-, second-, etc. neighbors (Figure 6-6); but the

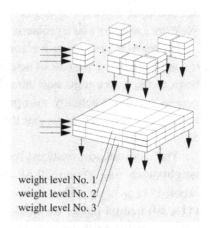

weight level No. 1
weight level No. 2
weight level No. 3

Figure 6-4: Neurons can as well be drawn as little boxes (above). The three inputs are coming from the side to all neurons at the same time. Schematically all neurons (small boxes) can be "packed" into a larger box or "brick" in which neurons are represented as columns (below).

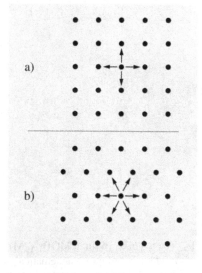

Figure 6-5: Square (a) and hexagonal (b) layout of neighbors.

net is a finite object, with edges. What about the neurons at the edges? Well, this wouldn't be a problem if the net were, say, a torus.

Figure 6-7 shows how a plane can be wrapped into a toroid: first, the upper edge (or the row of neurons located there) wraps around and links to the lower edge, and then the left edge joins the right one. Of course, we don't actually manipulate the net; we simply convert the indices of edge neurons so that they appear to wrap around (Figure 6-8).[1]

The coordinates (indices) of the neurons located in the k-th neighboring ring (Figure 6-6) around a particular neuron will be labeled $(x_{m1}, y_{m2})_k$. The following example shows how, for an (11×20)-neuron plane, we would calculate the coordinate pairs of all neurons in the forth neighboring ring around neuron $(3, 2)_c$ ("c" is used to designate the central neuron:

(In applications, the following algorithm for finding the neighbors at a certain topological distance from the center may be executed more efficiently because the entire neighborhood can be found by going through these loops.)

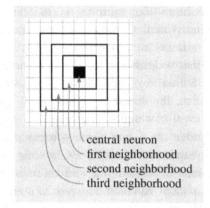

central neuron
first neighborhood
second neighborhood
third neighborhood

Figure 6-6: The square neighborhood is usually regarded as having 8, 16, 24, etc. neighbors in concentric neighborhoods.

1. The *modulus* (written MOD(A, M) in some programming languages, or "*a mod m*" in ordinary mathematical notation) is a function of two parameters *a* and *m*. It represents the remainder after division of the first parameter by the second. For example, 8 *mod* 3 (read "eight modulo three") = 2. Usually, the length or width of the neural network layout appears as the parameter *m*. Using the modulo function, any index no matter how large can be converted into what the position would be if the corresponding axis wrapped around to form a loop.

```
p = 11
q = 20
x =  3
y =  2
r =  4
for m1 = −r to r by 1
     for m2 = −r to r by 1
          xm1 = mod(x + m1 + p − 1, p) + 1
          ym2 = mod(y + m2 + q − 1, q) + 1
          if (m1 = r  or m2 = r) then
               the element (xm1, ym2) is in the r-th ring
          else
               the element is in the neighborhood ring
               determined by max(xm1, ym2)*
          end if
     next m2
next m1


*max(a, b) is the larger of the values a and b
```

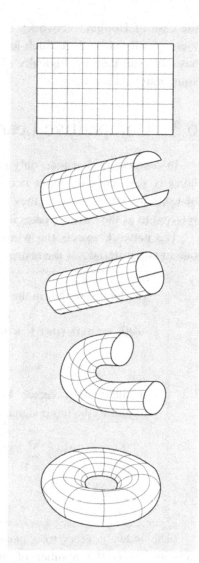

Figure 6-7: Wrapping a two-dimensional plane into a toroid.

In a Kohonen network, as always, we can speak of two layers: the input and the output layers. Only one layer is active. This active layer is usually arranged as a two-dimensional grid or "brick" of neurons (Figure 6-3 and Figure 6-4), but can also be arranged as a linear array (Figure 6-9). The toroidal wrapping in one-dimension is even simpler compared to the two-dimensional wrapping of neurons (Figure 6-7): the linear array becomes a circle.

Both mentioned layouts of neurons in a Kohonen layer – the one-dimensional (array) and the two-dimensional (plane) – are not the only possible solutions in this type of neural network application. Clearly, the layouts can be generalised to three- and higher-dimensional structures of neurons provided that the results from such complex networks can explain or solve the problem investigated in a more effective or plausible way.

All neurons in the active layer obtain the same multidimensional input. The most characteristic feature of the active layer in a Kohonen network is that it implements only a *local feedback*; that is, the output of each neuron is not connected to all other neurons in the plane (as in the case of Hopfield network), but only to a small number that are topologically close to it. Such local feedback of possible corrections has the result that topologically close neurons behave similarly when similar signals are input.

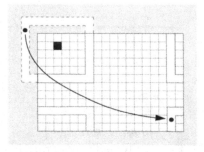

Figure 6-8: Elements of the fourth neighboring ring for the third neuron in the second row; ring segments found by "wrapping" (modulus function) are unhatched.

6.3   Competitive Learning

In *competitive learning*, only **one** neuron from those in the active layer is selected after input occurs; no matter how close the other neurons are to this best one, they are left out of that cycle. (This is also referred to as the "winner takes it all" method.)

The network selects the winner "*c*" (for "central") according to one of two criteria; *c* is the neuron having either:

The largest output in the entire network:

$$out_c \leftarrow \max(out_j) = \max\left(\sum_{i=1}^{m} w_{ji} x_{si}\right) \qquad (6.1)$$

$$j = 1, 2, ..., n$$

or the weight vector $W_j (w_{j1}, w_{j2}, ..., w_{jm})$ most similar to the input signal $X_s (x_{s1}, x_{s2}, ..., x_{sm})$:

$$out_c \leftarrow \min\left\{\sum_{i=1}^{m} (x_{si} - w_{ji})^2\right\} \qquad (6.2)$$

$$j = 1, 2, ..., n$$

(The index *j* refers to a particular neuron; *n* is the number of neurons; *m* is the number of weights per neuron; *s* identifies a particular input.) See also Section 7.4.

Index *j* that specifies the actual neuron in the Kohonen layer depends on the layout of the network. There is no problem if the neurons are arranged in a one-dimensional array consisting of *n* neurons. In such a case, the index *j* simply runs from 1 to *n*, while the closest neighbors to the selected neuron *c*, are the neurons with indices $j = c - 1$ and $j = c + 1$, the neurons of the second neighborhood have indices $j = c - 2$ and $j = c + 2$, and so on. In the case of a two-dimensional layout of the Kohonen network, index *j* has to be understood as describing the location of the particular neuron in a

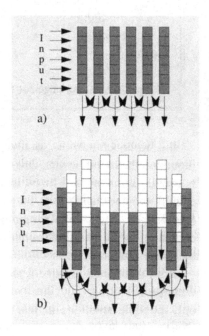

Figure 6-9: Kohonen network represented as a linear array of neurons (a). The local feedback connections are clearly seen: only the two neurons that are closest to each other receive the feedback when it occurs. The toroidal condition makes the array a circle (b).

two-dimensional plane. Usually, a two-dimensional location of neurons in the network is described by two indices: one describing the abscissa and the other one the ordinate axis of the position of the neuron. This means that the neuron j in a two-dimensional Kohonen layout having n neurons (ordered into an $n1 \times n1 = n$ network with $n1$ rows and $n1$ columns) can be found in the

$m1 = \lceil (j-1)/(n2) \rceil + 1$ column (x coordinate) runs from 1 to $n2$
    and in the

$m2 = j - (m1 n2)$ row (y coordinate) runs from 1 to $n1$

The mathematical expression $\lceil a \rceil$ means the largest integer not exceeding the value of a.

From now on, in all equations where index j occurs and whenever the two-dimensional Kohonen network architecture is applied (which is true in all examples) the index j will actually define the position of the j-th neuron in the $m1$-th column and $m2$-th row within the Kohonen network having $n = n1 \times n1$ neurons.

It must be added that rectangular neuron layouts with $n1 \neq n2$ are seldom used. In most cases the Kohonen layers are quadratic, i.e., $n1 = n2$. The quadratic layout minimizes the distortion of the 2D projection space that always occurs when projection from the multi-dimensional space of input vectors is applied.

After finding the neuron c, that best satisfies the selected criterion, its weights w_{ci} are corrected to make its response larger and/or closer to the desired one. This means that if a certain signal x_i coming to the weight w_{ci} has produced too large an output, the weight should be diminished, and vice versa.

The weights w_{ji} of neighboring neurons must be corrected as well. These corrections are usually scaled down, depending on the distance from c; for this reason, the scaling function is called a topology dependent function:

$$a(\cdot) = a(d_c - d_j) \qquad (6.3)$$

where $d_c - d_j$ is the *topological* distance between the central neuron c and the current neuron j, while the extent of the stimulation depends on the function $a(\cdot)$. Figure 6-10 shows some of the forms that this function can take; besides decreasing with increasing d_c, it decreases with each iteration cycle of the Kohonen learning process (see box on the opposite page explaining the algorithm). In Kohonen learning one can distinguish two different cases. The first one occurs when the number of objects is so large that each object X_s enters the Kohonen network only once and probably many more do not enter the learning procedure at all, while in the second case the number of objects for training the network is small, hence, it is necessary to input the entire set of objects again and again into the network, before it is properly trained.

In real world applications the second case occurs much more often. In order to describe the number of training cycles necessary for handling all objects by the network exactly once, the term called "one epoch" of training has been defined. Therefore, the duration of training is usually expressed in terms of epochs, meaning the number of times all objects have been processed by the network.

In view of this explanation and assuming the network is improving during the learning procedure Equation (6.3) is multiplied by another monotonically decreasing function $\eta(t)$:

$$f = \eta(t)\, a\, (d_c - d_j) \tag{6.4}$$

where t is the number of objects entered into the training process (if the number of objects is very large) or the number of epochs. The parameter t can easily be associated with time, since the time used for training is proportional to both - to the number of objects entering the network or to the number of epochs. $\eta(t)$ can be expressed as:

$$\eta(t) = (a_{max} - a_{min})\frac{t_{max} - t}{t_{max} - 1} + a_{min} \tag{6.5}$$

where t_{max} is either the total number of objects that will be input into the network until learning is completed or the maximum number of epochs predefined at the beginning of training. The two constants a_{max} and a_{min} define the upper and lower limit between which the correction $\eta(t)$ is decreasing from the beginning to the end of training.

Due to its destimulation of the border of the selected neighborhood, the "Mexican hat" function (Figure 6-10c)) enhances the "contrast" on the borders of the developing regions in the output plane. This makes it very useful, but if too much contrast is applied "empty" spaces ("no man's land") can develop in the map on the borders between the categories.

The size of a neighborhood for the scaling function need not be permanent; it may well be changed during the learning period. Usually, it shrinks, which means that fewer neurons will have their

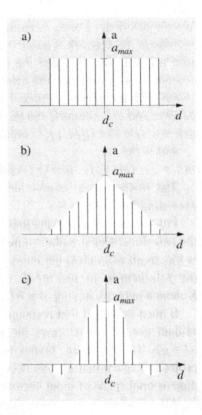

Figure 6-10: Typical functions for scaling corrections on neighbor-weights: constant (a), triangular (b), and Mexican hat (c).

In summary, the algorithm for one cycle of Kohonen learning is as follows:

- an *m*-dimensional object X_s enters the network;

- the responses of all neurons (each having *m* weights) are calculated;

- the position *c* is found for the neuron whose output is largest or most similar to the input;

- the weights of neuron *c* are corrected to improve its response for the same input *X* on the next cycle;

- the weights of all neurons in the (arbitrarily defined) neighborhood of the *c*-th neuron are corrected by an amount that decreases with increasing topological distance from *c*;

- the next *m*-variate object X_s is input and the process repeated.

weights corrected as the process goes on. Additionally, the maximum value of the scaling constant can be lowered.

The corrections of the weights w_{ji} of the *j*-th neuron lying within the region defined by the function f depend on the criterion used to select the central neuron *c*. We will describe in detail how to correct weights for one of these (Equation (6.2)):

$$w_{ji}^{(new)} = w_{ji}^{(old)} + \eta(t)\, a\, (d_c - d_j)\left(x_i - w_{ji}^{(old)} \right) \qquad (6.6)$$

(Here, x_i is a component of the input X_s; the central neuron is designated *c*, and the one being corrected is *j*; a particular weight of neuron *j* (and a particular input) is designated by *i*; *t* is (related to) which iteration cycle this is.)

Whether the difference $x_i - w_{ji}^{(old)}$ is positive or negative, i.e. whether x_i is greater or smaller than the weight $w_{ji}^{(old)}$, $w_{ji}^{(new)}$ will be closer to x_i than $w_{ji}^{(old)}$ was.

The correction function for the maximum signal criterion (Equation (6.1)) is evaluated similarly.

$$w_{ji}^{(new)} = w_{ji}^{(old)} + \eta(t)\, a\, (d_c - d_j)\left(1 - x_i w_{ji}^{(old)} \right) \qquad (6.7)$$

After the corrections have been made using Equation (6.6) or (6.7), the weights should be normalized to a constant value, usually 1:

$$\sqrt{\sum_{i=1}^{m} w_{ji}^2} = 1 \qquad (6.8)$$

Because of the specific architecture and learning algorithm of Kohonen networks, the outputs do not play as significant a **quantitative** role as in other networks. If the only significance of the output is to locate (topologically) the neuron with the largest output, then the actual magnitude of the output does not matter very much. Generally, our only concern is to keep the outputs within given limits in order to preserve the resemblance to actual biological neurons.

(In the case of criterion (6.2) and the associated correction (6.6), normalization does not improve the quality of the results.)

If the quantitative size of the output has little or no influence on the performance of the net, then normalizing the weights only disturbs the corrections. Besides, since the weights are corrected directly by comparison with the input signals (which presumably are normalized, or at least scaled to some reasonable values), the weights will be corrected to match them, and so will end up adjusted to normalized values anyway.

In any case, some kind of precaution must be taken to prevent the network output from "exploding". In our experience, providing an initial random distribution of weights (within the interval -0.1 to $+0.1$ or $-1/m$ to $+1/m$, where m is the number of weights) and scaling the input signals (and hence, the outputs) between -1 and $+1$ offers sufficient guarantee.

It must be clear from all of this that scaling or normalizing inputs must be performed very carefully.

> In most cases, it should be the nature of the problem rather than the method of solving it that dictates the criteria for normalizing (scaling and/or other transformations) of the input variables.

Though at first scaling the input may seem to be harmless at worst, this is not necessarily so. Improper scaling, especially across different variables, can change their internal relations and strongly influence the final results. Before applying normalization of any kind, check thoroughly that the transformed variables will still adequately describe your problem.

Always try the simplest transformations first, for example, simply dividing all the variables by a value approximately equal to the total system output; something like this may do the job quite as well as more complex procedures.

> Only one thing is more important than a complete understanding of the method you are using for handling data – a thorough knowledge and understanding of your data!

6.4 Mapping from Three to Two Dimensions

To provide you with a feeling for mapping from a higher to a lower dimension (what it looks like and how it can be done), we will work out a very simple example in detail: topological mapping will be used to transfer the entire surface of a three-dimensional sphere onto a square, planar (15 x 15) matrix. The problem is schematically shown in Figure 6-11.

A three-dimensional sphere of radius 1 is drawn around the coordinate system and divided into 8 spherical triangles (1 to 4 in the upper half and 5 to 8 in the lower half of the globe). 2000 points are generated randomly on the sphere's surface (approximately 250 on each triangular area) and labeled according to the eight possible combinations of coordinate signs (Table 6-1).

The Kohonen network used for this application is composed of three neurons in the input layer (one for each coordinate) and 225 neurons in the active output layer, arranged in a (15 x 15) matrix. Each of the three input neurons is connected to all 225 neurons on the Kohonen layer, which means that altogether 675 weights have to be trained for proper mapping. No bias weights are used.

Because the Kohonen network is square, the selection of square neighborhood rings (Figures 6-4 and 6-5) is the most natural choice. At the beginning, a neighborhood comprises seven rings of neighbors around the selected (central) neuron, which ensures that the neighborhoods are adjacent (no "no man's land")[1]. The number of

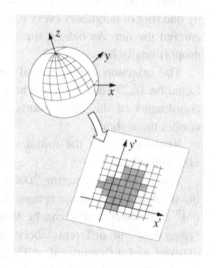

Figure 6-11: Topological mapping from a three- into a two-dimensional space.

	coordinates		label
x	*y*	*z*	
+	+	+	1
–	+	+	2
–	–	+	3
+	–	+	4
+	+	–	5
–	+	–	6
–	–	–	7
+	–	–	8

Table 6-1:   Labeling eight triangular areas on the sphere.

corrections declines linearly from the inner to the outer rings of the neighborhood. The maximum correction at the central neuron, $a^{(0)}$ (Equation (6.4)) is 0.3, while in the 7th layer it is 0.5% of the central value (= 0.005 x 0.3 = 0.0015). No decrease of the central value $a^{(0)}$ is made during the training; in other words, $\eta(t) = constant$.

During training, the outer border of the neighborhood is reduced by one ring of neighbors every time a fifth (400) of the total points has entered the net. As before, the more distant neighbors are corrected proportionally less.

The selection of the central neuron is made by the criterion of Equation (6.2) (the neuron whose weights are most similar to the coordinates of the input points); hence, Equation (6.6) is used to correct the weights.

Remember, that the initial weights are random values between –0.1 and +0.1.

The result after entering 2000 points is shown in Figure 6-12. All the triangle labels on the sphere cluster into irregular areas of the net ("4" is shaded). These can be thought of as groups of neurons that "specialize" in different labels; in other words, the weights have evolved and differentiated so that, for example, all points from area number "3" produce their best matches in the area numbered "3" in the planar pattern in Figure 6-12.

You can get a feeling for how the weights change in the Kohonen network by referring to Figure 6-13. Let's say that a certain neuron has weights of (0.5, –0.7, 0.9); after the signal from one point (input:

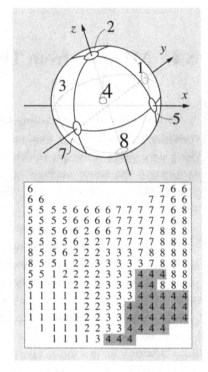

Figure 6-12: Mapping a sphere into two dimensions (see text).

---

1. Each ring is two neurons wider than the previous one. Hence, the width of *n* neighborhoods is 2*n*+1, in this case 2 x 7 + 1 = 15, the (topological) width of the network.

(0.8, −0.5, 0.33)) is received, the weight corrections are calculated assuming that this is a central neuron and that $a^{(0)}$ (given by (6.1) with $d_c − d_j = 0$) equals 0.2. The changes of the weights are calculated from the differences (input minus old weights: (0.3, 0.2, −0.57)) and the scaling value $a^{(0)}$ to give: 0.06, 0.04, and −0.11, respectively; hence, the new weights are (0.56, −0.66, 0.79). It is obvious that all three weights are closer to the input values, and that the changes are proportional to the error. The largest correction is calculated for the third weight, where the error is the largest of the three.

The *Kohonen map* shown in Figure 6-14 indicates that the topology of the surface of the sphere is preserved in the planar map. Figure 6-12 clearly shows that area 4 is adjacent to 1, 3 and 8, and shares corners with 5, 7, 2; it has no contact with 6, which is on the opposite "side" of the sphere. These features also appear in Figure 6-14 (if you remember to wrap it so that opposite edges come together).

The circles in Figure 6-14 correspond to the points where the axes penetrate the globe (Figure 6-13); the circle at the bottom of Figure 6-14 is the mapping of the north pole.

We can present our results more clearly by actually extending the (15 × 15) network with redundant neurons (instead of "wrapping" the array 1, 2, 3, ..., 15, we write it out as 1, 2, 3, ..., 15, 1, 2, 3, ..., 15, etc.). That is, we will use the original (15 × 15) neural network (Figure 6-14) as a tile to cover an area nine times larger than before (Figure 6-15). (Close examination of Figure 6-15 shows that the pattern repeats every 15 units.)

The topological relations (links) among the numbered areas can be expressed as a connection table or as a connectivity matrix. Both forms are shown below in Table 6-2. (The row or column for area 4 shows that it is adjacent to 1, 3 and 8.)

By comparing the topologies found on the sphere and on the (15 × 15) neural network, we can see that they are identical, whichever presentation form we use.

In addition to the longer borders between the numbered areas, shorter borders (as short as one or two neurons) can be observed as well. A closer look at the regions where such short borders are found shows that at these locations **four** areas converge in all cases. Such areas correspond to the places where the coordinate axes enter or leave the sphere (white circles, Figures 6-11 and 6-14). As expected, there are six such areas in the obtained map.

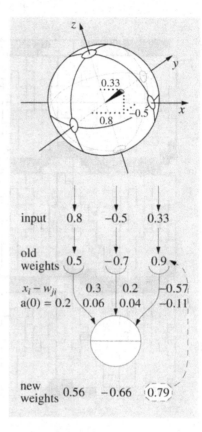

Figure 6-13: Correction of weights in the Kohonen network (cf. Equation (6.6)).

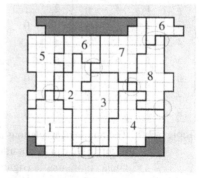

Figure 6-14: Topology of the Kohonen map of the sphere, indicating the border lines and common points of the spherical triangles.

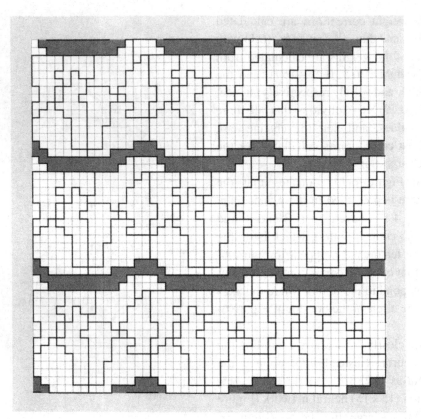

Figure 6-15: The plane can be tiled by repetition of patterns obtained on the toroid. In a periodic pattern the most important points can be seen more easily.

area	neighboring area no.			area	1	2	3	4	5	6	7	8
1	2	4	5	1	0	1	0	1	1	0	0	0
2	1	3	6	2	1	0	1	0	0	1	0	0
3	2	4	7	3	0	1	0	1	0	0	1	0
4	1	3	8	4	1	0	1	0	0	0	0	1
5	1	6	8	5	1	0	0	0	0	1	0	1
6	2	5	7	6	0	1	0	0	1	0	1	0
7	3	6	8	7	0	0	1	0	0	1	0	1
8	4	5	7	8	0	0	0	1	1	0	1	0

Table 6-2:   The topological relations among the numbered areas shown in Figure 6-12 can be recorded as a connection table (left) or as a connectivity matrix (right).

Another notable feature of Figure 6-12 is the "empty region", which separates the vertex where areas 1, 2, 3 and 4 come together (the "north pole" of Figure 6-12) from the areas numbered 5, 6, 7 and

8, which are in the "southern hemisphere". This same region can be thought of as separating the south pole from the triangles in the northern hemisphere; that is, there is only one such "empty region", a fact which is evident from the extended (tiled) representation of Figure 6-15 (see Reference 6-9).

The actual results of the Kohonen topology preserving mapping procedure always depend on the initial choice of weights (chosen randomly in the region from -0.1 to 0.1 for this experiment), the selected neighborhood, the correction function $a(d_c - d_j)$, and the initial value of correction $a^{(0)}$. However, the topology of the obtained map (as expressed above) will generally be preserved, i.e., the procedure is very robust or stable regarding small changes in the procedure.

6.5   Another Example

The mapping of two- and three-dimensional objects into a two-dimensional plane of neurons is very instructive, because you can visualize the results. However, in physics, chemistry, technology, sociology, economics, and other disciplines, we have data sets composed of more than three variables.

For example, wine producers analyze each wine for at least ten components, and they have thousands of sets of data. Engineers monitor a given technological process for even larger numbers of parameters (temperature, pressure, flow-rate, etc.).

All these sets of data can be regarded as m-tuplets or m-element vectors, each component in a given set being the value of one variable describing a certain wine, say, or a technological process. Such data can be analyzed in many ways and by many methods. If large datasets must be processed, Kohonen mapping can turn out to be quite effective for at least the prescreening of data. For more information on mapping, see Section 9.4.

In Part IV of this book, where some examples are worked out in detail, a set of a few hundred chemical bonds is described using seven different variables: electronic, energy, etc. (Chapter 11); that is, each chemical bond in this collection is represented in a seven-dimensional space.

There are quite a number of things that chemists would like to know about these chemical bonds, such as whether or under what conditions the bond is breaking.

This is a typical problem for complex statistical analysis, cluster analysis, or any other standard data processing method. Unfortunately, most of these standard methods are too inefficient for handling tens of thousands of datasets. Kohonen's competitive learning method offers a very simple and efficient (though maybe not well understood) method that can at least shed some light on the problem.

For the next example, we have selected 200 chemical bonds (seven-element vectors). For 94 of these bonds we know that they either break very easily (58) under specified conditions or with great difficulty or not at all (36), and the rest are bonds for which we do not have the relevant information. The questions we would like to answer are:

1) Can the easily breakable bonds and those that are hard to break be separated by this method?

2) If yes, how many variables are really needed to achieve this?

3) What can be said about the bonds that will overlap or "excite" the same neuron even though they have different representations?

4) How stable or robust is the method?

There are of course many more questions that can be raised, but not all of them are as important as those above. The Kohonen map of this dataset is shown in Figure 6-16.

Because the map shown in Figure 6-16 has only 121 neurons, it is clear that more than one of the 200 bonds must excite the same neuron. At the same time, some neurons that are not excited at all will co-exist with them in the network. The neurons excited by bonds that are "easy" or "difficult" (to break) are marked as "+" and "–", respectively, while the neurons excited by the bonds for which we did not have data are marked with asterisks (*).

As can be seen, only two neurons are excited by all three types of bonds; a detailed analysis has shown that the parameters of these bonds are actually very similar. There are of course many more interesting details in this study, which may be found in Chapter 11.

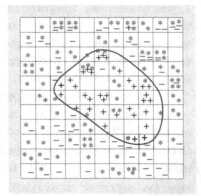

Figure 6-16: Kohonen network of (11 x 11) neurons that clusters breakable and non-breakable bonds on the basis of seven-dimensional data vectors.

6.6   Remarks

The Kohonen network has a very serious computational drawback that affects the performance of large scale applications running on parallel (but not serial) computers. In order to find out which neuron (and neighborhood) is to be stimulated, the program must check **all** n neurons; this is a serious restriction when large nets are to be trained. Even on a parallel computer, this involves $n/2$ parallel comparisons, which requires at least $\log_2 n$ steps; the relative advantage of (expensive) parallel computers is thus compromised. On ordinary machines, all calculations have to be done sequentially anyway, and keeping track of the largest output does not affect the overall performance very much.

The Kohonen network or a Kohonen layer can be built into a more complex network as one of its constituent layers (Reference 6-4) or implemented in combination with some other techniques (Reference 6-5). As we will see in the next chapter, Kohonen competitive learning can be combined with the counter-propagation corrections to form multilevel networks (Reference 6-7).

6.7   Essentials

- the main goal of Kohonen neural networks is to map objects from m-dimensional into n-dimensional space

- this mapping preserves the essential **topological** features of the data

- the primary neuron for weight modification is chosen by competition ...

- ... the algorithm modifies the weight of the neuron with the most intense output, or whose weights are most similar to the input signal ...

- ... and smooths the map by also making modulated changes to neurons in a defined "neighborhood" of that one

- the Kohonen learning procedure is unsupervised learning

- the topology of the planar network can vary considerably from application to application and there are different types of neighborhood relations in the network

• **finding the best neuron**

$$out_c \leftarrow \max(out_j) = \max\left(\sum_{i=1}^{m} w_{ji} x_{si}\right) \tag{6.1}$$

$$j = 1, 2, ..., n$$

$$out_c \leftarrow \min\left\{\sum_{i=1}^{m} (x_{si} - w_{ji})^2\right\} \tag{6.2}$$

$$j = 1, 2, ..., n$$

• **corrections of weights**

$$w_{ji}^{(new)} = w_{ji}^{(old)} + \eta(t)\, a\, (d_c - d_j)\left(x_i - w_{ji}^{(old)}\right) \tag{6.6}$$

$$w_{ji}^{(new)} = w_{ji}^{(old)} + \eta(t)\, a\, (d_c - d_j)\left(1 - x_i w_{ji}^{(old)}\right) \tag{6.7}$$

6.8   References and Suggested Readings

6-1.   T. Kohonen, "Self-Organized Formation of Topologically Correct Feature Maps", *Biol. Cybern.* **43** (1982) 59 – 69.

6-2.   T. Kohonen, *Self-Organization and Associative Memory*, Third Edition, Springer Verlag, Berlin, FRG, 1989.

6-3.   R. Hecht-Nielsen, "Counterpropagation Networks", *Appl. Optics* **26** (1987) 4979 – 4984; D. G. Stork, "Counterpropagation Networks: Adaptive Hierarchical Networks for Near Optimal Mappings", *Synapse Connection* **1** (1988) 9 – 17.

6-4.   P. D. Wasserman and T. Schwartz, "Neural Networks, Part 2: What are They and Why is Everybody so Interested in Them Now?", *IEEE Expert*, Spring 1988, 10 – 15.

6-5.   J. Zupan, *Algorithms for Chemists*, Paragraph 11.1.6: "Neural Network Algorithm A30 for Kohonen Learning", John Wiley, Chichester, UK, 1989, pp. 257 – 261.

6-6.   H. Ritter, T. Martinetz and K. Schulten, *Neuronale Netze, Eine Einführung in die Neuroinformatik selbstorganisierender Netzwerke*, Addison-Wesley, Bonn, FRG, 1990.

6-7.   J. Dayhoff, *Neural Network Architectures, An Introduction*, Van Nostrand Reinhold, New York, USA, 1990.

6-8.   K. L. Peterson, "Classification of Cm Energy Levels Using Counterpropagation Neural Networks", *Phys. Rev. A* **41** (1990) 2457 – 2461.

6-9.   X. Li, J. Gasteiger and J. Zupan, "On the Topology Distortion in Self Organizing Feature Maps", *Biol. Cybern.* **70** (1993) 189 – 198.

6-10. J. Gasteiger and J. Zupan, "Neuronale Netze in der Chemie", *Angew. Chem.* **105** (1993) 510 – 558; J. Gasteiger and J. Zupan, "Neural Networks in Chemistry", *Angew. Chem. Int. Ed. Engl.* **32** (1993) 503 – 527.

6-11. J. Zupan, "Introduction to Artificial Neural Network (ANN) Methods: What They Are and How to Use Them", *Acta Chim. Slov.* **41** (1994) 327 – 352.

6-12. J. Lozano, M. Novic, F. X. Rius and J. Zupan, "Modelling Metabolic Energy by Neural Networks", *Chemom. Intell. Lab. Syst.* **28** (1995) 61 – 72.

6-13. J. Novic and J. Zupan, "Investigation of IR Spectra-Structure Correlation Using Kohonen and Counterpropagation Neural Network", *J. Chem. Inf. Comp. Sci.* **35** (1995) 354 – 466.

6-14. N. Majcen, K. Rajer-Kanduc, M. Novic and J. Zupan, "Modeling of Property Prediction from Multicomponent Analytical Data Using Different Neural Networks", *Anal. Chem.* **67** (1995) 2154 – 2161.

6-15. J. A. Remola, J. Lozano, I. Ruisanchez, M. S. Larrechi, F. X. Rius and J. Zupan, "New Chemometric Tools to Study the Origin of Amphorae Produced in the Roman Empire", *Tr. Anal. Chem.* **15** (1996) 137 – 151.

6-16. I. Ruisanchez, P. Potokar, J. Zupan and V. Smolej, "Classification of Energy Dispersion X-Ray Spectra of Mineralogical Samples by Artificial Neural Networks", *J. Chem. Inf. Comput. Sci.* **36** (1996) 214 – 220.

6-17. J. Zupan, M. Novic and I. Ruisanchez, "Tutorial: Kohonen and Counter-propagation Artificial Neural Networks in Analytical Chemistry", *Chemom. Intell. Lab. Syst.* **38** (1997) 1 – 23.

6-18. I. Ruisanchez, J. Lozano, M.S. Larrechi, F.X. Rius and J. Zupan, "On-line Automated Analytical Signal Diagnosis in Sequential Injection Analysis Systems Using Artificial Neural Networks", *Anal. Chim. Acta* **348** (1997) 113 – 128.

Part III
Multilayer Networks

7 Counter-Propagation

learning objectives

- the first example of a network having more than one layer

- how supervised learning is used when you have a set of "correct" answers

- how data flows through a "counter-propagation" network: the correct answers flow backwards

- content-dependent data storage and retrieval ("associative memory")

- lookup tables and models, and how neural nets can serve as either

- how input data are prepared by "normalization"

- application of counter-propagation to simulate a simple tennis match

- usefulness of counter-propagation for creating lookup tables

7.1 Transition from One to Two Layers

The neural networks we have discussed up to now have had only one layer of neurons. The input signal is transferred to the active neuron layer via the input net, but this layer has more or less only a formal meaning. From a design point of view, it is easy to connect one neuron layer to a lower one, this one to still another one and so on. What we need is only to determine the number of neurons in one layer and the way the outputs of the neurons in this layer are connected to the synapses (weights) in the layer below.

The neurons of two layers can be fully, partially or randomly connected (Figure 7-1). *Full* connection means that each neuron in one layer is connected to all the neurons in the layer below. This is the

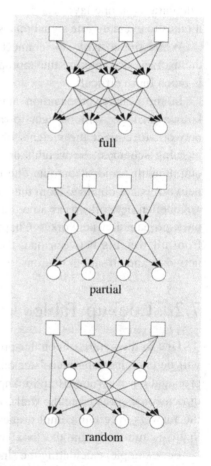

Figure 7-1: Connection of neurons among layers: full, partial and random.

most common scheme used in artificial neural networks. *Partial* and *random* connection are self-explanatory. The partial scheme is used when some aspect of the problem suggests it. When the neurons are connected randomly, the concept of layers becomes meaningless; the designer can determine only the average number of neurons to be connected between the two layers.

Once the links are made and the weights are determined, the signals can flow from the input through many layers towards the output very quickly. This calculation is a natural application for parallel computers, because each neuron processes data independently of all the others; hence (if enough processors are available), the calculations in one layer can be done simultaneously, i.e., in the time it takes to evaluate **one** signal on a sequential computer.

While it is easy to connect the layers of neurons among themselves, determining the appropriate weights for a given problem is much more difficult.

In the counter-propagation approach to modifying weights, the known answers will be sent towards the inputs back through the network to correct the weights of the neuron. As in all supervised learning schemes, the weights are adapted by comparing the actual output with an ideal output. The output of the counter-propagation network is not obtained from the weights of **one** output **neuron** (as in a Kohonen network) or as an output vector from **all neurons** (as in a back-propagation network – Chapter 8). Rather, the output is taken from **all weights** between one, the winning neuron of the Kohonen network, and all output neurons.

7.2   Lookup Table

Due to the nature of counter-propagation network learning, which will be explained in detail later on, we can treat the trained counter-propagation network as a lookup table (we can call it a multi-dimensional spreadsheet as well), which is an area of memory where the answers to certain complex questions are ordered in such a manner that we can readily find the "box" with the right answer.

For example, calculating the sine or cosine of an angle can be very time-consuming when it occurs within deeply nested loops; scientific programs often store values of the sine, tabulated at some convenient interval, and use the angle argument as an **index** to retrieve the proper value from this lookup table.

Thus, we calculate an **address** rather than a value from the input data. There are other situations, too, where lookup table is useful, for example, cases where the value to be retrieved is only a weak function of the input (i.e., where there is a wide degree of tolerance in the answers) or (which is equivalent to this) when corrupted input has to be used. We have already discussed the problem of retrieving the answer to a corrupted input in Chapters 4 and 5; the procedure for finding the sought answer on the basis of imperfect or fuzzy data is called content-dependent retrieval. It requires that **only similar** inputs cause a given box to be selected (the definition of "similar" being up to the user).

There are other *content-dependent retrieval* methods (e.g., the hash algorithm: Reference 7-8, or three-distance clustering: Reference 7-9). See the literature for a detailed discussion.

The concept of the lookup table is distinct from the concept of a *model*, where the goal is to obtain a "function" or a procedure that will yield an answer that is different for each different set of variables; for example, instead of obtaining the sine of an angle by lookup table, we might call the sine subprogram, which approximates (models) the actual value by a series expansion. (Figure 7-2).

These two methods differ not only in application goals, but in the realm of data to which they can be applied. Modeling typically requires only a few data (the sine, for example, requires only an angle argument and the coefficients of the series expansion), whereas a lookup table requires much more (depending on the resolution, maybe several hundred sine values).

That is, in constructing (or training) a model, the number of experiments needs to be larger than the number of parameters describing the model. However, a lookup table requires **at least one** experiment for each possible or expected answer. In order to fill all the boxes with information, a relatively dense distribution of input data (experiments) over the entire variable space is required.

The counter-propagation network is not well suited for modeling; other techniques, such as "back-propagation of errors" (see the next Chapter), can be employed much more efficiently. Counter-propagation is best used to generate lookup tables, where **all** the required answers (within some limits) are known in advance.

The larger the lookup table or the better the model, the smaller are the differences between the ideal and the supplied answers. However, a model can, in principle, yield an infinite number of answers, while a

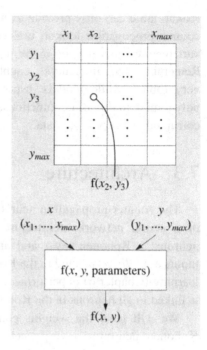

Figure 7-2: The difference between a lookup table and a model. *X* can have values between x_1 and x_{max} and *y* between y_1 and y_{max}.

lookup table can only provide as many answers as it has boxes. The counter-propagation lookup table is optimum if some of the input variables are missing, or if the input data are vague or very fuzzy. Remember that, in general, counter-propagation lookup tables are very robust, while models, especially those simulating high-order polynomial multivariate functions, can show unstable behavior for corrupted or fuzzy input data.

7.3 Architecture

The counter-propagation neural network will be our first example of a neural network architecture which has **two** active layers of neurons: a Kohonen layer and an output layer (Figure 7-3.). The inputs are *fully* connected to the Kohonen network, where competitive learning (Chapter 6) is performed; that is, each unit in the input layer is linked to all neurons in the Kohonen layer.

We will label the weights connecting the input unit i with the Kohonen neuron j as w_{ji}; each neuron in the Kohonen layer is described by a weight vector $\boldsymbol{W}_j$.

The neurons of the Kohonen layer are in turn connected to the neurons in the output layer. In principle, this is complete (full) connection. In practice, however, after each input, only a certain neighborhood of a given neuron is connected to the output neurons (Figure 7-4), and only the weights linking these neurons are allowed to change. The weights connecting the j-th neuron in the Kohonen layer with the k-th neuron in the output layer are labeled r_{kj}, and the weight vector belonging to a given output neuron is labeled $\boldsymbol{R}_k$.

The goal of constructing a lookup table immediately sets requirements on the architecture of the counter-propagation network. A good way to store numerous sets of answers is as the weights of the output neurons that receive signals from the Kohonen layer (Figure 7-4).

A given answer is **not** stored as a set of weights in one neuron, but holographically, as **one** component of the weights of **all** the output neurons.

Such an organization requires the number of neurons in the Kohonen layer to be equal to the number of answers we would like to

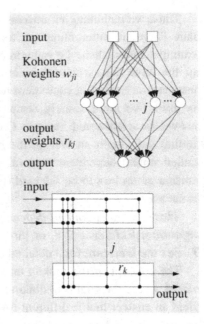

Figure 7-3: The layout of a counter-propagation network.

store, and the number of neurons in the output layer to be equal to the number of variables comprising the output answer.

For example, one thousand answers, each consisting of four variables, requires a Kohonen network with one thousand neurons, and an output layer with four. The input layer should have the same number of units as there are input variables.

Because the counter-propagation network is described in the literature as requiring normalized input data, the standard architectures are shown with an additional normalizing neuron, which provides a position for an extra input (a normalizing variable comparable to the bias); this is determined in such a way that the magnitude of the new, augmented input vector X is equal to unity (Figure 7-5):

$$X = (x_1, x_2, ...,x_m) \rightarrow X'(x_1, x_2, ...,x_m, x_{m+1})$$

and

$$\|X'\| = \sqrt{\left(x_1^2 + x_2^2 + ... + x_m^2 + x_{m+1}^2 \right)} = 1$$

from which it follows that x_{m+1} should be:

$$x_{m+1} = \sqrt{1 - \left(x_1^2 + x_2^2 + ... + x_m^2 \right)} = \sqrt{\left(1 - \|X\|^2 \right)} \qquad (7.1)$$

The normalization procedure, however, is not strictly required, unless the learning strategy selects the neuron with the **largest** output; if it selects the neuron **most similar** to the input vector, the counter-propagation algorithm works without any normalization or renormalization procedures. This is explained below, using an example.

The normalizing variable is in some respects similar to the bias described in Chapter 2. The purpose of both is to change the actual input vector X into an X' so that the calculated net input values (see Section 2.6 for details) lie within the range where the methods work best. Such an adjustment requires the additional weight at each neuron as well as the normalization of weights at the beginning and renormalization at the end of each cycle of training.

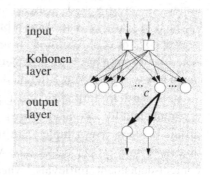

Figure 7-4: The answer (response) to a given input is stored as a weight vector connecting the selected neuron in the Kohonen network and all output neurons (bold). W_c is the vector of weights of the winning neuron c.

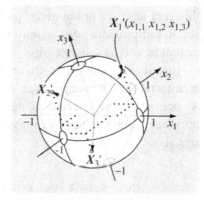

Figure 7-5: If the third variable x_3 is added to the two-dimensional objects $X_s(x_{s1}, x_{s2})$, this third variable is easily determined in such a way that all three-dimensional points X'_s (x_{s1}, x_{s2}, x_{s3}) will lie on the surface of a three-dimensional plane.

7.4   Supervised Competitive Learning

Supervised learning requires sets of pairs (X_s, Y_s) for input: the actual input into the network is the vector X_s, and the corresponding target, or prespecified answer, is labeled Y_s. The goal of any supervised learning is to form a black box that will give the correct answers Y_s for each vector X_s from the training set (see Figure 5-2). After the training has been completed successfully, it is hoped that the black box will give correct predictions for any new object X.

It is hard to formalize the types of predictions which can be accomplished by a counter-propagation network; they can be of very different types. The simplest are those classifying multidimensional objects X into proper categories. More complex predictions involve content-dependent retrievals, where incomplete or fuzzy data are entered and the originals are recovered. For this kind of retrieval, counter-propagation is the optimum method. Still another type is modeling a complex multivariate, nonlinear function yielding a low-dimensional answer (usually 1D or 2D).

The problem with the counter-propagation network is that it needs large quantities of data covering all possible answers. Also the number of **different** answers the counter-propagation method can yield is limited by the size of the network; if there are not too many different answers in the given problem domain, the network may be small, but problems that require large numbers of different solutions cannot be solved satisfactorily.

Recall that the first active layer in the counter-propagation architecture is a Kohonen layer. After each input of an m-variate input X, one neuron is selected as the "winner" exactly as shown in the previous chapter, either by choosing the neuron with the largest output signal out_c:

$$out_c = \max\,(out_j) = \max\!\left(\sum_{i=1}^{m} w_{ji} x_{si}\right) \tag{6.1}$$

$$j = 1, 2, ..., n$$

or by choosing the neuron j with the corresponding weight vector W_j ($w_{j1}, w_{j2}, ..., w_{jm}$) most similar to the input X ($x_1, x_2, ..., x_m$):

$$out_c \leftarrow \min\!\left[\sum_{i=1}^{m} (x_{si} - w_{ji})^2\right] \tag{6.2}$$

$$j = 1, 2, ..., n$$

The choice between these strategies is more or less a matter of personal preference, because they show almost identical performance; neither has a decisive advantage over the other.

However, remember that when Equation (6.1) is used, the input vector X and the weights must be normalized so that their magnitudes are equal to 1. The normalization has two effects on the economy of the method; first, it requires an additional weight on each Kohonen neuron; and second, it requires an additional loop inside the correction algorithm in order to renormalize the corrected weights.

Since the method requires a large number of neurons in the Kohonen layer, an additional weight at each of them significantly increases the memory requirements. Moreover, large Kohonen networks require large neighborhoods of neurons to be corrected at the beginning of each pass; since this loop occurs in the innermost part of the algorithm, the time requirements are increased as well.

The benefit of normalization is that a numeric overflow of the weights is not possible; there is a significant danger of this occurring when inputs and weights are not normalized.

After the winning neuron has been selected, **two** types of corrections are made:

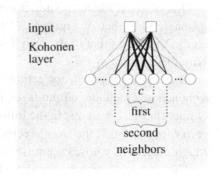

Figure 7-6: The first type of weight corrections affects the weights between the input and the Kohonen layer.

first,   the correction of weights w_{ji} within the neurons of the Kohonen layer

second,   the correction of weights in the output layer.

Corrections of the first type (Figure 7-6) are made using Equation (6.6):

$$w_{ji}^{(new)} = w_{ji}^{(old)} + \eta\,(t)\,a\,(d_c - d_j)\left(x_i - w_{ji}^{(old)}\right) \qquad (6.6)$$

The neighborhood-dependent function $a(d_c - d_j)$ and the monotonically decreasing function $\eta(t)$ are discussed in Chapter 6 in connection with competitive learning in the Kohonen network.

After the corrected weight vector $W_j^{(new)}$ ($w_{j1}^{(new)}$, $w_{j2}^{(new)}$, ..., $w_{jn}^{(new)}$) has been obtained, it has to be renormalized. If the selection of the winning neuron is made according to the largest output criterion (6.1), the renormalization (Equation (7.1)) is mandatory:

$$W_j^{\prime\,(new)} = \frac{W_j^{(new)}}{\|W_j\|^2} = \frac{W_j^{(new)}}{\sqrt{\displaystyle\sum_{i=1}^{m}\left(w_{ji}^{(new)}\right)^2}} \qquad (7.2)$$

Otherwise (if the most similar neuron criterion, (6.2), is used) it is optional. Note that in Equation (7.1), the dimension m of the input vector X and thus that of the weight vector W_j should be larger by one than if (6.2) is used.

The second type of correction affects the weights between the Kohonen layer and the output layer. (In Section 7.3 we used "r" as the symbol for these weights. In the following discussion, R_k is the weight vector for neuron k; the (row) vectors R_k together make up the output weight matrix R, whose columns will be designated C_j.)

The answers are not stored in R_k. Because only one neuron is activated in the Kohonen layer, per input component, the result vector must consist of **one** weight from each output neuron. This is equivalent to a **column** C_j of the output weight matrix R (the "j" reminds us that the vector C_j is actually associated with the Kohonen neuron j (Figure 7-7)).

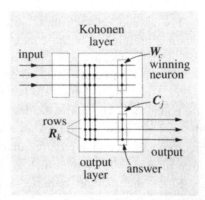

Figure 7-7: The outputs of the counter-propagation network are stored in the columns C_j and not in the rows R_k of the output weight matrix.

> Remember, when correcting output weights, that the weights are corrected within the column vector C_j and **not** within the row vector R_k!

Equation (7.3) shows how output weights are to be corrected:

$$c_{ji}^{(new)} = c_{ji}^{(old)} + \eta\,(t)\,a\,(d_c - d_j)\left(y_i - c_{ji}^{(old)}\right) \qquad (7.3)$$

As before, c is the index of the winning Kohonen neuron; j is the index of the neighboring neuron being corrected; i runs over all the weights linking the Kohonen neuron j with the output neurons i. Each of the n output neurons represents one component (variable) of the target vector $Y = (y_1, y_2, ..., y_n)$.

Figure 7-8 shows the weights and neurons to which the weights are connected in the entire procedure. The extent of the correction depends on the topological distance of the corrected neuron from the winning one, and the thickness of the lines used to draw them is in proportion to the degree of correction.

Therefore, the corrected weights c_{ji} need not be normalized, because they represent the result that should be used later as the output from the counter-propagation network.

Recall that, unlike the other networks we have studied before, the output from a counter-propagation network is obtained from **one** Kohonen network.

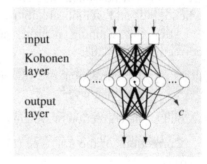

Figure 7-8: The thickness of the lines symbolizes the extent of the correction of the weights.

A counter-propagation network serving as a lookup table can be visualized as two boxes, one on top of the other (Figure 7-9). The two boxes may be of different height but they have the same length and width. The upper box represents the Kohonen layer and the lower one the output layer. The neurons are packed as columns in the boxes and are accessed by the indices j' and j'' describing the top plane of the box. The position of the Kohonen neuron most excited or most similar to the input object X is referenced by these coordinates j' and j''. The position (j', j'') also determines the position of the vertical column in the output box where the corresponding answers are stored.

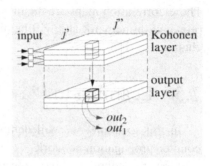

Figure 7-9: Counter-propagation network shown as two boxes. The upper box contains the weights W of the Kohonen network, while the lower one contains the weights C of the output layer.

In Figure 7-9 we have labelled the winning neuron c by the indices j' and j'' to indicate its position in the box and still retain the image of a two-dimensional Kohonen network. (In the equations in this Chapter we have combined the two indices j' and j'' into one index j, as is usually done.)

The Kohonen network is usually implemented as this box implies, as a three-dimensional array, $w_{j'j''i}$, where j' runs from 1 to the length of the map, j'' from 1 to the width of the map, and i indicates the number of input signals.

You may wonder why the results that are stored as weights on the output neurons are not just replaced by the actual targets instead of undergoing the tedious iteration involved in Kohonen mapping. This would be a valid observation if only binary vectors were involved, since in this case there is no need to adapt the output weights according to any scheme; they are either zero or one and can be substituted directly:

$$c_{ci}^{(new)} = y_i \qquad (7.4)$$

where y_i is the desired target answer. There is no need to look at the neurons in the neighborhood of the winning one; the substitution is made only for the weights connecting the winning neuron c with all n output neurons (index c instead of j).

When nonbinary values are used as components of the target vector, we have to be aware that different inputs with slightly different targets may excite the same winning neuron and thus yield the same answer. Thus, the similarity of inputs shows in the similarity of outputs. The correlation between the input and the output variables as well as correlations within each group can easily be obtained by inspecting the resulting weights after the lookup table is generated.

These correlation maps are the most outstanding result of the counter-propagation network and will be discussed in more detail in the next chapter.

7.5   Learning to Play Tennis

In this example, we will demonstrate various capabilities of the counter-propagation network.

First, the counter-propagation network is capable of dealing with vectors (objects) representing real values on the input side as well as on the output (target) side.

Second, a lookup table can be formed from a comparably large set of experimental data and can be utilized as a substitute for a mathematical model (that is, an explicit functional relationship).

Third, the correlations among different variables can be obtained from the resulting layer of output weights.

In real life, learning tennis requires a court and a lot of practice. The more different strokes you learn, the better your play. The two prime factors in tennis are getting to the right part of the court (in time), and returning the ball to an exact spot on your opponent's side.

Obviously, there are a lot of technical details, such as the speed and spin of the ball, or analyzing your opponent's weaknesses; but for now, let's assume that only these two issues (where you are, and how you hit the ball) are involved.

To "compute" your response, you need two data: your opponent's position and how he swings the racket. From these two data to estimate the trajectory of the ball (where it will land) and how to hit it.

For simplicity, we will assume that both players can move only on the service lines of the tennis court (bold lines in Figure 7-10). The players will be labeled as X and Y (trainer and trainee). Their positions on the service line are marked as x and y.

The input data that the trainee Y obtains for learning are the position x of the trainer X and the angle β at which the ball is flying against her. Therefore, the input X can be regarded as a two-dimensional vector:

$$X = (x, \beta)$$

The trainee must position herself at position y (where the ball will cross the service line), holding her racket at the angle γ which will return the ball towards position z. We will simplify further by

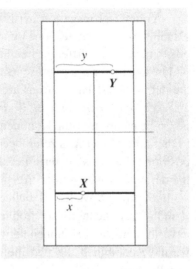

Figure 7-10: Model of a tennis court: x and y are the positions of players X and Y, who move only along their service lines.

assuming that z is fixed at the rightmost point of X's service line (Figure 7-11). The desired learning target is thus a two-dimensional vector Y:

$$Y = (y, \gamma)$$

The position of the trainer will be selected randomly between 0 and a (the width of the court). The range of the angle β at which he can send the ball depends on his position x; if he is at the point $x = 0$, then $\tan\beta$ can vary between 0 and a/b (b is the length of the court), while at the position $x = a$, $\tan\beta$ can range between $-a/b$ and 0. For a general position x of the trainer, $\tan\beta$ varies between $-x/b$ and $(a-x)/b$, so

for any $x \in [0, a]$

$$\tan\beta \in [-x/b, (a-x)/b]$$

If the trainee is a complete newcomer, her answers will be random; but if her best answer is always corrected towards even better performance, she will tend to select better and better answers. (Unlike a real tennis instructor, we must calculate the "right" answers using appropriate equations and then compare the trainee's answers with them.).

Figure 7-12 illustrates how the answer $Y(y, \gamma)$ is calculated for each $X(x, \beta)$. First we can see that:

$$\tan\beta = (y-x)/b$$

with which the first output y of the target $Y(y, \gamma)$ can easily be calculated:

$$y = x + b\tan\beta$$

Next, the angle γ between the racket plane and the service line should be estimated. From Figure 7-12 we see that:

$$\gamma + \varphi = 0.5\,(\beta + \varphi)$$

This gives us:

$$\gamma = 0.5\,(\beta - \varphi)$$

With:

$$\tan\varphi = (z-y)/b$$

we obtain:

$$\gamma = 0.5\,\{\beta - \arctan[(z-y)/b]\} \tag{7.5}$$

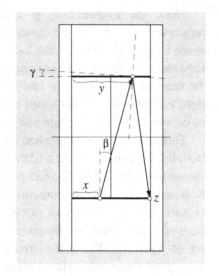

Figure 7-11: The variables (x, β) and (y, γ) representing the input (trainer's stroke) and desired output (trainee's response).

Thus, the output $Y(y, \gamma)$ can be calculated from the input parameters $X(x, \beta)$ and the dimensions of the tennis court a, b, and z using Equation (7.5) (z is the location to which the trainee wants to send the ball). Since in this example z is always equal to a, the goal of the training is, in effect, to teach the trainee how to hit the extreme right corner of the trainer's part of the court, regardless of where she is, or where the ball is coming from.

This can all be accomplished by the counter-propagation network. Initially, we construct a (2 x 625 x 2) counter-propagation network, as shown in Figure 7-13. (The 625 neurons come from a (25 x 25) neuron Kohonen layer used for competitive learning.) Then a number of input vectors $X_s = (x_s, \beta_s)$, let us say 4000, are randomly selected, and to each of them a theoretically correct answer $Y_s(y_s, \gamma_s)$ is assigned. The set of pairs (X_s, Y_s) is then ready to be presented to the counter-propagation network. After having all weights set to small random numbers, the following procedure is launched:

– input a vector $X_s = (x_s, \beta_s)$

– evaluate n sums in all n neurons in the Kohonen layer:

$$out_j = (x_s - w_{j1})^2 + (\beta_s - w_{j2})^2$$

$$j = 1, 2, ..., n$$

– select the winning neuron c as the one having the minimum out_j:

$$out_c = \min \{ out_1, out_2, ..., out_n \}$$

$$j = 1, 2, ..., n$$

– correct both weights in each neuron from a given neighborhood around the winning neuron c in the Kohonen layer (see Equation (6.6)):

$$w_{j1}^{(new)} = w_{j1}^{(old)} + \eta (t) \, a \, (d_c - d_j) \left(x_s - w_{j1}^{(old)} \right)$$

$$w_{j2}^{(new)} = w_{j2}^{(old)} + \eta (t) \, a \, (d_c - d_j) \left(\beta_s - w_{j2}^{(old)} \right)$$

– at the beginning of the learning procedure, the product $\eta(t) \, a \, (d_c - d_j)$ is about 0.5 (Equation (6.5) with $a_{max} = 0.5$ and $a_{min} = 0.01$); for each other neuron this product decreases as a function of which

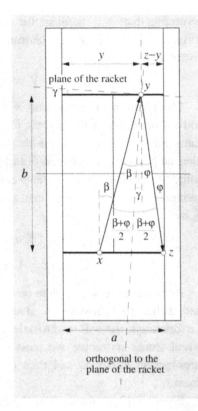

Figure 7-12: The calculation of the response $Y(y, \gamma)$ for each $X(x, \beta)$.

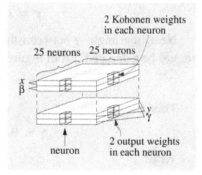

Figure 7-13: 625-neuron network for learning tennis.

neighbor-ring it is in, and how many iteration cycles have already occurred.

– correct the two weights in the output layer leading from all neurons in a given neighborhood of the winning neuron c towards both outputs (see Equation (7.3)):

$$c_{j1}^{(new)} = c_{j1}^{(old)} + \eta(t)\, a\, (d_c - d_j) \left(y_s - c_{j1}^{(old)} \right)$$

$$c_{j2}^{(new)} = c_{j2}^{(old)} + \eta(t)\, a\, (d_c - d_j) \left(\gamma_s - c_{j2}^{(old)} \right)$$

In the above correction, exactly the same value of the product $\eta(t)\, a\, (d_c - d_j)$ can be employed as is used during the correction of the Kohonen weights. But if you have some good reason, the correction factor $\eta(t)\, a\, (d_c - d_j)$ can be changed later on:

– change the factor $\eta(t)$ and the neighborhood range to which the function $a\, (d_c - d_j)$ is applied, and go to the first step; repeat until all pairs have been sent through the network.

The product $\eta(t)\, a\, (d_c - d_j)$ is a kind of adjusting "dial" in the method; you have to find by trial and error which values work best for a given application. In our example, the entire (25 x 25) Kohonen layer has to be encompassed initially. Of course, the weights of neurons lying in the 25th neighbor's ring away from the winning neuron will be changed very little; the correction can amount to only one twentyfifth of the correction applied to the winning neuron's weights (Figure 7-14).

If there are 4000 input vectors, the neighborhood can be shrunk by one ring after each 160 inputs; hence, in the last 160 inputs only the weights of the winning neurons will be corrected and nothing else. Of course, these corrections will be considerably smaller than at the beginning because the factor $\eta(t)$ decreases as we continue to provide inputs. In the present example, by the 4000th input, η had decreased to 0.01 compared to the initial value of 0.5 (Figure 7-14).

Because the corrections of the output weights are implemented in the same loop as the corrections of the Kohonen weights, the learning procedure is quite efficient.

After 4000 inputs, the answers obtained by the (25 x 25) network (which is able to store 625 answers) were quite encouraging. The root-mean-square (RMS) error between the targets and the results given by the counter-propagation network is 0.33.

The RMS error is calculated from the general equation:

$$\text{RMS} = \sqrt{\frac{\displaystyle\sum_{s=1}^{n_i}\sum_{i=1}^{n}(y_{si}-out_{si})^2}{n_i n}} \qquad (7.6)$$

where y_{si} is the i-th component of the desired target Y_s, out_{si} is the i-th component of the output produced by the network for the s-th input vector, n_i is the number of inputs, and n is the number of output variables.

In our case, this takes the following form:

$$\text{RMS} = \sqrt{\frac{\displaystyle\sum_{k=1}^{4000}\left[(y_s-c_{c1})^2+(\gamma_s-c_{c2})^2\right]}{8000}}$$

c_{c1} and c_{c2} are the weights in the output layer selected by the winning Kohonen neuron (c) for the s-th input vector.

An RMS error of 0.33 corresponds to a tolerance of ±0.025 in the position y and a tolerance of ±1 arc degree in the angle γ. It is interesting to see what the correspondence is between this and the player's success rate. Table 7-1 shows the number of unsuccessful hits for different RMS errors (an unsuccessful hit means that her/his response misses the calculated value by more than the tolerance limits).

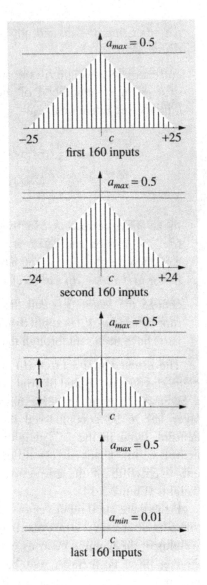

Figure 7-14: How the neighborhood and η change during the training.

RMS error	number of unsuccessful hits per 1000
1.52	110 – 160
1.10	47 – 76
0.77	8 – 18
0.62	1 – 8
0.45	1 – 5
0.35	0 – 5
0.26	0 – 2
0.14	0

Table 7-1:   Simulation of playing tennis; translation of RMS error into number of unsuccessful hits.

This requires us to create many counter-propagation networks yielding different degrees of RMS error; for each one, a test of one thousand randomly selected inputs was repeated five times. Each net gave almost identical RMS errors in the five repeated tests, but different numbers of unsuccessful hits were recorded; hence, the range of unsuccessful hits in Table 7-1.

Additional information can be found in Section 8.6, where the tennis example is worked out more thoroughly.

7.6 Correlations Among the Variables

One of the most valuable properties of the counter-propagation network is that the final values of the output-layer weights contain information about the correlations (lack of functional independence) among the input variables.

Let us consider both "planes" of output weights that were generated during our tennis simulation experiment (Figure 7-15). The architecture of the counter-propagation network for this example can be represented as several square double pyramids. The two upper ones represent the Kohonen weights: one receiving the positions x and the other one the angles β, while the two upside down pyramids represent the output weights: one yielding the positions y and the other one the racket angles γ.

Computationally, all the pyramids can easily be separated and written in the form of a square matrix. The position of each weight is the position of the Kohonen neuron with which it is involved. In the case of Kohonen weights, these matrices represent the already familiar Kohonen maps (Section 6.4), showing topological relationships among the input variables; there is one for each variable.

Similarly, in the case of the output weights, the upside-down pyramids generate matrices containing the Kohonen maps of the output variables. Figure 7-16 shows all four Kohonen maps obtained in the present example. The upper two maps show the lines that connect points with the same x-position (*iso-x-position-lines*) and the lines with the same β-angle, while the lower two maps show the output's *iso-y-position* and *iso-γ-angle* lines.

These maps tell us that for a given position of x only a certain range of angles β is possible, and vice versa, that a specific angle is allowed only at a certain range of positions. The same is true for the position y and the angle γ for the trainee. That is, these variables are

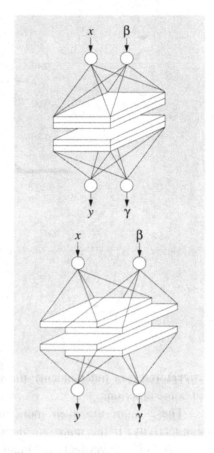

Figure 7-15: The counter-propagation network architecture for the tennis simulation (2 inputs, 2 outputs) visualized as four pyramids.

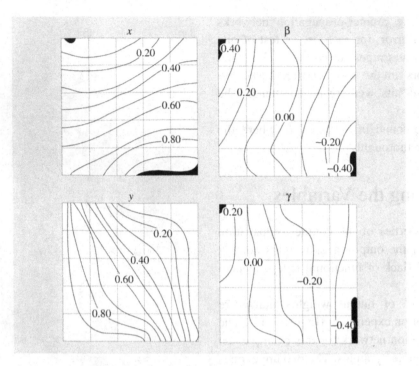

Figure 7-16: The four Kohonen maps for the tennis simulation problem obtained on a (25 x 25) matrix after 5000 random inputs (positions x and y between 0 and 1, angles β and γ in radians). The RMS error of the answers is 0.35.

correlated (not independent): the values of one depend on the value of some other one.

These maps are even more useful because of their "vertical" connectivity; if the maps are drawn on transparencies and overlaid (Figure 7-17), a vertical line going through all maps gives the output variables for a given set of input variables. By recording all values at the intersections of the Kohonen maps with the vertical line **travelling along** a selected *iso*-value line on the map of one variable, all the plots of the remaining variables are obtained at the constant value of the selected one. It is easy to program this *intersection retrieval* once the complete output weights have been obtained.

The procedure of "following the vertical intersections" enables us to obtain the answer to inverse questions, that is, questions relating to an **inverse** model – very hard problems to solve by standard analytical methods.

In the tennis example, such an inverse question would be: does a (trainer's) position x exist which would force the trainee standing at $y = 0.6$ to hold the racket at an angle of 0 degrees in order to hit the corner $x = 1$? To obtain the answer, the vertical line is put on the beginning of the γ = 0 line in the γ angle map (Figure 7-16d)). Then the vertical line is moved along the γ = 0 line until it encounters the

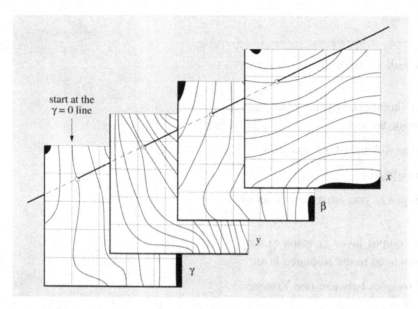

start at the
$\gamma = 0$ line

Figure 7-17: With the help of a vertical intersection line, the overlap of all four maps enables the retrieval of answers to *inverse questions*.

0.6-*iso*-y-line in the map above (Figure 7-16c)). Then the intersections of this line with Figures (a) and (b) give the values $x = 0.2$ and $\beta = 0.20$. This means that if the trainer at $x = 0.2$ sends the ball at angle 0.20 radians, the trainee should be at $y = 0.6$ holding the racket with $\gamma = 0$ in order to hit the corner $x = 1$ on the trainer's side of the court.

Additionally, this method allows you to determine whether some conditions can be met at all (and still yield an acceptable solution) – a quite important piece of information for many types of problems.

7.7   Essentials

- the counter-propagation network is basically a two-layer network

- it consists of a Kohonen layer (influenced by the inputs) and an output layer (influenced by the targets)

- it is trained very similarly to the Kohonen-type networks

- the input data are usually normalized

- it employs supervised learning, i.e. you must have a set of "correct" answers

- the answers are stored in the output layer as maps exactly corresponding to the maps generated in the Kohonen layer

- the output is taken from all weights between one Kohonen neuron and all the output neurons

- counter-propagation networks are used as lookup tables

- there is a one-to-one correspondence between the neurons in the Kohonen map and those in the output map

- **normalization**

$$X = (x_1, x_2, ..., x_m) \rightarrow X'(x_1, x_2, ..., x_m, x_{m+1})$$

$$x_{m+1} = \sqrt{\left(1 - x_1^2 + x_2^2 + ... + x_m^2\right)} = \sqrt{\left(1 - \|X\|^2\right)} \tag{7.1}$$

- **finding the best neuron c**

largest output:

$$out_c = \max(out_j) = \max\left(\sum_{i=1}^{m} w_{ji} x_{si}\right) \\ j = 1, 2, ..., n \tag{6.1}$$

best agreements with the weights:

$$out_c \leftarrow \min\left[\sum_{i=1}^{m} (x_{si} - w_{ji})^2\right] \quad j = 1, 2, ..., n \tag{6.2}$$

corrections of weights:

- **Kohonen weights**

$$w_{ji}^{(new)} = w_{ji}^{(old)} + \eta(t)\, a\, (d_c - d_j)\left(x_i - w_{ji}^{(old)}\right) \tag{6.6}$$

$$W_j^{(new)} = \frac{W_j^{(new)}}{\|W_j\|^2} = \frac{W_j^{(new)}}{\sqrt{\sum_{i-1}^{m}\left(w_{ji}^{(new)}\right)^2}} \tag{7.2}$$

correction of output weights

$$c_{ji}^{(new)} = c_{ji}^{(old)} + \eta(t)\, a\, (d_c - d_j)\left(y_i - c_{ji}^{(old)}\right) \tag{7.3}$$

RMS error

$$RMS = \sqrt{\frac{\sum_{s=1}^{n_i}\sum_{i=1}^{n}(y_{si} - out_{si})^2}{n_i n}} \tag{7.6}$$

decreasing learning rate

$$\eta(t) = (a_{max} - a_{min})\frac{t_{max} - t}{t_{max} - 1} + a_{min} \tag{6.5}$$

7.8   References and Suggested Readings

7-1.   R. Hecht-Nielsen, "Counterpropagation Networks", *Proceedings of the IEEE First International Conference on Neural Networks*, 1987 (II) 19 – 32.

7-2.   R. Hecht-Nielsen, "Counterpropagation Networks", *Appl. Optics* **26** (1987) 4979 – 4984.

7-3.   R. Hecht-Nielsen, "Applications of Counterpropagation Networks", *Neural Networks* **1** (1988) 131 – 140.

7-4.   T. Kohonen, *Self-Organization and Associative Memory*, Springer Verlag, Berlin, FRG, 1989.

7-5.   G. A. Carpenter, "Neural Network for Pattern Recognition and Associative Memory", *Neural Networks* **2** (1989) 243 – 257.

7-6.   H. Ritter, T. Martinetz and K. Schulten, *Neuronale Netze, Eine Einführung in die Neuroinformatik selbstorganisierender Netzwerke*", Addison-Wesley, Bonn, FRG, 1990.

7-7.   J. Dayhoff, *Neural Network Architectures, An Introduction*, Van Nostrand Reinhold, New York, USA, 1990.

7-8.   D. E. Knuth, *The Art of Computer Programming*, Addison-Wesley, Reading, USA, 1975, Vol. *3*, p. 506.

7-9.   J. Zupan, *Clustering of Large Data Sets*, Research Studies Press, Chichester, UK, 1982.

7-10. J. Gasteiger and J. Zupan, "Neuronale Netze in der Chemie", *Angew. Chem.* **105** (1993) 510 – 558; J. Gasteiger and J. Zupan, "Neural Networks in Chemistry", *Angew. Chem. Int. Ed. Engl.* **32** (1993) 503 – 527.

7-11. J. Zupan, "Introduction to Artificial Neural Network (ANN) Methods: What They Are and How to Use Them", *Acta Chim. Slov.* **41** (1994) 327 – 352.

7-12. J. Zupan, M. Novic and I. Ruisanchez, "Kohonen and Counter-propagation Artificial Neural Networks in Analytical Chemistry", *Chemom. Intell. Lab. Syst.* **38** (1997) 1 – 23.

7-13. I. Ruisanchez, J. Lozano, M. S. Larrechi, F. X. Rius and J. Zupan, "On-line Automated Analytical Signal Diagnosis in Sequential Injection Analysis Systems Using Artificial Neural Networks", *Anal. Chim. Acta* **348** (1997) 113 – 128.

7-14. J. Zupan and M. Novic, "Counter-propagation Learning Strategy in Neural Networks and its Application in Chemistry", in *Further Advances in Chemical Information*, Ed. H. Collier, Royal Soc. Chem., Cambridge, UK, 1994, pp. 92 – 108.

7-15. J. Zupan, M. Novic and J. Gasteiger, "Neural Networks with Counter-propagation Learning Strategy Used for Modelling", *Chem. Intell. Lab. Syst.* **27** (1995) 175 – 188.

7-16. J. Lozano, M. Novic, F.X. Rius and J. Zupan, "Modelling Metabolic Energy by Neural Networks", *Chemom. Intell. Lab. Syst.* **28** (1995) 61 – 72.

7-17. N. Majcen, K. Rajer-Kanduc, M. Novic and J. Zupan, "Modeling of Property Prediction from Multicomponent Analytical Data Using Different Neural Networks", *Anal. Chem.* **67** (1995) 2154 – 2161.

7-18. F. Ehrentreich, M. Novic, S. Bohanec and J. Zupan, "Bewertung von IR-Spektrum-Struktur-Korrelationen mit Counter-propagation-Netzen", in *Software Development in Chemistry 10*, Ed.: J. Gasteiger, GDCh, Frankfurt am Main, FRG, 1996, pp. 271 – 292.

8   Back-Propagation of Errors

learning objectives:

- the differences between "back-propagation" and "counter-propagation"

- why the back-propagation algorithm (a scheme for correcting weights) is the most widely used method

- how two empirical factors have been introduced into the weight-correction equations to overcome some problems

- how a back-propagation net can be used to "model" the simple tennis game introduced in Chapter 7

8.1   General

Back-propagation of errors is not the name of a specific neural network architecture, but the name of a learning method, a strategy for the correction of weights. First introduced by Werbos and later on intensely popularized in connection with neural networks by Rumelhart and coworkers, back-propagation has become the most frequently used method in the field.

In fact, it has become so popular that for many authors the term "neural networks" simply means the back-propagation method. A recent study made by the authors (Reference 8-11) discovered that almost 90 percent of all publications using neural networks in chemistry had used the back-propagation method. Even more interestingly, a number of applications had used this method for clustering, as a lookup table, or as an associative memory – tasks for which other already described methods seem to be far more suitable.

The attractiveness of the back-propagation method comes from the well-defined and explicit set of equations for weight corrections.

These equations are applied throughout the layers, beginning with the correction of the weights in the last (output) layer, and then continuing backwards (hence the name!) towards the input layer (Figure 8-1).

The weight-correction procedure in the back-propagation algorithm does probably not resemble the real process of changing the weights (synaptic strengths) in the brain. However, it must be said that the back-propagation algorithm most closely follows the description of artificial neurons given in Chapter 2.

The back-propagation method, shown schematically in Figure 8-1, is a *supervised* learning method; therefore, it needs a set of **pairs** of objects, the inputs X_s and the targets Y_s, (X_s, Y_s). Because the objects and targets can be represented by sets of real variables, X_s $(x_{s1}, x_{s2}, ..., x_{sn})$ and Y_s $(y_{s1}, y_{s2}, ..., y_{sm})$, the resulting network can be regarded as a *model* yielding an *m*-variate answer for each *n*-variable input (Figure 8-2).

Compared to standard statistical and pattern recognition methods for supervised learning, three things have to be stressed. First, almost all features known in standard model-generating techniques (choice of variables, representation of objects, experimental design, etc.) play an important role in the back-propagation procedure as well: the troublesome ones as well as the desirable ones.

Second, neural networks trained by back-propagation of errors have one very important advantage: there is no need to know the exact form of the analytical function on which the model should be built. This means that neither the functional type (polynomial, exponential, logarithmic, etc.) nor the number and positions of the parameters in the model-function need to be given.

In order to test the influence of weights on the final output, we would need to be able to determine the effect of **each input** variable on **each weight** separately. In a fully-connected multilayer network, however, **each** input influences **all** weights. Since the influence of a given input on any weight is virtually impossible to predict in advance, our only hope is to input as many objects as possible and try to observe how the weights act. (With neural networks having a million weights and more this is almost impossible.)

Nor can much information be obtained from the inspection of final weights. In any more or less complex back-propagation neural network, a **large** number of weights are trained to yield the correct answers. Therefore, it is very hard, if not impossible, to establish exactly what each weight is responsible for.

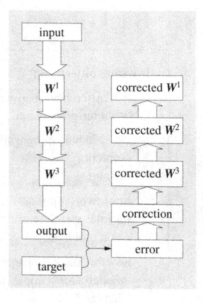

Figure 8-1: Schematic presentation of weight correction by back-propagation of errors.

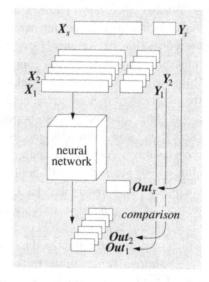

Figure 8-2: Supervised learning requires pairs of data: inputs X_s and targets Y_s.

Thus, third, a back-propagation neural network acts as a black box, allowing no physical interpretation of its internal parameters.

8.2   Architecture

The architecture of the network is the main feature influencing the flexibility of the model it generates; that is, the number of layers, the number of neurons in each layer, and the way the neurons are connected.

Although the back-propagation algorithm was primarily designed for use in multilayer neural networks, it can also be applied to neural networks having only one layer. Such a network has, aside from the input units, only one layer of active neurons, the output layer whose weights are to be corrected via backward propagation.

The layers of neurons are usually fully connected. Figure 8-3 shows an architecture consisting of one input and three active layers of neurons (two hidden layers and the output layer). The connections to the biases and the bias weights are indicated by heavier lines and black squares, respectively.

The number of layers as well as the number of neurons in each layer depends on the application for which the neural network is set up, and is, as a rule, determined by trial and error. In the applications reported in the literature, as many as one million weights and as few as ten have been used.

In most cases, neural networks consisting of two active layers – one hidden and one output layer – are used. Only seldom have more than three active layers been linked together. However, quite a number of authors have used separate neural networks linked into a decision hierarchy rather than one large network making all decisions simultaneously (See Chapter 18).

Such designs are used either because of inadequate computer resources, or because not enough data are available to cover the entire variable space.

At any given time during learning by back-propagation of errors, a considerable number of interlayer calculations may occur, involving as many as **three different** layers of neurons. Great care must therefore be taken to make clear which layer is involved in a given operation. Therefore, we will need a comprehensive system of notation.

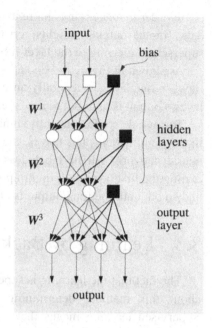

Figure 8-3: Architecture of a back-propagation network with three active layers.

In order to make things more understandable, each specific item of data (input, output, weights, errors and corrections) will bear a superscript referring to the layer it belongs to.

As shown in Figure 8-4 and as already discussed in Chapter 3, the input to one layer is generally an output from the layer above, and a layer's output is an input to the layer of neurons below.

To avoid confusion, all signals will be labeled as outputs throughout this chapter; that is, the actual input signal that enters the neural network will be labeled as Out^0, it will produce Out^1 after exiting the first layer, will in turn produce Out^2 from the second layer, and so on, until the final output is obtained and labeled Out^{last}.

8.3   Learning by Back-Propagation

The fact that the learning is supervised is the most important thing about this method, determining all of its other characteristics. Supervised learning means that the weights are corrected so as to produce prespecified ("correct") target values for as many inputs as possible.

The correction of weights, the most important step in the learning process, can be made **after each individual new input** (*immediate correction*), or **after all inputs have been tested** (*deferred correction*). In the first case, the correction is made immediately after the error is detected; in the second, the individual errors for all data pairs are accumulated, and then the accumulated error of the entire training set used for the correction.

Most applications use immediate correction; deferred correction is more rare and does not offer any apparent advantage. Therefore, we will focus on the former.

During learning, the object X (input vector) is presented to the neural network and the output vector Out is immediately compared with the target vector Y (y_1, y_2, ..., y_m), which is the correct output for X.

Once the actual error produced by the network is known, we have to figure out exactly how to use this to correct weights throughout the entire neural network. Before going into the details, here is the final result in condensed form:

$$\Delta w_{ji}^{l} = \eta \delta_j^l out_i^{l-1} + \mu \Delta w_{ji}^{l\,(previous)} \qquad (8.1)$$

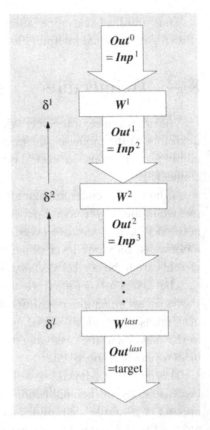

Figure 8-4: Notation used in discussing back-propagation networks: superscripts label the neural levels to which data refer.

Remember that l is the index of the current layer, j identifies the current neuron, and i is the index of the input source, i.e. the index of the neuron in the upper layer. In this equation, δ_j^l, the error introduced by the corresponding neuron, is calculated in two ways, depending on whether the *last* (output) layer or one of the hidden layers is under consideration. In the following two equations, (8.2) and (8.3), a sigmoidal transfer function is assumed (see also Table 8-1).

According to Equation (8.1), the correction of weights in the l-th layer is composed of two terms, which pull in opposite directions: the first one tends towards a fast "steepest-descent" convergence, while the second is a longer-range function that prevents the solution from getting trapped in shallow local minima. The constant η (the same one as $\eta(t)$ in Section 6.3 and Section 7.4, and serving the same purpose) is called the learning rate and μ is called the momentum constant. By taking into account the correction made on the previous cycle, μ can (to the degree you specify) prevent sudden changes in the direction in which corrections are made: this is particularly useful for damping oscillations. The magnitudes of these constants determine the relative influence of the two terms.

For the output layer ($l = last$) the error δ_j^l is expressed as:

$$\delta_j^{last} = \left(y_j - out_j^{last} \right) out_j^{last} \left(1 - out_j^{last} \right) \tag{8.2}$$

For all other layers l ($l = last - 1$ to 1) the error δ_j^l is calculated by:

$$\delta_j^l = \left(\sum_{k=1}^{r} \delta_k^{l+1} w_{kj}^{l+1} \right) out_j^l \left(1 - out_j^l \right) \tag{8.3}$$

Substituting (8.3) into (8.1) gives the full expression of the weight correction in a hidden layer:

$$\Delta w_{ji}^l = \eta \left(\sum_{k=1}^{r} \delta_k^{l+1} w_{kj}^{l+1} \right) out_j^l \left(1 - out_j^l \right) out_i^{l-1} + \mu \Delta w_{ji}^{l \, (previous)} \tag{8.4}$$

This equation shows that values from three layers influence the correction of weights in any one layer: values from the current layer, l, and the ones above ($l - 1$) and below ($l + 1$).

Section 8.4 below shows how the expressions (8.2) and (8.3) can be derived from the delta-rule (Section 2.3, Equation (2.12)), and how the gradient descent method is used to evaluate these terms. You may want to skip this section the first time you read this book, but you should probably study it later in order to understand the essentials of this widely-used method.

8.4   The Generalized Delta-Rule

Learning by back-propagation sends the data through the network in one direction, and scans through it, changing weights, in the opposite direction. The correction of the i-th weight on the j-th neuron in the l-th layer of neurons is defined as:

$$\Delta w_{ji}^{l} = w_{ji}^{l\,(new)} - w_{ji}^{l\,(old)} \tag{8.5}$$

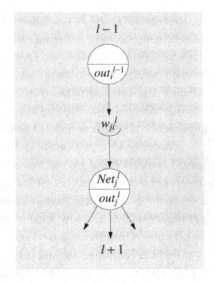

Figure 8-5: The weight w_{ji}^{l} within the layer l is linked to the next-higher and -lower layer via the i-th input and j-th output.

Within the j-th neuron in the l-th layer, the weight w_{ji}^{l} links the i-th **input** with the j-th **output** signal. These two links, one with the upper and one with the lower layer, (Figure 8-5), reflect the fact that the error originates partly on the input and partly on the output side.

A well known way to consider both influences was discussed in paragraph 2.3; it is called the delta-rule and is expressed as follows (cf. Equations (2.22) and (2.22a)):

$$\Delta parameter = \eta \; g\,(output\; error)\, f\,(input) \tag{8.6}$$

In its most general form, the delta-rule states that the change of any parameter in an adapting process should be proportional to the input signal **and** to the error on the output side. The proportionality constant η (learning rate) determines how fast the changes of this parameter should be implemented in the iteration cycles.

In order to give Equation (8.6) a more familiar look, it is rewritten by substituting into it the terms used in the neural network approach:

$$\Delta w_{ji}^{l} = \eta \; \delta_{j}^{l} \, out_{i}^{l-1} \tag{8.7}$$

Formally, Equations (8.6) and (8.7) are identical. The parameter which causes the error is the weight w_{ji}^{l}; its correction Δw_{ji}^{l} is proportional to the term δ_{j}^{l}, corresponding to the function g above (Equation (8.6)). Although the term out_{i}^{l-1} is labeled as the output of the $(l-1)$-st layer, it is at the same time the input to the l-th layer, which is consistent with Figure 8.4.

The function f is simply the input itself; therefore, the remaining problem is the estimation of the function δ_{j}^{l}.

In the back-propagation algorithm the change δ_j^l needed in the correction of the weights is obtained using the so-called *gradient descent* method, the essence of which is the observation that an error ε plotted against the parameter that causes it **must** show a minimum at some (initially) unknown value of this parameter. By observing the slope of this curve, we can decide how to change the parameter in order to come closer to the sought minimum. In Figure 8-6, the value of the parameter to be changed, i.e., the weight of the neuron, is to the right of the minimum; if the derivative $d\varepsilon/dw$ is **positive**, the new value of the parameter should be **smaller** than the old one and vice versa. In other words, we can write:

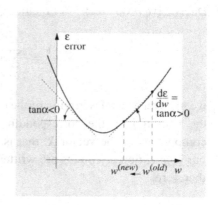

Figure 8-6: The error ε as a function of the weight.

$$\Delta w = w^{(new)} - w^{(old)} = -\kappa d\varepsilon/dw \qquad (8.8)$$

where κ is just a positive numerical scaling factor; note the minus sign.

For a specific weight w_{ji}^l in the layer l, the corresponding equation is:

$$\Delta w_{ji}^l = -\kappa \partial \varepsilon^l / \partial w_{ji}^l \qquad (8.9)$$

For the sake of clarity of presentation we defer the calculation of the error ε^l to a later part of this section and first discuss the dependence of the error on the weight.

This error function represents that part of the error caused by this particular weight at the output of layer l. Because the error function is a complicated and rather indirect function of the parameters w_{ji}^l, we can evaluate the derivative $\partial \varepsilon^l / \partial w_{ji}^l$ in a stepwise manner employing the chain rule:

The chain rule:

$$F = f(z(y(x)))$$

$$\frac{\partial F}{\partial x} = \left(\frac{\partial F}{\partial z}\right)\left(\frac{\partial z}{\partial y}\right)\left(\frac{\partial y}{\partial x}\right)$$

$$\Delta w_{ji}^l = -\kappa \frac{\partial \varepsilon^l}{\partial w_{ji}^l} = -\kappa \left(\frac{\partial \varepsilon^l}{\partial out_j^l}\right)\left(\frac{\partial out_j^l}{\partial Net_j^l}\right)\left(\frac{\partial Net_j^l}{\partial w_{ji}^l}\right) \qquad (8.10)$$

The derivatives of the error function ε^l are calculated consecutively with respect to the values of out_j^l, Net_j^l and w_{ji}^l. Because all derivatives in Equation (8.10) play an important role in the back-propagation algorithm, we will take a closer look at each of them separately, starting with the last one.

Derivative $\partial Net_j^l / \partial w_{ji}^l$. In Equation (3.5) of Section 3.7:

$$\boxed{\partial Net_j^l / \partial w_{ji}^l}$$

$$Net_j^l = \sum_{i=1}^m w_{ji}^l x_i^l \qquad (8.11)$$

we have an exact description of the dependence of the net input Net_j^l of the neuron j on the corresponding set of weights. The x_i^l are the components of the vector X^l that is input to the level l. According to our convention, all inputs are written as outputs from the layer above; therefore:

$$x_i^l = out_i^{l-1} \qquad (8.12)$$

If Equation (8.11) is written as a sum of products, the derivative of Net_j^l with respect to the particular weight can clearly be seen:

$$\frac{\partial Net_j^l}{\partial w_{ji}^l} = \frac{\partial\left(w_{j1}^l out_1^{l-1} + ... + w_{ji}^l out_i^{l-1} + ... + w_{jm}^l out_m^{l-1} \right)}{\partial w_{ji}^l} \qquad (8.13)$$

$$= out_i^{l-1}$$

By inserting Equation (8.13) into Equation (8.10) for the corrections of weights, we obtain:

$$\Delta w_{ji}^l = -\kappa \left(\frac{\partial \varepsilon^l}{\partial out_j^l} \right) \left(\frac{\partial out_j^l}{\partial Net_j^l} \right) out_i^{l-1} \qquad (8.14)$$

Now we see a correspondence of terms between Equations (8.14) and (8.7) representing the delta-rule correction in its expanded form:

$$\Delta w_{ji}^l = -\kappa \left(\frac{\partial \varepsilon^l}{\partial out_j^l} \right) \left(\frac{\partial out_j^l}{\partial Net_j^l} \right) out_i^{l-1} \qquad (8.15)$$

$$\Delta w_{ji}^l = \eta \qquad \delta_j^l \qquad out_i^{l-1}$$

The above comparison leads directly to the delta-term:

$$\delta_j^l = -\left(\frac{\partial \varepsilon^l}{\partial out_j^l}\right)\left(\frac{\partial out_j^l}{\partial Net_j^l}\right) \qquad (8.16)$$

This result is of major importance in the back-propagation model.

Derivative $\partial out_j^l/\partial Net_j^l$. Because the form of the transfer or squashing function (see Section 2.4, Figure 8-7) of the neurons is usually known explicitly, the derivative $\partial out_j^l/\partial Net_j^l$ is not difficult to obtain.

The relationship between out_j^l and Net_j^l was discussed extensively in Section 2.5, where we mentioned the hard-limiter and threshold functions. Although very convenient for evaluation, these two squashing functions are nevertheless often passed over in favor of the more complex sigmoidal transfer function:

$$out_j^l = \frac{1}{1 + exp\left(-Net_j^l\right)} \qquad (8.17)$$

The main reason why (8.17) is used instead of simpler ones is because its derivative can be obtained analytically. As shown in Section 2.5 (Equation (2.32)), function (8.17) can not only be differentiated easily, but its derivative can be expressed in terms of the function itself:

$$\frac{\partial out_j^l}{\partial Net_j^l} = out_j^l\left(1 - out_j^l\right) \qquad (8.18)$$

This property is very convenient for use on computers: the derivative is, so to speak, bundled free with the function. The sigmoidal function (8.17) is not the only one whose derivative can be expressed in terms of the function itself; some other squashing functions having this particular property are discussed in Section 8.5.

Derivative $\partial \varepsilon^l/\partial out_j^l$. This is the last derivative remaining from Equation (8.10). For this derivative, we have to distinguish two cases, depending on whether or not ε^l is explicitly known; in other words, whether the correction is calculated for:

- the last (output) layer, or
- the hidden layers.

$$\boxed{\partial out_j^l/\partial Net_j^l}$$

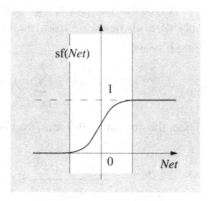

Figure 8-7: Squashing function.

$$\boxed{\partial \varepsilon^l/\partial out_j^l}$$

The first case: correction on the last layer. This case is much simpler to handle; since the back-propagation method is a supervised learning method, the error ε^l at the very last (output) level l is always known:

> The error ε^l in the output is the difference between the expected output (target), $\boldsymbol{Y}$ ($y_1, y_2, ..., y_m$), and the actual output $\boldsymbol{Out}^l$ ($out_1^{\,l}, out_2^{\,l}, ..., out_m^{\,l}$).

Obviously, the error ε^l can be expressed by subtracting the output $out_j^{\,l}$ of each neuron j from the corresponding component y_j of the target vector $\boldsymbol{Y}$:

$$\varepsilon^l = \sum_{j=1}^{n} \left(y_j - out_j^{\,l} \right)^2 \qquad (8.19)$$

Then, the derivative $\partial \varepsilon^l / \partial out_j^{\,l}$ can be obtained easily:

$$\frac{\partial \varepsilon^l}{\partial out_j^{\,l}} = \frac{\partial \left(y_1 - out_1^{\,l} \right)^2}{\partial out_j^{\,l}} + ... + \frac{\partial \left(y_j - out_j^{\,l} \right)^2}{\partial out_j^{\,l}} + ... = \qquad (8.20)$$

$$= -2 \left(y_j - out_j^{\,l} \right)$$

As only the j-th component in this expansion is dependent on $out_j^{\,l}$, only this component gives a nonzero value in the derivation.

The final expression for the evaluation of the weight corrections in the **last** layer of a neural network is obtained by collecting all three derivatives, $\partial Net_j^{\,l} / \partial w_{ji}^{\,l}$, $\partial out_j^{\,l} / \partial Net_j^{\,l}$, and $\partial \varepsilon^l / \partial out_j^{\,l}$ in the above procedure from Equations (8.13), (8.18) and (8.20), and inserting them into the expression for the delta-rule, (8.10).

Because the only error we know exactly (Figure 8-8) comes from the last layer, ε^{last}, the superscript l indicating the neuron's layer must be changed from l to *last*: Furthermore, we substitute η for 2κ.

$$\frac{\partial \varepsilon^{last}}{\partial out_j^{\,last}} = -2 \left(y_j - out_j^{\,last} \right)$$

$$\frac{\partial out_j^{\,last}}{\partial Net_j^{\,last}} = out_j^{\,last} \left(1 - out_j^{\,last} \right)$$

$$\frac{\partial Net_j^{last}}{\partial w_{ji}^{last}} = out_i^{last-1}$$

resulting in:

$$w_{ji}^{last} = \eta \left(y_j - out_j^{last} \right) out_j^{last} \left(1 - out_j^{last} \right) out_i^{last-1} \qquad (8.21)$$

The second case: corrections on the hidden layers. The expression we have not yet evaluated is the case where the explicit relationship between the error function ε^l and the output out_j^l is not known, which is the case in the **hidden** layers l.

In a hidden layer l, the actual output error ε^l cannot be calculated directly, because the "true" values of their outputs are not known (even in supervised learning). Therefore, the derivative $\partial \varepsilon^l / \partial out_j^l$ can be calculated only if we make some assumptions.

One simple, defensible assumption is that the error ε^l produced by the forward process at a given layer l has been distributed **evenly** over all neurons r in the lower layer $l + 1$:

$$\varepsilon^l = \sum_{k=1}^{r} \varepsilon_k^{l+1} \qquad (8.22)$$

The summations run over all r neurons in level $(l + 1)$ (Figure 8-9). Therefore, the error at a level l (which is needed to calculate the corrections of weights on the same level) can be obtained by collecting the errors from the level $l + 1$ below it.

Assuming hypothesis (8.22), the derivative $\partial \varepsilon^l / \partial out_j^l$ is not hard to figure out; by application of the chain rule and use of Equation (8.22), it follows that:

$$\frac{\partial \varepsilon^l}{\partial out_j^l} = \sum_{k=1}^{r} \left(\frac{\partial \varepsilon_k^{l+1}}{\partial Net_k^{l+1}} \right) \left(\frac{\partial Net_k^{l+1}}{\partial out_j^l} \right) \qquad (8.23)$$

The rightmost derivative $\partial Net_k^{l+1} / \partial out_j^l$ is obtained similarly to the derivative described by Equations (8.11) and (8.13). The net input Net_k^{l+1} is written as in Equation (3.5):

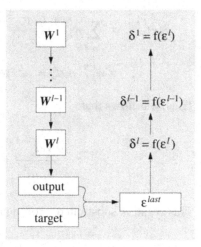

Figure 8-8: For each layer in the network ε^l and $\partial \varepsilon^l / \partial out_j^l$ have to be evaluated. Only the error in the last layer, ε^{last}, is known explicitly.

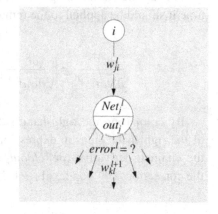

Figure 8-9: Diffusion of error over all lower weights.

$$Net_k^{l+1} = \sum_{j=1}^{m} w_{kj}^{l+1} x_j^{l+1} = \sum_{j=1}^{m} w_{kj}^{l+1} out_j^l =$$
$$= w_{k1}^{l+1} out_1^l + \dots + w_{kj}^{l+1} out_j^l + \dots + w_{km}^{l+1} out_m^l \tag{8.24}$$

Thus, it follows that:

$$\frac{\partial Net_k^{l+1}}{\partial out_j^l} = w_{kj}^{l+1} \tag{8.25}$$

This result is substituted into Equation (8.23):

$$\frac{\partial \varepsilon^l}{\partial out_j^l} = \sum_{k=1}^{r} \left(\frac{\partial \varepsilon_k^{l+1}}{\partial Net_k^{l+1}} \right) w_{kj}^{l+1} \tag{8.26}$$

Due to the differentiation over out_j^l, all terms of the expanded sum (8.24) have vanished except for the j-th term.

The last step in this long story is to apply the chain rule again. This time it should be applied to the remaining derivative $\partial \varepsilon^{l+1}/\partial Net_k^{l+1}$:

$$\frac{\partial \varepsilon^{l+1}}{\partial Net_k^{l+1}} = \left(\frac{\partial \varepsilon^{l+1}}{\partial out_k^{l+1}} \right) \left(\frac{\partial out_k^{l+1}}{\partial Net_k^{l+1}} \right) \tag{8.27}$$

By comparing the right-hand side of (8.27) with the middle parts of the right-hand sides in expressions (8.15), it is easy to deduce that the chained derivative $(\partial \varepsilon^{l+1}/\partial out_k^{l+1})$ $(\partial out_k^{l+1}/\partial Net_k^{l+1})$ is equal to the corrections δ_k^{l+1} on level $l+1$. Hence:

$$\frac{\partial \varepsilon^{l+1}}{\partial Net_k^{l+1}} = \delta_k^{l+1} \tag{8.28}$$

By inserting the derivative $\partial \varepsilon^{l+1}/\partial Net_k^{l+1}$ into Equation (8.23), the following expression is obtained:

$$\frac{\partial \varepsilon^l}{\partial out_j^l} = \sum_{k=1}^{r} \delta_k^{l+1} w_{kj}^{l+1} \tag{8.29}$$

As when correcting the weights in the output layer, the three derivatives $\partial Net_j^l / \partial w_{ji}^l$, $\partial out_j^l / \partial Net_j^l$ and $\partial \varepsilon^l / \partial out_j^l$ are collected from Equations (8.13), (8.18) and (8.29), and inserted into Equation (8.10) for the delta-rule:

$$\frac{\partial Net_j^l}{\partial w_{ji}^l} = out_i^{l-1}$$

$$\frac{\partial out_j^l}{\partial Net_j^l} = out_j^l \left(1 - out_j^l \right)$$

$$\frac{\partial \varepsilon^l}{\partial out_j^l} = \sum_{k=1}^{r} \delta_k^{l+1} w_{kj}^{l+1}$$

resulting in:

$$\Delta w_{ji}^l = \eta \left(\sum_{k=1}^{r} \delta_k^{l+1} w_{kj}^{l+1} \right) out_j^l \left(1 - out_j^l \right) out_i^{l-1} \qquad (8.30)$$

In Equation (8.30) the learning parameter η has the same meaning and the same function as the parameter η mentioned in Chapters 6 and 7 in Equations (6.5) - (6.7) and (7.3), respectively. In the error back-propagation algorithm η is mostly held constant, however, it can be linearly diminishing during learning according to Equation (6.5), already explained in Chapter 6:

$$\eta(t) = (a_{max} \quad a_{min}) \frac{t_{max} - t}{t_{max} - 1} + a_{min} \qquad (6.5)$$

The result obtained in Equation (8.30) distinctly shows how values from **three** different layers are involved in the calculation of the weight correction in the hidden layer l:

– the output out_i^{l-1} of the layer above acting as the input i to the l-th layer,

– the out_j^l of the j-th neuron on the current layer l,

– the correction δ_k^{l+1} of the weight w_{kj}^{l+1} from layer $l+1$.

8.5   Learning Algorithm

Now that we are familiar with the equations used in back-propagation learning, the procedure for the weight correction will be described algorithmically. (For understanding this method, the flow of data in the network and the timing of weight corrections are just as important as the equations.)
The learning procedure involves the following steps:

– input an object X $(x_1, x_2, ..., x_m)$

– label the components x_i of the input object X as out_i^0 and add a component 1 for bias; the input vector thus becomes: $\boldsymbol{Out}^0$ $(out_1^0,$ $out_2^0, ..., out_m^0, 1)$

– propagate $\boldsymbol{Out}^0$ through the network's layers by consecutively evaluating the output vectors $\boldsymbol{Out}^l$; for this, we use the weights w_{ji}^l of the l-th layer and the output out_i^{l-1} from the previous layer (which acts as input to layer l):

$$out_j^l = f\left(\sum_{i=1}^{m} w_{ji}^l out_i^{l-1} \right)$$

where f is the chosen transfer function, e.g. the sigmoidal function.

– calculate the correction factor for all weights in the output layer δ_j^{last}, by using its output vector $\boldsymbol{Out}^{last}$ and the target vector Y:

$$\delta_j^{last} = \left(y_j - out_j^{last} \right) out_j^{last} \left(1 - out_j^{last} \right)$$

– correct all weights w_{ji}^{last} on the last layer:

$$\Delta w_{ji}^{last} = \eta \delta_j^{last} out_i^{last-1} + \mu \Delta w_{ji}^{last\,(previous)}$$

– calculate consecutively layer by layer the correction factors δ_j^l for the hidden layers from $l = last - 1$ to $l = 1$:

$$\delta_j^l = \left(\sum_{k=1}^{r} \delta_k^{l+1} w_{kj}^{l+1} \right) out_j^l \left(1 - out_j^l \right)$$

– correct all weights w_{ji}^l on the layer l:

$$\Delta w_{ji}^l = \eta \delta_j^l out_i^{l-1} + \mu \Delta w_{ji}^{l(previous)}$$

– repeat the procedure with a new input-target pair (X, Y).

Due to the widespread use of back-propagation learning, it is important to comment on some of the steps listed in the learning algorithm above, and to point out some problems which may arise when applying it.

Before the actual learning begins, three things have to be done:

– initial choice of the neural network architecture

– randomization of initial weights, and

– selection of the learning rate η and momentum constant μ.

The initial architecture of the neural network (the number of layers, neurons and weights) is only a starting guess. You may want to modify the architecture after you see how the network performs during the learning or testing phase; this will be discussed later on in this Chapter.

The weights are initialized by setting them to small random numbers. Be sure that not all weights are equal to zero. Usually this is done automatically by the program package, but you will probably be asked to specify the interval within which the weights should be randomized. A typical choice for a layer l is the interval between $\{-1/n, 1/n\}$, n being the number of all weights in that layer. Because the weights will be changed anyway, the exact starting values have no particular significance.

The most important of these choices, apart from the neural network architecture, is the *learning rate constant* η (Equations (8.1) – (8.3)), which determines the speed at which the weights change; if they change too quickly, the procedure may end up in a local minimum (the steepest way downhill does not necessarily lead to the global minimum). The gradient descent method is justified for continuous functions and requires infinitesimally small corrections of weights. This is not possible in the back-propagation approach. The trick is to find a reasonable tradeoff between fast learning and converging to the lowest minimum.

The learning rate constant η is generally obtained by trial and error; good starting values are between 0.3 and 0.6.

The *momentum constant* μ determining the size of the momentum term has a close connection with η. Figure 8-10 shows its influence. In a real multivariate system, of course, the situation is much more complex than shown in this one-dimensional picture; the paths among the local minima are very hard to follow or predict accurately.

As can be seen from Equation (8.1), the momentum term takes into account the **most recent** correction of weights; this is how μ gets its power to prevent sudden changes in the direction in which the solution is being moved.

Let's assume for simplicity that μ = η; if, further, the two terms of Equation (8.1) are equal, then:

$$\delta_j^l out_i^{l-1} = -\Delta w_{ji}^{l\,(previous)} \tag{8.31}$$

This would cause the weight w_{ji}^l **not to be changed at all**, even though the current cycle taken by itself recommends a change equal to $\eta\delta_j^l\,out_i^{l-1}$.

Hence, a value of μ larger than η tends to suppress oscillations, but possibly at the price of overlooking some narrow ways to the global minimum. The learning rate η and the momentum constant μ may need to be systematically changed during the learning process. Then, they are usually decreased during the iteration process.

The inclusion of the momentum term considerably increases the need for computer memory, since in addition to the current values of the weights, we must store their values on the previous cycle as well.

Augmenting the input vectors by including the bias is generally an automatic program feature. Normalization of input vectors, a desirable if not always mandatory procedure, is included in standard packages for neural network calculations by means of algorithms from linear scaling in the required interval, to a statistic auto-scaling that ensures that each variable will have a mean value of zero and a standard deviation of one.

It should be noted that the expression $out_j^l\,(1 - out_j^l)$ (Equations (8.2) and (8.3)) results from the derivative $\partial out_j^l/\partial Net_j^l$ given by Equation (8.18). In general, the squashing function does not always need to have the form of (8.17); if a different squashing function is used, the derivative $\partial out_j^l/\partial Net_j^l$ will have a different form from the one given by Equation (8.18) (see Table 8-1). Therefore, in Equations (8.2) and (8.3), the expression $out_j^l(1 - out_j^l)$ can be different. Sometimes it is advisable to write the derivative $\partial out_j^l/\partial Net_j^l$ in a shorter form:

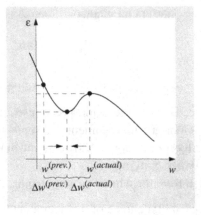

Figure 8-10: Escape from a local minimum is possible if the momentum term is large enough to "push" the weight over the barrier.

$$\frac{\partial out_j^l}{\partial Net_j^l} = \text{f}'\!\left(Net_j^l\right) \qquad\qquad (8.32)$$

Table 8-1 gives a few examples of functions that can be used in the back-propagation evaluation of weight corrections.

As can be seen from Figure 8-11, the second function in Table 8-1, the hyperbolic tangent, tanh(x), is especially attractive. It has its inflection point at the origin and two asymptotes at ±1. This makes it useful in applications where the input data lie between +1 and −1.

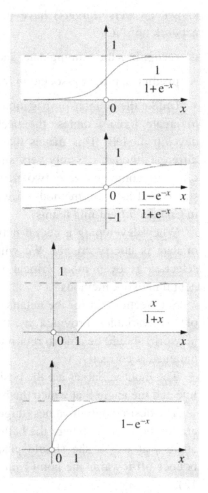

f(x)	f'(x)	f'(f(x))
$\dfrac{1}{1+e^{-x}}$	$\dfrac{-e^{-x}}{\left(1+e^{-x}\right)^2}$	$\text{f}(x)(1-\text{f}(x))$
$\dfrac{1-e^{-x}}{1+e^{-x}}$	$\dfrac{2e^{-x}}{\left(1+e^{-x}\right)^2}$	$1-(\text{f}(x))^2$
$\dfrac{x}{1+x}$	$\dfrac{1}{(1+x)^2}$	$(1-\text{f}(x))^2$
$1-e^{-x}$	e^{-x}	$1-\text{f}(x)$

Table 8-1:   Functions that can be used as transfer functions in artificial neurons; these have the property that the first derivative is expressible in terms of the function itself.

Figure 8-11: Functions for which the derivative can be expressed in terms of the function itself. The hyperbolic tangent, tanh(x), is especially attractive because it is antisymmetric with respect to the coordinate origin.

The last two functions are used only for argument values larger than or equal to 1; for argument values smaller than 1 they are regarded as equal to zero.

In the back-propagation of errors learning scheme, one pass of all objects through the network is called one iteration *cycle* or one *epoch*. As a rule, many hundreds or even thousands of cycles are necessary to achieve convergence in this learning scheme. Because the number of weights is large even for a medium sized net, the convergence can be quite a lengthy procedure, and may not even be achieved at all.

8.6   Example: Tennis Match

There are many different applications where the back-propagation method can be useful in neural networks. Since back-propagation is usually a supervised learning method, training and test sets with known answers (targets) have to be selected and run through the network until it:

– **recognizes** the training data, and

– **predicts** a proper association for new input data.

Unlike the counter-propagation method (Chapter 7), which is used to create lookup tables, the back-propagation method is used to develop models. This means that for each different input vector, a different (though possibly very similar) output vector is obtained. In order to compare these two neural networks (both of which are supervised learning methods), we will use the same example here as in Chapter 7 (learning tennis).

When developing a neural network application, careful selection of data is utterly crucial. We will see from this example that data selection is even more critical in back-propagation than in other neural network methods.

See Section 7.5 for the details of the tennis problem. On the basis of two variables provided by the trainer, the trainee (the neural network) should be able to return the stroke by choosing her own two parameters correctly.

The input variables are the position, x, of the trainer on his service line and the angle β at which the ball moves towards the trainee.

The desired output values (targets) that the trainee is trying to learn are the position, y, where the ball will cross her service line, and the angle γ at which she should place her racket so that the ball will bounce off towards the point z on the trainer's service line (Figure 8-12).

8.6.1   Choice of Data

Before beginning the training, we must construct a set of data that will be representative of all possible cases; here, we will produce sixteen such training pairs, using trigonometry rather than resorting to an actual tennis court.

Theoretically, the pairs of input data and desired outputs (targets), can be expressed as follows (Figure 8-12):

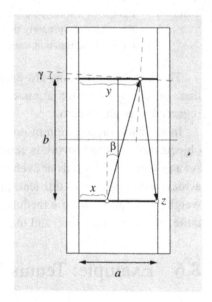

Figure 8-12: Tennis example. Input values are pairs of (x, β), while outputs are pairs (y, γ).

$$y = x + b \tan \beta$$

$$\gamma = 0.5 \left(\beta - \arctan\left(\frac{z - y}{b} \right) \right) \tag{8.33}$$

The tangent function of small angles can be approximated with good precision by the angles themselves; hence, the outputs y and γ can be approximated by:

$$y = x + b\beta$$

$$\gamma = 0.5 \left(\beta - \frac{z - y}{b} \right) \tag{8.34}$$

Regardless of whether the data are simulated or real, the question we are immediately faced with in any type of modeling is: how many data (pairs of input and target variables) do we need to obtain a good model? Ten, a hundred, a thousand?

In standard modeling techniques where an analytical function (say, a quadratic polynomial of two variables) is to be used as a model, the answer is not complicated: there must be at least one more data point than there are parameters in the model (Figure 8-13). In the case of neural networks, however, where there is no analytical form to fall back on, the answer to this question is much harder to come by.

Because the output variables do not seem to be very complicated, let us start by taking sixteen data points evenly distributed over the entire variable space (the tennis court). Table 8-2 lists all sixteen data points (two inputs and two targets each) chosen carefully to represent all of the conditions that can occur.

The angles are given in radians; positive and negative angles indicate clockwise and counterclockwise directions with respect to the perpendicular axis.

8.6.2 Architecture and Parameters of the Network

Once we have the data, we then must decide on the network architecture to be used: how many layers, how many neurons, how many weights?

Because the numbers of input and output neurons are known (two inputs and two outputs – see Figure 8-14), the input and output layers are defined; but the "inner" architecture of the hidden layers is still entirely up to us.

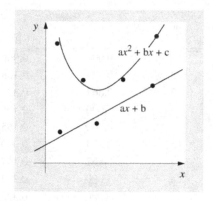

Figure 8-13: Standard models, such as the straight line $(ax + b)$ or parabola $(ax^2 + bx + c)$, need at least one more data point than there are parameters to be determined.

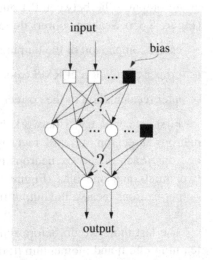

Figure 8-14: The architecture of the network is not yet defined.

no.	inputs		targets	
	x	β	y	γ
1	0.000	0.000	0.000	−0.232
2	0.333	−0.165	0.000	−0.314
3	0.667	−0.322	0.000	−0.393
4	1.000	−0.465	0.000	−0.464
5	0.000	0.165	0.333	−0.078
6	0.333	0.000	0.333	−0.161
7	0.667	−0.167	0.333	−0.243
8	1.000	−0.322	0.333	−0.322
9	0.000	0.322	0.667	0.078
10	0.333	0.165	0.667	0.000
11	0.667	0.000	0.667	−0.083
12	1.000	−0.165	0.667	−0.165
13	0.000	0.465	1.000	0.232
14	0.333	0.322	1.000	0.161
15	0.667	0.165	1.000	0.083
16	1.000	0.000	1.000	1.000

Table 8-2: The sixteen calculated input-target pairs for back-propagation learning in the tennis problem.

In a back-propagation network, the output is obtained directly from the neurons in the output layer. Therefore, it is advisable to scale each component of the target to lie between 0 and 1. Due to the nonlinear character of the transfer function, it is better to scale the entire output to lie between 0.1 and 0.9 or even between 0.2 and 0.8 (Figure 8-15). Scaling confers three advantages:

– easier comparison of the output and target data,

– proper calculation of RMS (root-mean-square) error,

– later recalculation of the correct answer from the output neuron.

First, we will try a network having one hidden layer with six neurons. Later on we will investigate different numbers of hidden neurons. Each of the six neurons has three weights to accommodate two inputs and one bias (Figure 8-16); the seven weights on both output neurons receive the output from each of those six neurons plus a bias.

The last thing we do before running the example is to choose the learning rate η and momentum μ; our first choice will be 0.5 and 0.9, arbitrarily. We shall soon see that this choice is not a very good one.

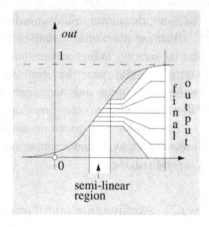

Figure 8-15: Scaling of the output into the linear response region.

Now we can run the first learning experiment with sixteen data points. These are randomly mixed and then input to the above (2 x 6 x 2) – or with biases, (3 x 7 x 2), – network for 1000, 2000, 3000 and 4000 epochs.

Recognition (Recall)

After each input (not after each epoch), all the weights are changed according to Equations (8.2) to (8.4). At the end of each thousand epochs of learning, the RMS error (Equation (7.6)), is calculated. The results of learning the training patterns – what we call the **recognition** or **recall** of the objects – are very good at first glance. The RMS error is 0.008, or slightly less than 1%. Table 8-3 explicitly shows the recall results of the sixteen target pairs. Table 8-3 also indicates what an error of less than 1% actually looks like, in terms of actual data.

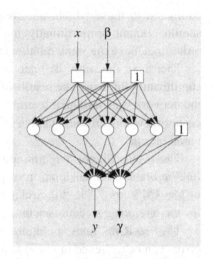

Figure 8-16: The first network used for the application of back-propagation learning to the tennis problem.

no.	y			γ		
	target	output	error	target	output	error
1	0.000	0.003	0.003	−0.232	−0.236	0.004
2	0.000	0.001	0.001	−0.314	−0.307	0.007
3	0.000	0.001	0.001	−0.393	−0.392	0.001
4	0.000	0.001	0.001	−0.464	−0.461	0.003
5	0.333	0.320	0.013	−0.078	−0.073	0.005
6	0.333	0.326	0.007	−0.161	−0.156	0.004
7	0.333	0.315	0.018	−0.243	−0.235	0.008
8	0.333	0.312	0.021	−0.322	−0.325	0.003
9	0.667	0.665	0.002	0.078	0.091	0.013
10	0.667	0.659	0.008	0.000	−0.006	0.006
11	0.667	0.651	0.016	−0.083	−0.082	0.001
12	0.667	0.661	0.006	−0.165	−0.158	0.007
13	1.000	0.998	0.002	0.232	0.223	0.009
14	1.000	0.999	0.001	0.161	0.168	0.007
15	1.000	0.998	0.002	0.083	0.078	0.005
16	1.000	0.996	0.004	0.000	0.009	0.009

Table 8-3:   The output from sixteen training data after 4000 epochs of learning on (2 x 6 x 2) back-propagation network. The error column contains the absolute differences between the targets and the outputs.

It can be seen clearly that the average absolute error is about 0.006 and that it can be as small as 0.001 or as large as 0.021. If the relative errors are important for a given application, one has to be aware that the results at lower values always have **larger** relative errors compared to those which are close to 1. Therefore, the training set should contain proportionally more objects giving low outputs in order to achieve the same relative error over the entire interval.

The next question is: can the learning rate and momentum significantly influence the results of learning? To answer this question the network was retrained several times with the same sixteen objects, each time with a different choice of the learning rate η and momentum parameter μ.

The result (Figure 8-17) is quite interesting. Each *iso*-RMS error line on the two-dimensional η vs. μ chart represents a constant value of the RMS error. The full circles represent the RMS values obtained for the learning procedures actually performed as described above.

The *iso*-RMS lines in Figure 8-17 tell us that reasonable RMS errors can be achieved in our example with higher learning rates, even if the momentum is zero. Using the information obtained from these preliminary runs, the best combination of the learning rate and momentum constant can be selected.

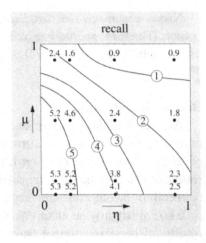

Figure 8-17: Plot of RMS error (in %) in training versus learning rate (η) and momentum parameter (μ), showing lines of constant error.

Prediction

After testing the recall or recognition ability of the network comes the fun part: checking the network's ability to predict output from new inputs.

To do this, we took 1000 random (x, β) pairs and calculated the correct answer pairs, (y, γ), using the "tennis equations" (Equations (8.33)). The calculated values were then compared to the values output by the neural network. See Table 8-4 for a small sample of these results.

Table 8-4 gives a completely different impression from Table 8-3; the errors can be larger by almost an order of magnitude than when simply recalling the training inputs. In the worst case (no. 2 in Table 8-4), the correct position for the player is 0.1 units from the side line, while the program puts her/him almost at the line, 0.008. And the errors in the angle, although at first glance smaller than the errors in position, are quite large, especially since the error in the racket angle is doubled when we consider both the trajectories towards the racket and away from it (after the stroke).

> Always evaluate errors in view of your actual application and needs.

Now let's examine the influence of the learning constant η and the momentum parameter μ on the RMS error. Again, the *iso*-RMS-error lines are plotted against η and μ. The same network as for Figure 8-17 is used for Figure 8-18.

Figure 8-18 shows that the high values of η and μ that yield the network having the **best recall** (recognition) are associated with the **worst prediction** ability.

From both Figures 8-17 and 8-18, we can conclude that for the tennis problem the best choice of these parameters lies in the shaded area of Figure 8-19.

> The sum of η and μ should be more or less equal to one:
>
> $$\eta + \mu \approx 1 \qquad (8.35)$$

It has to be stressed that this relationship will certainly depend on the kind of data being investigated and therefore might not necessarily be the best one in all cases.

Table 8-4 gave us a qualitative indication that our system for modeling tennis is not very good; Figure 8-19 confirms this: with the present training data set and present size of the neural network (2 x 6 x 2) we cannot expect better results than about 4% RMS error. How can we make our "tennis brain" smarter?

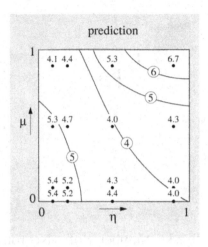

Figure 8-18: Plot of RMS error (in %) in prediction output versus learning rate and momentum, showing lines of constant error.

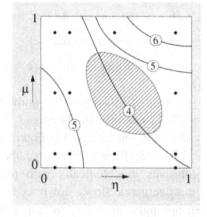

Figure 8-19: The optimum region for the learning rate η and momentum μ for the tennis problem.

no.	y calc.	y output	y error	γ calc.	γ output	γ error
1	0.074	0.020	0.054	−0.207	−0.208	0.001
2	0.097	0.008	0.089	−0.414	0.439	0.025
3	0.348	0.337	0.011	−0.301	−0.296	0.005
4	0.412	0.385	0.073	−0.048	−0.044	0.004
5	0.671	0.665	0.006	−0.011	−0.017	0.006
6	0.755	0.833	0.078	0.108	0.125	0.013
7	0.902	0.985	0.083	0.110	0.111	0.001
8	0.937	0.993	0.056	0.187	0.201	0.014

Table 8-4: Output from new test data, presented after training: 4000 epochs in the (2 x 6 x 2) back-propagation network with $\eta = 0.5$ and $\mu = 0.9$. The error column contains absolute differences between the targets and the outputs.

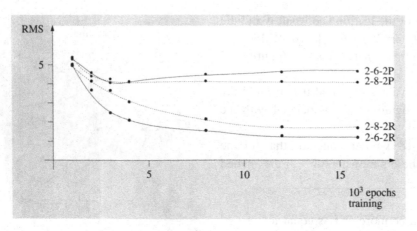

Figure 8-20: Recall (R) and prediction (P) error vs. number of training epochs.

8.6.3 Number of Neurons in the Hidden Layer: Overtraining

There are three things that can still be tried to improve the results:

- enlarge the number of learning steps (epochs), i.e. train the network longer,

- change the network design,

- expand the training data set.

(In our further attempts, we will hold η and μ at the values 0.5 and 0.6.)

First, let's increase the number of training steps from 4000 to 8000, 12000, and 16000 epochs. The resulting RMS errors for recall and prediction are given in Figure 8-20.

The recall performance of the original network (bottom curve) improves monotonically with longer training (though the rate of improvements slows down considerably after 4000 epochs); but, surprisingly, its ability to handle new data (top curve) actually gets worse!

This is known as the *overtraining effect*. Overtraining can be explained as a consequence of parameter redundancy; that is, the system has more parameters than are needed for the solution of the problem. In curve-fitting, we might see this in a polynomial with too many terms: it can make a "better" fit to a set of data by adapting to, rather than smoothing out, the "wiggles" caused by noise.

For example, suppose we are fitting three points in the xy-plane to a quadratic function $ax^2 + bx + c$ instead of to a straight line $ax + b$ (Figure 8-21). In spite of the fact that the straight line cannot go

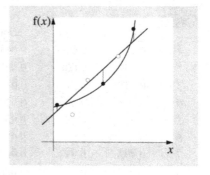

Figure 8-21: In constructing a model, more adjustable parameters are **not** necessarily better. In this example, a three-parameter curve fits the data better than a two-parameter (linear) one — but it may be fitting noise. The "worse" fit may make better predictions.

through all three points (full circles) exactly, as the quadratic function can, it is possible that the prediction of other points (not taken into account before; open circles) will be much better if based on the straight line than based on the quadratic function.

Based on the overtraining data, it appears that learning with about 4000 epochs should be optimum in our case.

The next thing to consider is changing the neural network design and/or the training data set. While the network design can be changed at any time, getting more data is sometimes not as easy as it may seem. In the tennis example, it is easy to calculate as many as we like; but when the data come from an expensive experiment, it's a different story!

> If your problem just does not involve enough different data, you might do well to consider some other, more standard approaches and forget about neural networks.

If, in spite of having a small data collection, you still insist on using the neural network approach as a solution to your problem, then your best chance is to find the **smallest** adequate design!

In the tennis problem, we will try both things: changing the design of the network, and training the network with larger numbers of objects.

For the first, we select four designs: (2 x 4 x 2), (2 x 6 x 2) (the original), (2 x 8 x 2), and (2 x 10 x 2) (Figure 8-22); for the second, we select three training sets containing 16, 36, and 100 objects at specific locations in the variable space, and three more containing 100, 200, and 300 randomly chosen ones. These six data sets are used to test all four networks to see which combination gives the best results.

The RMS errors for recall and predictions for these twenty-four cases are very informative. First of all, increasing the number of neurons neither improves the recall nor the prediction when only sixteen training data are used. The same behavior can be observed in all other cases too, although to a lesser extent; when there are 300 training data, the situation is not so clear, because with more data in the training set we are allowed to have more weights to adjust (Table 8-5 and Figures 8-23 and 8-24).

Nevertheless, it can be seen from Figure 8-24 that the best prediction ability is achieved when the number of weights is

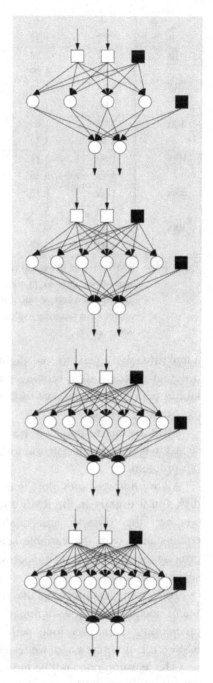

Figure 8-22: Four networks (with 22, 32, 42, and 52 weights, resp.) used for back-propagation learning in the tennis example.

h	4	6	8	10	
o \ w	22	32	42	52	
16	2.91	2.10	3.05	3.25	R
	4.66	4.09	4.11	4.08	P
36	3.26	1.79	1.93	2.19	R
	4.06	3.45	3.42	3.51	P
100	2.07	1.60	1.25	1.34	R
	3.50	3.37	3.24	3.51	P
100r	1.60	1.11	0.95	0.93	R
	3.66	3.36	3.29	3.29	P
200r	1.49	1.12	0.80	0.70	R
	3.46	3.31	3.23	3.23	P
300r	1.44	0.77	0.70	0.70	R
	3.41	3.17	3.17	3.18	P

Table 8-5:   RMS error of the recall (R) and prediction (P), as a function of the size of the network (h = number of hidden neurons; w = total number of weights) and the size of the training set (o = number of objects; r = randomly selected). In all cases, the training period was 4000 epochs.

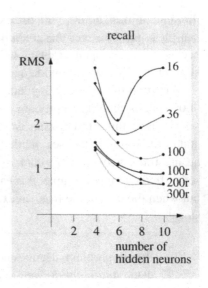

Figure 8-23: The RMS error of recall as a function of the number of neurons in the hidden layer for different sizes of the training sets. The numbers labeled with r (e.g. 100r) refer to randomly chosen training sets.

approximately equal to or slightly smaller than the number of different objects in the training set. When the number of weights is larger than the number of **different** objects in the training set, overtraining appears.

After all this testing, we are finally in a position to select what seems to be the appropriate training set and the best suited network for our problem.

All training sets with more than 100 objects generate networks that differ only slightly in the RMS error of the predictions, regardless of whether the objects are randomly selected or are sampled equidistantly over the variable space. More than 200 objects would prolong the learning time unnecessarily; thus, a choice of 200 objects seems optimum.

Concerning architecture, the (2 x 8 x 2) design with 52 weights seems adequate: it shows a remarkable stability in the RMS error for predictions, even over long periods of training, and gives the best RMS error of all networks we tested (Figure 8-24).

(The learning rate and the momentum were 0.5 and 0.6. Any other values within the ±0.1 interval would do equally well).

The network obtained after 4000 epochs is shown in Figure 8-25.

The predictive results of the network shown at the right are not overwhelmingly good (RMS error of 3.1). Nevertheless, they show very well what can be expected from back-propagation learning: what the problems are and how to tackle some of them.

Additional examples, with practical hints and descriptions of problems, are given in Part IV.

Modeling is not the most favorable example of back-propagation learning in neural networks; for example, sorting multivariate objects into classes might be more interesting or even more exciting, certainly more "successful"; but this example better illustrates the method's limitations.

In classification problems, the output layer consists of as many neurons as there are classes of objects, and the RMS error at each output neuron need not be as small as in the tennis example. For applications producing binary/bipolar outputs, it is adequate for the "1" output to lie between 1.0 and 0.6, and "0", at 0.4 to 0.0. Modeling a system with continuous output values, on the other hand, requires a precision of at least a few percent.

A number of authors claim that their back-propagation networks are able to **generalize** the solution of the problem; by this, they mean that the network yields reliable answers even outside the area of the variable space from which the objects for the training are taken. Unless it is shown beyond a reasonable doubt that the network has not been overtrained (i.e., that it does not fit the training data so closely that its predictive ability is compromised), such claims are, to put it mildly, exaggerated.

> The most important aspect of designing a network for back-propagation learning is to ensure that it will not become overtrained.

See Section 16.5 for a detailed comparison of the back-propagation and counter-propagation learning techniques.

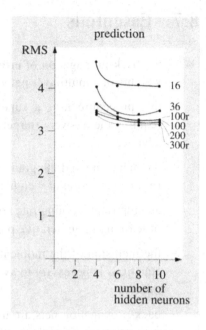

Figure 8-24: The RMS error of predictions as a function of the number of neurons in the hidden layer and the sizes of the training sets.

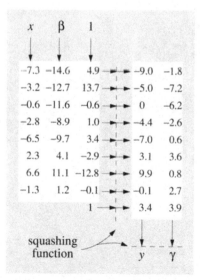

Figure 8-25: The (2 x 8 x 2) neural network obtained after 4000 epochs.

8.7   Essentials

- the back-propagation of errors enables the correction of all weights in a multilayer network

- this procedure uses a supervised learning method, i.e. it requires the answers (targets) to the inputs to be known in advance

- the delta-rule and the gradient descent method are the basis for the correction of weights

- an empirical learning rate constant, η, determines the speed of learning in an iterative procedure

- the inclusion of the momentum term, μ, into the corrective equations is necessary to avoid being trapped in small local minima

- back-propagation nets are used as models, especially when the analytic form of the model relationship is not known

- finding the proper network design (number of layers, number of neurons, and number of weights) is usually a trial and error procedure

- **total weight change:**

$$\Delta w_{ji}^{l} = \eta \delta_{j}^{l} out_{i}^{l-1} + \mu \Delta w_{ji}^{l\,(previous)} \qquad (8.1)$$

error in the last layer

$$\delta_{j}^{last} = \left(y_{j} - out_{j}^{last} \right) out_{j}^{last} \left(1 - out_{j}^{last} \right) \qquad (8.2)$$

error in the hidden layer

$$\delta_{j}^{l} = \left(\sum_{k=1}^{r} \delta_{k}^{l+1} w_{kj}^{l+1} \right) out_{j}^{l} \left(1 - out_{j}^{l} \right) \qquad (8.3)$$

- **required weight change:**

in the last layer

$$\Delta w_{ji}^{last} = \eta \left(y_{j} - out_{j}^{last} \right) out_{j}^{last} \left(1 - out_{j}^{last} \right) out_{i}^{last-1}$$

$$(8.21)$$

in the hidden layer

$$\Delta w_{ji}^{l} = \eta \left(\sum_{k=1}^{r} \delta_{k}^{l+1} w_{kj}^{l+1} \right) out_{j}^{l} \left(1 - out_{j}^{l} \right) out_{i}^{l-1} \qquad (8.30)$$

if changing of learning rate is required:

$$\eta(t) = (a_{max} - a_{min}) \frac{t_{max} - t}{t_{max} - 1} + a_{min} \qquad (6.5)$$

8.8   References and Suggested Readings

8-1.   P. Werbos, "Applications of Advances in Nonlinear Sensitivity Analysis", in *System Modeling and Optimization: Proc. of the Int. Federation for Information Processes*, Eds.: R. Drenick, F. Kozin, Springer Verlag, New York, USA, 1982, pp. 762 – 770.

8-2.   D. E. Rumelhart, G. E. Hinton and R. J. Williams, "Learning Internal Representations by Error Propagation", in *Parallel Distributed Processing: Explorations in the Microstructures of Cognition*, Eds.: D. E. Rumelhart, J. L. McClelland, Vol. *1*, MIT Press, Cambridge, MA, USA, 1986, pp. 318 – 362.

8-3.   W. P. Jones and J. Hoskins, "Back-Propagation, A Generalized Delta-Learning Rule", *Byte*, October 1987, 155 – 162.

8-4.   L. B. Elliot, "Neural Networks – Conference Update and Overview", *IEEE Expert*, Winter 1987, 12 – 13.

8-5.   R. P. Lippmann, "An Introduction to Computing with Neural Nets", *IEEE ASSP Magazine*, April 1987, 4 – 22; P. D. Isserman and T. Schwartz, "Neural Networks, Part 2: What are They and Why is Everybody so Interested in Them Now?", *IEEE Expert*, Spring 1988, 10 – 15.

8-6.   T. Kohonen, "An Introduction to Neural Computing", *Neural Networks* **1** (1988) 3 – 16.

8-7.   D. S. Touretzky, D. A. Pomerleau, "What's Hidden in the Hidden Layers?", *Byte*, August 1989, 227 – 233.

8-8.   W. Kirchner, "Fehlerkorrektur im Rückwärtsgang, Neuronales Backpropagation-Netz zum Selbermachen", *c't*, Heft 11 (1990) 248 – 257.

8-9.   J. Dayhoff, *Neural Network Architectures, An Introduction*, Van Nostrand Reinhold, New York, USA, 1990.

8-10.   H. Ritter, T. Martinetz and K. Schulten, *Neuronale Netze, Eine Einführung in die Neuroinformatik selbstorganisierender Netzwerke*, Addison-Wesley, Bonn, FRG, 1990.

8-11.   J. Zupan and J. Gasteiger, "Neural Networks: A New Method for Solving Chemical Problems or Just A Passing Phase?", *Anal. Chim. Acta.* **248** (1991) 1 – 30.

Part IV
Applications

9 General Comments on Chemical Applications

learning objectives:

- the types of problems to which neural networks can be applied

- multivariate (multiple input) and multiresponse (multiple output) systems

- how to choose between supervised and unsupervised learning

- classification of objects into categories and hierarchies of categories

- why modeling is simpler with neural nets than with classical techniques ... and how that very simplicity can be a disadvantage

- how mapping can be used in chemistry

- how neural nets can be used in process feedback and control systems

- the crucial role of data representation in neural network applications

- what is a moving window approach

- overview of the examples presented in Chapters 10 to 20

9.1 Introduction

In chemistry, as in all natural sciences, we would like to learn new methods that can improve, shorten, or bring new insight into old ways of handling experimental data. Part IV will show how different neural network architectures and learning strategies can be applied to some of the problems encountered in the everyday practice in chemistry.

While we can't show all possible applications, we will discuss certain **types** of problems that can be dealt with by neural networks.

The neural network approach is basically a method for handling multivariate and multiresponse data. *Multivariate* data are used to describe **one** object with **several** variables: for example, analysis of air samples for the pollutants NO_x, SO_2 and CO would comprise three-variate data.

If such multicomponent data are studied with respect to several factors (responses) such a problem is generally also described as *multiresponse* (Figure 9-1).

The relation between a multivariate object $X_s = (x_{s1}, x_{s2}, ..., x_{sm})$ and the putative factors Y_s $(y_{s1}, y_{s2}, ..., y_{sn})$ can be written in the following way:

$$(y_{s1}, y_{s2}, ..., y_{sn}) = A (x_{s1}, x_{s2}, ..., x_{sm}) \qquad (9.1)$$

Here, A is a $(m \times n)$-variate matrix that linearly transforms vector X_s into vector Y_s. The index s, according to the convention of Section 1.4, indicates different samples: $s = 1, 2, ..., r$.

Equation (9.1) is the linear (and therefore simplest) version of the multivariate multiresponse problem (A can actually be a very complex operator).

In many applications, A is not known; it may be that all you have is a set of carefully collected m-variate data $\{X_s\}$ accompanied by a set of n-variate responses $\{Y_s\}$. And in some applications, even the responses $\{Y_s\}$ are not known and have still to be figured out. From such sets of carefully measured multivariate data, conclusions are sought that are valid for unknown samples.

The key to finding these answers is that the sought information is already hidden in the multivariate data. For example, an infrared spectrum measured with a resolution of 1 cm^{-1} between 4000 and 200 cm^{-1} comprises 3800 intensity points; these data contain the complete information about the structure of the compound – say, which 15 groups make up the molecule, out of a pool of 200 possibilities. The essence of this 3800-variate 200-response problem is to find out how to confirm or exclude any of the 200 functional groups based on each of the 3800-point spectra (Figure 9-2).

Problems of this type are usually not solved by *ab initio* (theoretical) calculations; instead, they require various statistical, pattern recognition, and, as we want to show, neural network methods.

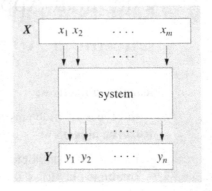

Figure 9-1: Multivariate input X producing multiresponse output Y.

As has been stressed previously, learning can be either supervised or unsupervised. In supervised learning, the system must adapt itself to yield the known correct answers to all query objects in the training set, while unsupervised learning just maps the objects according to some internal criterion, such as similarity, into the "virtual" area defined by the architecture of the network.

The choice of a supervised or an unsupervised approach depends on the problem and the data available to solve it. In both cases, objects with known answers are needed. In supervised learning, the answers are directly used to influence the learning system, i.e., to calculate changes in the weights of the neural network; in unsupervised learning, the answers are needed only to identify and label the output neurons.

The basic question to consider when deciding on the learning method is: do you want to force the system to adapt itself to an already selected representation of objects and classes, or do you want to keep your options open? An unsupervised neural network method is more flexible due to its many possible outputs. Thus, unsupervised learning can be used as a screening step, allowing you to inspect the behavior of the "response space". After enough knowledge is accumulated, you might be able to switch to supervised learning.

For supervised learning, the multivariate objects should be divided into three sets:

- a training set

- a control set for determining when to stop training (see overtraining, Section 8.6.3), and

- a test set for checking the achieved predictive ability.

For unsupervised learning, we don't need a control set, since learning has to continue until the network stabilizes. (This is achieved in a Kohonen network when the neighborhood of the corrections shrinks to a single neuron.)

The next decisive factor for the choice of the appropriate learning strategy is the number of available objects. The number of objects within reach for all three sets of data (training, control, and test) is very relevant to the efficiency of the method. A supervised learning procedure can take tens or even hundreds of thousands of epochs (an epoch is one pass through all the objects in the input set). Furthermore, for each object, the corrections of thousands of weights might be required.

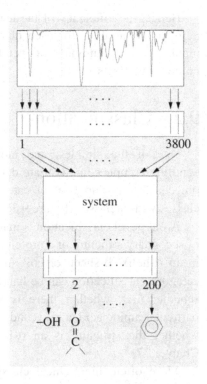

Figure 9-2: A 3800-variate spectrum is input to the system, and a 200-variate output points to specific chemical fragments.

Hence, the efficiency of the learning process is influenced by the number of variables for each object (which defines the size of the input layer), the number of weights in the hidden layers, and the number of objects.

9.2   Classification

One of the simplest and most frequently used operations in handling complex multivariate data is *classification*, which may sort objects into a simple set of categories, or into a hierarchy of sub-classes within classes (Figure 9.3).

A one-level classification is made, for example, in finding the type of secondary structure of a protein: whether a particular amino acid is in an α-helix, β-sheet, or coil structure. Even such a simple three-category classification can be handled by a hierarchical classification scheme: first, whether there is any defined structure around the particular amino acid, and second, if a structure is confirmed, to decide whether the structure is an α-helix or β-sheet (Figure 9-3) (see Chapter 17).

More often, hierarchical classification involves many levels of decisions, classifying an object into the proper group, sub-group, category, sub-category, or class: for example, the prediction of different structural features in an unknown compound on the basis of its spectrum, or the selection of the proper chromatographic method for a given analysis.

In many chemical applications, the object belongs to several **different classes simultaneously**, so that the result of the classification is a product of two or more decisions.

The spectrum-structure correlation problem is a typical multidecision classification: the "object" is the spectrum, while the output is a list of fragments (classes) present in the structure that produced the spectrum. Figure 9-4 shows the hierarchy for deciding which of the five structural fragments A, B, C, D, and E are present in a structure represented by its spectrum. On the first level, a four-category decision is made: whether either, both or neither of the fragments A and B are present in the structure ("$\varnothing$" = $\overline{A} \wedge \overline{B}$ = neither; "AB" = $A \wedge B$ = A and B).

For each of theses decisions, an analogous second-level decision is made regarding the fragments C and D, while at the final level a two-category decision about E is made.

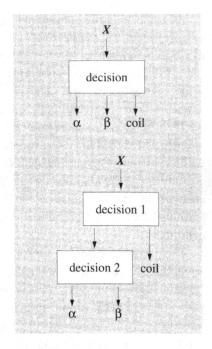

Figure 9-3: One-level and hierarchical classification.

Figure 9-4 shows one of many possible decision hierarchies that can be set up to solve the problem.

Usually, when faced with the task of setting up a classification hierarchy for a complex decision, you should begin with a statistical study of the relationship between the representation of objects and their class membership. You should consult the literature on this matter (Section 9.8: Massart et al., *Chemometrics*, or Zupan, *Algorithms for Chemists*).

For our purpose, it is enough to say that setting up the hierarchy of decisions has to be done **before** the learning procedure is started; if the final results are not satisfactory, it is advisable to look at the representation of the objects and/or the way the decision hierarchy was set up, change one or both, and try the entire procedure again.

9.3 Modeling

Some classifications systems produce a **discrete** answer, such as yes or no, or an integer identifying the input object with one of several classes. However, *modeling* requires a system that yields a **continuous** answer for each input. In modeling, a relatively small number of data are used to build a model that can give predictions for **all** possible objects. Curve-fitting (Figures 8-13, 8-15, 9-5) falls into this category; in most applications, it is a one-variable/one-response procedure:

$$y = f(x, a, b, ..., p) \qquad (9.2)$$

Here, the function f contains a set of unknown parameters a, b, ..., p (in the case of polynomial models, the number and types of terms are chosen to reproduce the degree of curvature anticipated to be in the data). These parameters have to be determined to minimize the discrepancy between the set of experimentally determined answers $\{y_s^{experimental}\}$ and the set of answers obtained by the model $\{y_s^{model}\}$ at the **same** values of x. The following criterion is often used:

$$\left(y_s^{experimental} - y_s^{model}\right)^2 \rightarrow minimum \qquad (9.3)$$

Modeling can also be a multivariate/one-response problem, or even a multivariate (x)/multiresponse (y) problem; in the latter case, we need to find a **different** model for **each** response y_i, i.e., a different function form f_i with a different parameter set.

In any kind of modeling, a tacit assumption is always made:

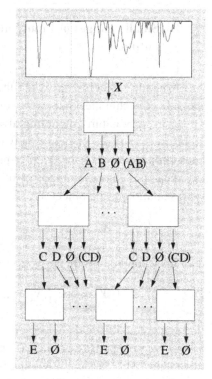

Figure 9-4: Hierarchical classification of an infrared spectrum according to five classes of functional groups, A, B, C, D, and E. A means that only group A is present, in (AB) both groups are present, in Ø both groups are absent.

> **Small changes of input** will cause **small changes in output**.

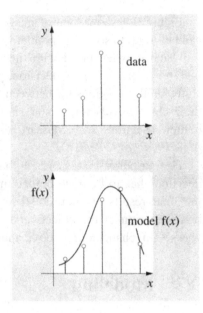

Figure 9-5: One-variable/one-response model.

("Chaotic" systems, in which this requirement is relaxed, require specialized techniques that are beyond the scope of this book.)

By its very nature, modeling always requires supervised learning; therefore two types of neural networks are generally applicable: the back-propagation and the counter-propagation methods. In modeling, the data are always composed of two parts. The first part is the representation of the object X_s:

$$X_s = (x_{s1}, x_{s2}, ..., x_{sj}, ..., x_{sm}) \qquad (9.4)$$

and the second one is the answer or target, Y_s:

$$Y_s = (y_{s1}, y_{s2}, ..., y_{sj}, ..., y_{sn}) \qquad (9.5)$$

that is expected from the object X_s.

In multiresponse problems, neural networks have a major advantage over classical modeling by analytical functions:

In neural networks the targets Y_s ($y_{s1}, y_{s2}, ..., y_{sj}, y_{sn}$) may be multi-dimensional!

What does this mean? Simply that a **multiresponse** answer can be modeled without bothering with different analytical functions or coefficients (Figures 9-6 and 9-7).

> Classical models require a separate predefined analytical function for each response, while neural networks can model a multiresponse vector without any *a priori* knowledge!

$$y_1 = f_1(x_1, ..., x_m, a_{11}, a_{12}, ..., a_{1p})$$
$$y_2 = f_2(x_1, ..., x_m, a_{21}, a_{22}, ..., a_{2r})$$
$$\vdots$$
$$y_n = f_n(x_1, ..., x_m, a_{n1}, a_{n2}, ..., a_{ns})$$

The functions
$f_1(\cdot), f_2(\cdot), ..., f_n(\cdot)$
are valid.

All parameters
a_{ij}
must be determined.

Figure 9-6: Classical modeling.

Of course, as a wise man said, for every silver lining there's a cloud; in classical modeling, the coefficients of the model function can often be given a physical interpretation (e.g., the virial coefficients of a gas). There is no such thing as the coefficient of a given variable in the neural network approach because **all** variable values are shared among **all** weights (albeit not to the same extent). Great effort has been expended (so far, with little success) to find methods that would associate specific weights or paths through the network with given variables or combinations of them.

In many applications, especially when the experimental error is comparatively large, we can simplify the above-mentioned proportionality between input and output changes. In such cases, a lookup table (Figure 9-8) is adequate, even though it gives the same answer for a range of inputs.

Modeling is mainly used for objects having a few (one to ten) variables, which produces neural networks much smaller and computationally less demanding than those used for classification problems.

In the selection of a learning strategy, the distribution of the objects within the variable space has to be considered. If the objects for training are few and more or less evenly spaced (some experimental techniques make this possible), then the back-propagation method is suitable. On the other hand, if the objects are plentiful and scattered irregularly through the variable space, they should first be reduced to cover the space evenly and then the counter-propagation method is used. (The reduction of objects can be carried out either by a Kohonen network (Chapter 6, and the examples in Chapters 10 and 11), or with a statistical evaluation of the intervals of the variables.)

Basically, a counter-propagation method used for modeling needs more objects than back-propagation.

9.4 Mapping

Mapping is the transformation of an *m*-dimensional space into a space of lower (often 2 or 3) dimension (see Section 6.4) in order to display some features that cannot be shown in a higher-dimensional space. All kinds of mappings can be achieved with Kohonen networks.

To some extent, mapping is similar to the clustering or classification of objects. The difference lies not so much in the method itself as in the interpretation of the results. In classification, we want to identify which cluster or class a given object belongs to, while in mapping we focus on the entire map and, in effect, **derive** the clusters from the data.

The essence of mapping with neural networks is this:

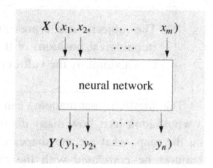

Figure 9-7: Modeling with neural networks.

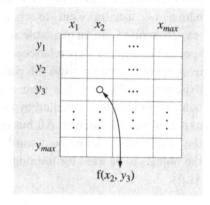

Figure 9-8: A lookup table acting as a model.

> The objects are represented by the **coordinates** (topological position) of the central neuron (Section 6.3) and **not** by the **values** of the output.

The questions that mapping can answer are: what does the map as a whole look like, how many distinguishable areas can be identified on the map, what are the shapes of these areas, how can the map's features be correlated with the objects from which the map was obtained.

A mapping can be either linear or nonlinear, as illustrated in Figure 9-9.

One of the useful applications of mapping involves a preprocessing method (usually called "experimental design"), by which we usually want to choose the most appropriate objects (experiments) from an available set, for example, selecting spectra for spectrum-structure correlations, or eliminating redundant (or overlapping) combinations of parameters to be used in a series of experiments. In the latter case, we would monitor how often each neuron in the plane is excited by an object (Figure 9-10), e.g., a set of experimental parameters. All but one of the objects that excite or fire the same neuron should be discarded. (This is also useful for selecting the objects to be used for training a back-propagation net; see Section 11.4).

9.5 Associations; Moving Window

Among neural network applications not yet mentioned, the following should be used more often in chemistry:

– auto- and hetero-associations

– prediction of time-dependent events

The problem of *association* is to find a target object associated with an input, even if the input is corrupted or incompletely known, for example, when identifying the baseline type in a recorded spectrum, or the type of spectrometer malfunction responsible for a particular bad spectrum. In a sense, then, auto- and hetero-associations can be regarded as classifications.

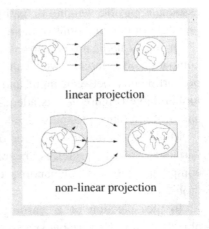

Figure 9-9: Mapping the world globe into a plane.

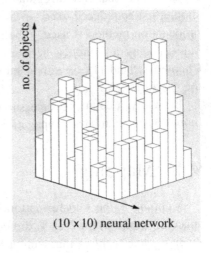

Figure 9-10: Distribution of objects that excite specific neurons on the mapping plane.

In both examples, neither the patterns X (spectra) nor the target responses Y (baselines or malfunction types) can be defined exactly. For example, it is easy to see whether the baseline of an infrared spectrum is concave or convex, but hard to describe the difference mathematically. It's even worse in the second example, where the "target" may be anything from a badly prepared sample to drift in the electronic circuits.

Because associations can be learned only through examples, i.e., through input-target pairs, we are restricted to neural networks capable of performing supervised learning.

Investigation of *time-dependent processes* is one of the main topics in process control research (see Section 16.4). Here, the user wants a model that will predict the behavior of a multiresponse system based on a series of data taken over time.

In time-dependent modeling, the input and output variables are basically the same: the only difference is that the input consists of **present and past** values of process variables, while the output predicts the **future** values of the same variables (Figure 9-11, for more details see Chapter 16, particularly Table 16-1). The consecutive sets of variables taken as one input vector and the consecutive sets of future values on the output side are called the *past* and the *future horizons of learning*.

In principle, these horizons can be of arbitrary length, but in process control the future horizon usually covers only one time step. The *moving window* shown in Figure 9-11 encompasses five consecutive time events; the past horizon is of length four: the input contains all process variables at time t_{-3}, t_{-2}, t_{-1}, and t_0 (the current value), while the process variables beginning with t_1 (in the future horizon) are taken as targets. The "future" data used for training are obtained either from theoretical models or from actual past observations.

Let's construct a vector P_t containing the values at time t of some process variables, say x_1, the flow rate; x_2, the input temperature and x_3, the output temperature:

$$P_t = (x_{1t}, x_{2t}, ..., x_{mt}) \qquad (9.6)$$

Then, taking into account **three** consecutive steps in the process, the input vector X looks like this:

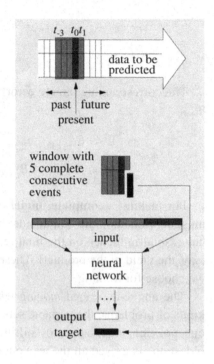

Figure 9-11: Present, past and future data as input/output pairs in time-dependent models.

$$X = (x_{1t}, x_{2t}, ..., x_{mt}, \underbrace{x_{1, t+1}, x_{2, t+1}, ..., x_{m, t+1}}, \underbrace{x_{1, t+2}, ...})$$
$$\quad\quad\quad \underbrace{}_{P_t} \quad\quad\quad\quad\quad \underbrace{}_{P_{t+1}} \quad\quad\quad \underbrace{}_{P_{t+2}} \quad (9.7)$$

The corresponding target vector Y can consist of the fourth vector P_{t+3}:

$$Y = (x_{1, t+3}, x_{2, t+3}, ..., x_{m, t+3})$$
$$Y = \quad\quad\quad P_{t+3} \quad\quad\quad\quad\quad\quad\quad (9.8)$$

For making a **complete model** of a process, **all** process variables must be trained on the output side; but not all variables in the process have equal influence on the final results, and only few are of interest (say, the yield of the product). These significant variables are the ones we choose for training.

The above-described *moving window* procedure is used in many kinds of problems; the variable sets under the window all shift **left** by one position (equivalent to shifting the window **right**), so that the oldest one is moved all the way out (and discarded), and what was the first "future" value becomes the "present" (time = zero) value.

This technique can be used for any problem that deals with a sequence of events or objects, from predicting environment parameters in chemical process plants, to deducing the secondary structure of a protein based on the sequence of amino acids.

We hope that these examples have prepared you to believe that:

> The applicability of neural networks is limited only by your imagination.

9.6 Overview of the Examples in Chapters 10 to 20

The diversity and number of applications of neural networks in chemistry has increased dramatically in the last few years. This is reflected in the number of publications dealing with the use of neural networks to solve chemical problems:

In the rest of the book we will present some of these applications. With such a large number to choose from, we can necessarily give only a fragmentary and personal overview.

year	no. of publications
1988	3
1989	5
1990	20
1991	105
1992	290
1993	441
1994	498
1995	743
1996	855
1997	927

Table 9-1:   Number of publications on neural networks in chemistry per year.

In the rest of the book we will present some of these applications. With such a large number to choose from, we can necessarily give only a fragmentary and personal overview.

While, in general, different types of problems require different neural network architectures and different learning strategies, it turns out that 90% of the problems described until now in the chemical literature have used one-hidden-layer neural networks, and the back-propagation learning strategy.

One of our goals is to encourage you to think about the **problem** first and then select the proper neural network, rather than trying to bend all your problems to the one neural network method you happen to be familiar with.

See Table 9-2 for some hints about the kinds of problems that can be dealt with using different neural network learning methods.

Our examples cover the area as broadly as possible and illustrate the diversity of potential applications. For example:

– various chemical disciplines

    – analytical chemistry (Chapters 10, 12, 18),
    – organic chemistry (Chapters 11, 13, 14, 18, 19),
    – pharmaceutical and biochemistry (Chapters 13, 17, 19, 20),
    – chemical engineering and chemical industry (Chapters 15 and 16).

168 General Comments on Chemical Applications

strategy / problem	back-propagation	counter-propagation	Kohonen network
classification	✻	✻	✻
modeling	✻	(✻)	
mapping	(✻)	(✻)	✻
association	✻	✻	
moving window	✻	✻	

Table 9-2: Neural network learning strategies and their application to different types of problems. An asterisk indicates whether this learning method can be applied to the indicated problem type.

- various types of applications

- different neural network learning methods

- different sizes of neural network architectures and datasets

Table 9-3 summarizes the essential features of the examples in the following Chapters.

chapter	problem	type of problem	size of the network	method
10	origin of olive oils	classification + mapping	medium	BPE + KL
11	bond reactivity	classification + mapping	small	KL + BPE
11	reaction classification	classification	medium	KL
12	HPLC separation	modeling	small	BPE
13	QSAR	modeling	small	BPE
13	QSAR	classification + modeling	medium	KL + BPE
13	QSAR	variable selection + modeling	medium	GA + CP
14	electrophilic aromatic substitution	modeling	small	BPE
15	paint coat recipe	modeling	small	BPE
16	fault detection, process control	classification + modeling	small	BPE + CP + MW
17	protein structure	classification	large	BPE + MW
18	infrared spectrum/structure correlation	classification + mapping	large	BPE + KL
18	infrared spectrum simulation	classification + modeling	large	CP
19	molecular surfaces	mapping	large	KL
20	chemical libraries	classification	large	KL

Table 9-3: Main features of the examples in Chapters 10 – 20; BPE: back-propagation of error, KL: Kohonen learning, CP: counter-propagation, MW: moving window, GA: genetic algorithm.

That's why we have intentionally not arranged the chapters according to the subdisciplines of chemistry. Rather, we want to help you develop an understanding and an eye for the types of problems –

classification, modeling, mapping – that can be tackled by neural networks.

> Before attempting to design a neural network solution to a problem, you should first **classify** the problem and only then attempt to identify the type of neural network and the learning method most appropriate to solve it.

We start with classification problems (Chapters 10 and 11) and then turn our attention to modeling problems (Chapters 12 – 16). However, Chapter 16 also contains a classification problem and classification problems are the main theme again in Chapters 17 and 18 (these were placed so late in the book because they involve such large network architectures and datasets). Chapter 19 deals with a mapping problem, where we will show that **artificial neural networks** can help us understand **biological neural networks**.

Mapping problems are also mentioned in Chapters 10, 11, 18, and 20. This brings up another important point:

> Quite often, a certain application contains aspects of several different types of problems and should therefore be attacked with **several different neural network methods simultaneously**.

In Chapters 10 – 20 we do exactly that – we look at many of the problems using multiple neural network methods: multilayer networks with learning by back-propagation of errors (BPE), counter-propagation network (CP), Kohonen learning (KL), and the moving window input scheme (MW).

A (perhaps the) most important issue in any application of neural networks is the representation of information that is fed into the network and/or obtained from it. Since neural networks are such general-purpose problem-solvers, it is all the more important to be very careful in choosing an appropriate representation of the information that goes into and leaves it. We therefore put strong emphasis on the proper representation of chemical information. The representation chosen has to be adapted to the problem.

As a case in point, the problem of how to represent the structures of organic compounds is addressed – and solved in different ways – in the examples shown in Chapters 11, 13, 14, 17, 18, 19, and 20.

Because of the overall importance of structure representation in many chemical applications we have added a special chapter (Chapter 21) that deals with this problem and collects the various methods that have been used in the different chapters.

Bernard Widrow stressed the importance of proper data representation this way:

"The three most important issues that must be addressed in the development of neural networks are:

1. representation

2. representation

3. representation"

9.7   Essentials

Classification

Either supervised or unsupervised learning may be used in creating a classification engine; in the former, one-hidden-layer neural network architectures with back-propagation learning are used, and Kohonen networks with the latter.

Modeling

Modeling applies only to systems in which changes in the output are proportional to the changes in the input. Modeling always requires supervised learning, and therefore mainly two types of neural networks are applicable: the back-propagation and the counter-propagation methods.

Neural net models are simpler to implement than with classical methods, because of having no need for adjustable parameters; but they also lack the potential for physical interpretation that those parameters possess.

Mapping

The essence of mapping is reduction of dimensionality. The objects are represented by the **coordinates of the position** of the "central" neuron and **not** by the **values of the output**.

9.8 References and Suggested Readings

9-1. D. L. Massart et al., Eds., *Chemometrics Tutorials*, Elsevier, Amsterdam, NL, 1990.

9-2. S. N. Deming and S. L. Morgan, *Experimental Design: A Chemometrics Approach*, Elsevier, Amsterdam, NL, 1987.

9-3. D. L. Massart, B. G. M. Vandeginste, L. M. C. Buydens, S. De Jong, P. J. Lewi and J. Smeyers-Verbeke, *Handbook of Chemometrics and Qualimetrics: Part A*, Elsevier, Amsterdam, NL, 1997.

9-4. B. G. M. Vandeginste, D. L. Massart, L. M. C. Buydens, S. De Jong, P. J. Lewi and J. Smeyers-Verbeke, *Handbook of Chemometrics and Qualimetrics: Part B*, Elsevier, Amsterdam, NL, 1998.

9-5. K. Varmuza and H. Lohninger, "Principal Component Analysis of Chemical Data", in *PCs for Chemists*, Ed.: J. Zupan, Elsevier, Amsterdam, NL, 1990, pp. 43 – 64.

9-6. K. Varmuza, *Pattern Recognition in Chemistry*, Springer-Verlag, Berlin, FRG, 1980.

9-7. J. Zupan and J. Gasteiger, "Neural Networks: A New Method for Solving Chemical Problems or Just a Passing Phase?", *Anal. Chim. Acta* **248** (1991) 1 – 30.

9-8. J. Zupan, *Algorithms for Chemists*, John Wiley, Chichester, UK, 1989.

9-9. J. Gasteiger and J. Zupan, "Neuronale Netze in der Chemie", *Angew. Chem.* **105** (1993) 510 – 536; "Neural Networks in Chemistry", *Angew. Chem. Int. Ed. Engl.* **32** (1993) 503 – 527.

9-10. J. Zupan, "Neural Networks in Chemistry", in *Encyclopedia of Computational Chemistry*, Eds.: P. v. R. Schleyer, N. L. Allinger, T. Clark, J. Gasteiger, P. A. Kollman, H. F. Schaefer III and P. R. Schreiner, Wiley, Chichester, UK, 1998, pp. 1813 – 1827.

9-11. D. J. Livingstone, G. Hesketh and D. Clayworth, "Novel Method for the Display of Multivariate Data Using Neural Networks", *J. Mol. Graphics* **9** (1991) 115 – 118.

9-12. B. Kocjancic and J. Zupan, "Application of a Feed-Forward Artificial Neural Network as a Mapping Device", *J. Chem. Inf. Comput. Sci.* **37** (1997) 985 – 989.

9-13. M. Novic and J. Zupan, "Computer-Aided Identification", in *Encyclopedia of Analytical Science*, Vol. 2, Ed.: A. Townshend, London, Academic Press, UK, 1995, pp. 817 – 825.

9-14. J. Zupan, M. Novic and I. Ruisanchez, "Tutorial: Kohonen and Counterpropagation Artificial Neural Networks in Analytical Chemistry", *Chemom. Intell. Lab. Syst.* **38** (1997) 1 – 23.

10 Clustering of Multi-Component Analytical Data for Olive Oils

learning objectives:

- how chemists analyze samples of olive oils for eight fatty acids, and from this can determine from which of nine regions in Italy the oils come

- the difference in classification ability between back-propagation and Kohonen learning

- how a Kohonen network can "associate" the analyses and the regions and learn to classify an oil sample as to region, given the analysis

- the way to select the appropriate Kohonen network architecture

- the significance of empty spaces in a Kohonen map, and how to deal with cases where objects are mapped into empty spaces.

10.1 The Problem

Monitoring the origins of goods is an important application of analytical chemistry in the food industry as well as among consumer protection groups.

The problem of determining geographical origin is not as hard as it sounds, because you are generally choosing from a limited set of possibilities; in fact it may be better if we rephrase the problem: how can we show that the object actually comes (or does not come) from the place named on the label. Hence, the problem is reduced to a standard **classification**.

In addition, there is the problem of how to present such results to the customer. Consumers, especially consumer activists, are not likely to accept a short answer like "yes" or "no".

In fact, there are lots of reasons for giving clear and easy-to-understand presentations of analytical results; so, the chosen classification method must show a clear picture, and must be easy to justify; it should be robust enough to allow an easy classification of unknown objects. Such a robust approach to the classification problem can be made by mapping the original multivariate objects into a two-dimensional plane and assigning (or trying to assign) the clusters of object projections formed on the map to the sought categories.

Before a procedure is accepted as a reliable classification technique, it still has to be tested with additional "unknown" objects. If the proposed mapping procedure does not provide a reliable classification, either the mapping method or the representation of the objects has to be changed.

Before the arrival of neural networks, the best method for mapping multivariate data into a two-dimensional plane was *Principal Component Analysis* (PCA). This method first calculates the correlation matrix, then diagonalizes it to obtain the eigenvalues and eigenvectors. Finally, it transforms the original data into new ones by using the matrix of eigenvectors as a transformation matrix. The map is obtained by plotting the transformed data against whatever two of the new components bring the largest portion of the information into the correlation matrix (Figure 10-1). Although the entire procedure can be made completely transparent to the user, such complex statistical calculations are hard to explain to the general public.

Therefore, a simpler method seems to be desirable. We will show here that this can be achieved by a Kohonen neural network.

Of course, if only the classification of objects is needed, the back-propagation method can be used just as well. In the following section, the same set of data is treated by both methods in a number of networks, each having a different architecture.

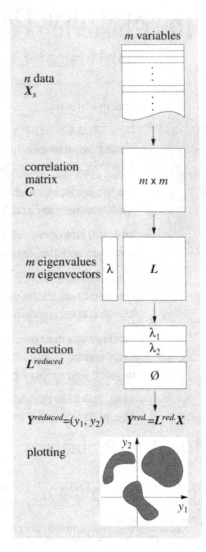

Figure 10-1: The flow of activities necessary to map data by a principal component analysis (PCA).

10.2 The Data

In order to show how problems of classifying multivariate objects can be treated by neural networks, we will use a dataset that has been extensively studied by various statistical and pattern recognition methods. This data set consists of analytical data from 572 Italian olive oils produced in nine different regions of Italy (Reference 10-1). For each oil, a chemical analysis determined the percentage of the following eight different fatty acids:

– palmitic

– palmitoleic

– stearic

– oleic

– linoleic

– arachidic (eicosanoic)

– linolenic

– eicosenoic

Figure 10-2: Italian regions used in this study.

This formidable analytical work produced a matrix of almost five thousand values (572 x 8 = 4576) (each of which a result of a careful analysis!).

Because the proportions of some fatty acids may differ by two orders of magnitude, all values belonging to a given variable were normalized.

Table 10-1 shows how the Italian regions were numbered, and how many different oils were analyzed from each part. Figure 10-2 shows the Italian regions and their numbers.

no.	region	no. of samples
1	North Apulia	25
2	Calabria	56
3	South Apulia	206
4	Sicily	36
5	Inner Sardinia	65
6	Coastal Sardinia	33
7	East Liguria	50
8	West Liguria	50
9	Umbria	51
total		572

Table 10-1: Origin and number of samples of each oil used in this study.

(The data, which are widely known among chemometricians, were kindly supplied to us by Professor Forina of the University of Genoa, Italy, for which we express our sincere thanks.)

Various groups of researchers have investigated this particular dataset using a number of statistical treatments and inspection methods, including various clustering methods, principal component and discriminate analysis. The results of these studies are mainly published in the journals of chemometrics and analytical chemistry. Some of these studies are listed in Section 10.6.

10.3 Preliminary Exploration of Possible Networks

We started by trying a simple classification of all 572 objects of four different back-propagation and four different Kohonen neural networks. We just wanted to see how different types of networks would react to these data. The results are summarized in Table 10-2; the back-propagation trials were aimed at finding the most promising architecture (i.e., the best recall), and the Kohonen to estimate the proper size for the resulting map.

network	type	dimension	weights	time [min]	errors or conflicts
1	BPN	8 x 5 x 1	51	15.0	138
2	BPN	8 x 10 x 1	101	20.0	71
3	BPN	8 x 5 x 9	108	45.0	23
4	BPN	8 x 8 x 9	153	75.0	15
5	Kohonen	10 x 10 x 8	800	1.0	24
6	Kohonen	15 x 15 x 8	1800	2.0	14
7	Kohonen	17 x 17 x 8	2512	2.5	14
8	Kohonen	20 x 20 x 8	3200	3.0	12
(Kohonen networks used the criterion of Equation (6.1))					

Table 10-2:   Characteristics of neural networks used in the preliminary step. (BPN: back-propagation network; time measured on IBM 387 compatible)

In the back-propagation network, "best recall" was defined as the smallest number of objects that cannot be learned after 400 epochs.

Two criteria were used to evaluate the Kohonen maps: the number of conflicts, and the amount of empty space. A *conflict* occurs when two objects from different classes trigger the same neuron. The number of conflicts is given in Table 10-2. An *empty space* is a neuron that is not triggered by any of the 572 oils. Too many empty spaces indicate either that the network is too large for the given set, or that the network did not spread the objects well enough across the projection plane.

In designing a back-propagation network, we have to decide; first, whether to use one or nine output lines; and second, how many neurons should be in the hidden layer. The answer to the first question is not hard: the networks with nine outputs perform far better than those with one (in Table 10-2, compare the number of errors in networks 3 and 4 against 1 and 2).

But the question about the number of neurons in the hidden layer is much harder. Fortunately, there is no need to fix the number of hidden neurons exactly at the optimum value in small networks like the ones used here. The time needed to learn the classification using the entire group of 572 objects was on the order of ten minutes on a SUN Sparc workstation. Hence, computational time is not a limiting problem. However, a network too large can increase the time considerably, with no payback in improved results; finally, we chose back-propagation networks with about 150 weights. (Don't forget to count the bias in the number of weights (see Section 2.5). Biases on Figure 10-3 are shown, as usual, as black squares.

Considering this particular problem, it can be said that the (20 x 20) neuron matrix is slightly too large for the given data set, since many of the objects are rather similar to each other and end up firing the same neuron. If we later separate the data into a training set of 250 and a test set of 322 objects, the (20 x 20) matrix would be populated too sparsely. Therefore (and for another reason that will be explained in the next section), a slightly smaller matrix was eventually chosen.

Another reason for selecting these particular back-propagation and Kohonen networks was the fact that other studies on the classification of these oils (including Principal Component Analysis (PCA), K-nearest neighbor technique (KNN), SIMCA, or three-distance clustering (3-DC)) gave a comparable number – 10 to 20 – of misclassifications or errors. The numbers of bad recalls and conflicts ("errors or conflicts" in Table 10-2) are comparable with the results from these other methods.

The objects that cause the errors or conflicts are either wrongly assigned or contain excessive experimental error. In any case, in a pool of almost 600 complex experiments, it is hard to make fewer than 2% errors in the analytical determinations.

Table 10-2 also contains the time each network needs to learn the 572 objects. The training time of the Kohonen networks is more than one order of magnitude smaller than for the back-propagation

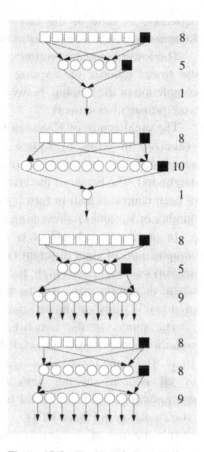

Figure 10-3: The back-propagation networks used in the preliminary study.

networks, in spite of the fact that the number of weights in the Kohonen networks is an order of magnitude larger (Figure 10-4).

The lower computation times of the Kohonen networks are due to the lower number of learning periods (epochs) required for the completion of the training: between 6 and 15, compared to 400 for the back-propagation network.

The architecture of Kohonen networks used in this study can be visualized in terms of square bricks; the base of such a brick accommodates the neurons (each of which has eight closest neighbors). The height of the brick depends on the number of weights in each neuron, which in turn is related to the number of inputs (the number of variables representing each object).

In our case, each object is represented by eight variables (the compositions of eight different fatty acids). Thus, all Kohonen neural networks are eight units high; they differ only by their bases, the areas where the future maps will be formed. All four Kohonen networks used in this study are shown schematically in Figure 10-4.

The inputs to the networks are treated as eight-dimensional column vectors (shown at the left-hand side of each network in Figure 10-4). The same input vector (complete analysis of an olive oil) goes to **all** neurons in the network simultaneously. However, **one component** of the input vector is connected to **one layer** of weights (shown darker in Figure 10-4).

NOTE: the map of oil samples was produced by training the Kohonen network **without** considering toroidal boundary conditions (see Section 6.2) when making the corrections. Thus, the projection was not made onto the surface of a torus but onto a normal two-dimensional plane.

Figure 10-5 shows the map obtained from 572 records of olive oil data with the (20 x 20 x 8) Kohonen network. With respect to geographical origin, regions which are topologically close to each other are mapped into areas of the map that are also close together.

For example, the Sardinian oils – both those from the inner and those from the coastal regions, classes 5 and 6 – are separated from the rest by an obvious empty region. The oils from the northern parts – the Liguria and Umbria regions, oils 7, 8, and 9 – form a tight cluster in the upper part of the map. Those from the southern parts of Italy – from Apulia, Calabria, and Sicily, oils 1, 2, 3, and 4 – are again clearly separated from the rest by an U-shaped region extending from left to right.

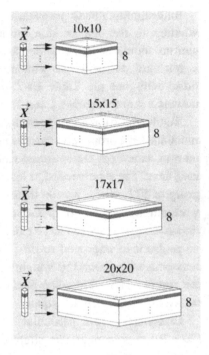

Figure 10-4: The Kohonen networks used in this study.

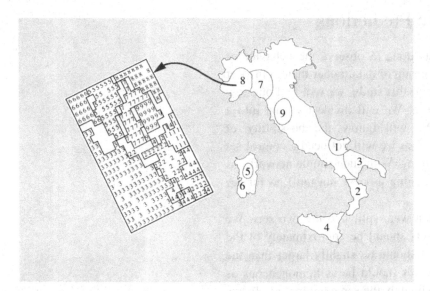

Figure 10-5: Mapping the data for olive oils on the (20 x 20) neural network.

The only significant inconsistency in the Kohonen map (compared with the actual map of oil growing regions in Italy) is the position of the South Apulian oils (class 3), which is in the lower left instead of the lower-right corner of the map. However, this correspondence between the geographical map and the Kohonen map in this example is purely fortuitous.

This clustering is remarkably good, considering the simple learning scheme used to produce it. Not only is there a clear gap between the southern and the northern oils, there is even a clear separation between the two types of Sardinian oils, and of those two types from the rest. We can safely say that there is a clear correlation between the topology of labeled regions in the Kohonen map and the actual positions of the oil-producing areas in Italy.

Two groups that are not as homogeneous as others are the Sicilian and Calabrian oils, which intersect each other. It is interesting to note that most of the errors in classification are produced by the oils from these two regions. This exception is quite instructive: even if excellent results are produced overall, we should remember that our input objects do not actually have any explicit geographical information embedded in their representation.

10.4 Learning to Make Predictions

All this preliminary work was done to observe the behavior of different networks with a certain group of data, rather than to inspect the data itself. In the second part of this study, we will find out how good our networks are at **learning**. We will divide the data into a training set and a test set. We will ignore the possibility of overtraining (Section 8.6.2); therefore we will not need a control set for signaling when to stop the training. With such simple networks as these, it is sufficient to let the learning process run until no further improvements can be detected.

Table 10-3 shows how the data were split into the two sets. We used two conditions: first, both sets should be approximately of the same size (if possible, the test set should be slightly larger than the training set); second, the training set should be as homogeneous as possible (it should contain approximately the same number of objects from each class). An good division is 270 objects for training and 302 for testing (the 270 comprise 30 objects from each of the 9 classes). But not all classes have equal numbers of data, so there has to be some tradeoff between smaller and larger groups. Table 10-3 shows the final distribution of the data objects.

no.	region	label	training	test
1	North Apulia	NA	15	10
2	Calabria	CA	35	21
3	South Apulia	SA	40	166
4	Sicily	SI	20	16
5	Inner Sardinia	IS	30	35
6	Coastal Sardinia	CS	20	13
7	East Liguria	EL	30	20
8	West Liguria	WL	30	20
9	Umbria	UM	30	21
total			250	322

Table 10-3:   Splitting the data into the training and test group.

The objects within each group were selected at random with no prior inspection of their behavior, or knowledge of how they group together.

The training set was first used with the four back-propagation networks described above (Table 10-2). Although it was clear from the beginning that the first two networks (with only one output neuron each) will yield worse predictions than those with nine, they were included in the experiment to show examples of bad design (Table 10-4).

All learning procedures were carried out by applying the set of equations given in Section 8.7 and using value of 0.2 and 0.4 for the learning rate η and the momentum μ.

network	dimension	errors	
		recall from 250	predictions from 322
1	8 x 5 x 1	104 (42)	121 (38)
2	8 x 10 x 1	89 (36)	117 (36)
3	8 x 5 x 9	5 (22)	25 or 27 (8)
4	8 x 8 x 9	1 (0.4)	31 (10)

Table 10-4:   Prediction ability of different back-propagation neural networks (percentages in parentheses).

The networks with only one output neuron gave almost 40% wrong predictions and thus clearly are inadequate for a nine-class identification. On the other hand, it is surprising that the network with eight neurons in the hidden layer is less successful in learning compared to the one having five (as demonstrated by the number of errors in predictions for the test set). Although the difference is not very large, this demonstrates that larger networks do not necessarily yield better performance.

Thus, the best performer is the (8 x 5 x 9) network, which gives only 25 wrong classifications for the 322 data in the test set. This 92% prediction ability can be considered quite good. The number 27 shown together with number 25 in Table 10-4 means that two more objects would be classified wrongly if a stricter criterion were selected for class membership. The issue of criteria for class membership will be explained soon.

Now let's compare the desired output (targets) and the actual output values.

For a neural network with nine output neurons, we have to provide nine-element target vectors *Y*. In the ideal case, only one output neuron should have an output signal equal to 1 (associated with a particular input class), while all other output neurons produce zero. The 9-variable targets are coded as follows:

class	NA	CA	SA	SI	IS	CS	EL	WL	UM	
Y (class 1) = (	1	0	0	0	0	0	0	0	0	)
Y (class 2) = (	0	1	0	0	0	0	0	0	0	)
Y (class 3) = (	0	0	1	0	0	0	0	0	0	)
Y (class 4) = (	0	0	0	1	0	0	0	0	0	)
Y (class 5) = (	0	0	0	0	1	0	0	0	0	)
Y (class 6) = (	0	0	0	0	0	1	0	0	0	)
Y (class 7) = (	0	0	0	0	0	0	1	0	0	)
Y (class 8) = (	0	0	0	0	0	0	0	1	0	)
Y (class 9) = (	0	0	0	0	0	0	0	0	1	)

The actual output from the networks was seldom like this; more often the output values of neurons that should be zero were around 0.1, and the values of the neurons signaling the correct class had values ranging from 0.95 to 0.5.

All of the 25 wrong predictions produced the largest signal on the **wrong** neuron. But when a stricter criterion is used, the two additional cases have to be considered. Both have their largest output on a neuron signaling the **correct** class, but in one case this largest signal was only about 0.45, with all other signals being much smaller; in the second case, the largest signal of the correct neuron was 0.9, but the second largest signal was well over 0.55.

Strictly speaking, the total error in both of the anomalous answers was more than 0.5; so, we could simply dismiss them as being wrong. But "right" and "wrong" are relative terms in science; if we consider only the neuron with the largest signal, regardless of its absolute strength, both can be regarded as correct answers.

The second learning experiment was set up on the Kohonen networks (see Table 10-2). These were trained by the same 250-member training set as used for the back-propagation networks, and then tested for predictions with the remaining 322 member test set.

Because a Kohonen network is usually not used for making predictions, we have to discuss in more detail how this can be done.

To review, the result of the Kohonen network is a **two**-dimensional map of assigned neurons, each of which carries a "tag" or "label" of the object that excited it at the final recall test. If more than one object excites one neuron, it is hoped that all of them belong to the same class. Even more so, it is expected that the neurons most excited by objects of the same class will form clusters or small regions on such a map.

If a test object falls into such a cluster, it can be classified as belonging to the group corresponding to this cluster. The region where the objects of a certain class excite the neurons can form tight borders with regions formed by other classes, or such regions can be separated by empty spaces, corresponding to neurons not excited by any object from the training set. Sometimes, empty spaces appear within the region of a class (Figure 10-6). An input object that maps into an empty space may still be classifiable, as we will now see.

Along tight borders of two or more regions, it can easily happen that the **same neuron is excited by two objects belonging to different classes**. Such cases are called *conflicts* and the neurons are called *conflicting neurons*. Conflicts can occur in the recall process, but much more often they happen during the later prediction phase. If the input object excites a neuron in the wrong region, this is clearly a conflict situation. On the other hand, if the excited neuron corresponds to an empty region, the class membership of the **neighboring** neurons can help us decide whether to consider it a conflict.

Before trying to settle this question, let's inspect the predictions made by different Kohonen networks. Table 10-5 shows the prediction abilities of eight Kohonen networks. Four were obtained from our preliminary investigation and four from additional investigations. This table contains the correct and wrong classifications, as well as the numbers of empty spaces.

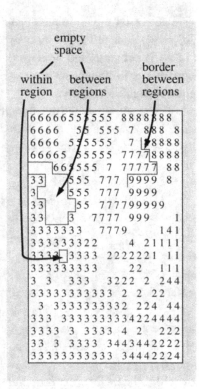

Figure 10-6: Borders between regions, and empty spaces between and within the regions representing classes.

no.	dimension	network map size	learning empty spaces	conflicts	predictions correct	hits into empty space	hits into wrong space
1	20 x 20 x 8	400	193	0	216	108	8
2	17 x 17 x 8	289	108	2	215	96	11
3	15 x 15 x 8	225	64	1	251	60	11
4	10 x 10 x 8	100	19	2	280	25	17
5	7 x 7 x 8	49	5	9	290	3	29
6	5 x 5 x 8	25	2	27	285	0	37
7	4 x 4 x 8	16	2	33	247	0	75
8	3 x 3 x 8	9	0	57	267	0	55

Table 10-5:   Performance of various Kohonen networks in learning and predicting geographical origins of olive oils.

All objects were mapped by exactly the same Kohonen learning procedure with the same training set of 250 objects, the only difference being the size of the network.

The **learning** produced only two conflicting neurons in the (10×10) and (17×17) networks, one conflicting neuron in the (15×15) network, and none in the (20×20) network. Thus, it can be said that a Kohonen network of adequate size has a good recall or good recognition ability.

The **prediction** ability of each network was tested with the same set of 322 oils, and it was found that an increase of the size of the Kohonen maps improves the prediction ability from 17 to only 8 mistakes; however, the number of hits into empty spaces increases at the same time by a factor of four: from 25 to 108.

The factor of four correlates with the increase the size in the Kohonen map from 100 to 400 neurons. It makes sense that the number of empty spaces ("unused" neurons) will increase, as the network size increases, and that the number of hits into empty spaces also depends on network size. As Figure 10-7 shows, the latter relationship is linear – within limits. For networks much larger than (20×20), there will be many more empty spaces, but because of the limited number of data the number of hits into empty spaces will level off. With very small maps, the number of empty spaces cannot be linearly related to the number of empty space hits because there wouldn't be any. Rather, the number of conflicts increases (Table 10-5).

Based on this discussion, we can see, that our network sizes were well chosen.

Because there always exist a certain number of repulsive interactions between certain classes, it is hard to generate a map completely covered by hits with no empty spaces. To investigate this, we generate four more Kohonen networks producing 49-, 25-, 16-, and 9-neuron maps. It was found that only the (3×3) map does not contain any empty spaces; the (4×4) and the (5×5) maps have two, while the (7×7) map has five empty spaces. So we see that, in designing Kohonen networks a compromise between the number of hits into empty spaces and the number of conflicts has to be found.

As Table 10-6 shows, when we test the network for prediction, there are quite a number of hits into empty spaces. It seems therefore worthwhile to explore these cases in more detail.

In order to make a guess about class membership for objects that map into empty spaces, we can try the K-nearest neighbor (KNN) technique, which determines the class by counting a number k of

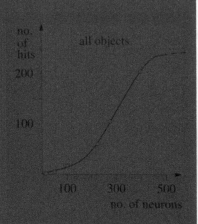

Figure 10-7: Number of hits into empty space as a function of map size.

closest neighbors, with the majority determining the class of the central object.

class	no. of objects	network						
		10 x 10		15 x 15			17 x 17	20 x 20
1	10	0	1	(1	–	–)	3	5
2	21	3	6	(4	2	–)	7	9
3	166	20	32	(32	–	–)	59	52
4	16	2	6	(3	–	3)	7	8
5	35	0	1	(1	–	–)	3	11
6	13	1	2	(–	–	2)	0	4
7	20	0	5	(3	2	–)	5	10
8	20	0	7	(7	–	–)	10	6
9	21	0	0	(–	–	–)	3	2
total	322	26	60	(51	4	5)	96	108

Table 10-6:   Number of hits into empty spaces in predictions on the 322-object test set.

A detailed count was made for all 60 empty-space hits of the (15 x 15) map. The numbers of correct, undecided and wrong classifications for each class are given in parentheses in Table 10-6.

It was found that 51 of the hits would be classified correctly based on the majority vote of the eight closest neighbors. Four were undecided, which means that an equal number of these neighbors belong to two different classes; and only five would be classified wrongly on this basis. Now, we can add these figures to the corresponding figures (Table 10-5) for correct and wrong answers, 251 and 11, respectively, predicted by the (15 x 15). Altogether, we infer that by considering the KNN decision for empty space hits **in addition** to the predictions made by hits in labeled areas, the (15 x 15) network produces 302 (= 51 + 251) correct, four undecided, and 16 (= 5 + 11) wrong classifications.

Figure 10-8 shows the case where an unlabeled neuron, i.e., an empty space, has two neighbors from class 2, two neighbors from class 3 and four additional blanks.

Comparing Figures 10-8 and 10-9, it appears that the neighborhoods of the wrong hits were quite different from those of the hits into "empty space". Working with a well-diversified test set (Table 10-3), we were able to check predictions involving a variety of topological configurations.

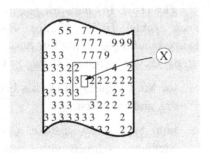

Figure 10-8: Out of eight neighbors of an unlabeled neuron, two are members of class 2 and two are members of class 4; the rest are blanks. The prediction cannot be made.

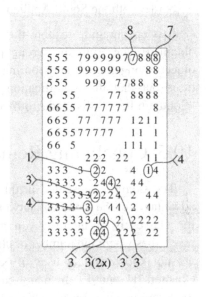

Figure 10-9: The 11 cases in the (15 x 15) network for which the predictions were wrong.

The (15 x 15) network makes a total of 11 wrong classifications (see Table 10-5); Figure 10-9 shows all neurons (circles) corresponding to wrong hits during the 322-object test.

In the eleven "wrongs" shown in Figure 10-9, it can be seen that

- the West Liguria oil (8) has been assigned as an East Ligurian one (7), and vice versa,

- both Sicilian oil mistakes (4) were predicted on the borders of Sicilian oils (once with North Apulia, (1), and once with South Apulia, (3)),

- all six South Apulia oil mistakes were made within (five cases) or in contact with (one case) the North Apulia oils, (4),

- only one case: the North Apulian oil, class (1), triggered a neuron far away from its own group. But even this "far away shot" places the North Apulian sample into the Calabrian region, (2), on the border with the South Apulian ones.

It is encouraging to note that some valuable information can be obtained even from the wrong predictions. All the errors involve objects being out into neighboring classes in the geographical sense; it turns out that the identification of origins as "Northern Italy" or "Southern Italy" was made 100% correctly for all samples.

10.5 Concluding Remarks

The example presented in this chapter was selected to show the advantages of the Kohonen network. It is useful when the topology of the classes is of interest; you may also use it in all preliminary researches where the number of clusters and the relations among them are not known. Other uses of Kohonen networks are discussed in Chapters 11 and 19. In many cases however, the back-propagation methods can give considerably better results than a Kohonen network.

10.6 References and Suggested Readings

10-1. M. Forina and C. Armanino, "Eigenvector Projection and Simplified Non-linear Mapping of Fatty Acid Content of Italian Olive Oils", *Ann. Chim. (Rome)* **72** (1982) 127 – 143; M. Forina, E. Tiscornia, ibid. 144 – 155.

10-2. M. P. Derde and D. L. Massart, "Extraction of Information from Large Data Sets by Pattern Recognition", *Fresenius' Z. Anal. Chem.* **313** (1982) 484 – 495.

10-3. M. P. Derde and D. L. Massart, "Supervised Pattern Recognition: the Ideal Method?", *Anal. Chim. Acta* **191** (1986) 1 – 16.

10-4. J. Zupan and D. L. Massart, "Evaluation of the 3-D Method with the Application in Analytical Chemistry", *Anal. Chem.* **61** (1989) 2098 – 2182.

10-5. D. L. Massart and L. Kaufmann, *Interpretation of Analytical Chemical Data by the Use of Cluster Analysis*, Wiley, New York, USA, 1983.

10-6. B. Everitt, *Clustering Analysis*, Heineman, London, UK, 1975.

10-7. D. L. Massart, B. G. M. Vandeginste, L. M. C. Buydens, S. De Jong, P. J. Lewi and J. Smeyers-Verbeke, *Handbook of Chemometrics and Qualimetrics: Part A*, Elsevier, Amsterdam, NL, 1997.

10-8. B. G. M. Vandeginste, D. L. Massart, L. M. C. Buydens, S. De Jong, P. J. Lewi and J. Smeyers-Verbeke, *Handbook of Chemometrics and Qualimetrics: Part B*, Elsevier, Amsterdam, NL, 1998.

10-9. J. Zupan, M. Novic, X. Li and J. Gasteiger, "Classification of Multicomponent Analytical Data of Olive Oils Using Different Neural Networks", *Anal. Chim. Acta* **192** (1994) 219 – 234.

10-10. X. Li, J. Gasteiger and J. Zupan, "On the Topology Distortion in Self-Organizing Feature Maps", *Biol. Cybern.* **70** (1993) 189 – 198.

10-11. B. Kocjancic and J. Zupan, "Application of a Feed-Forward Artificial Neural Network as a Mapping Device", *J. Chem. Inf. Comput. Sci.* **37** (1997) 985 – 989.

10-12. J. Zupan, M. Novic and I. Ruisanchez, "Tutorial: Kohonen and Counterpropagation Artificial Neural Networks in Analytical Chemistry", *Chemom. Intell. Lab. Syst.* **38** (1997) 1 – 23.

11 The Reactivity of Chemical Bonds and the Classification of Chemical Reactions

learning objectives:

- the electronic and energetic effects that determine the polar reactivity of a chemical bond

- the possibility of using a neural network to classify bonds by reactivity (susceptibility to heterolysis)

- the importance of properly choosing the training data

- a classical method of data selection: "experimental (factorial) design"

- using a Kohonen network to select data to be used to train a **different** network

- how the Kohonen classifying map can be interpreted to reveal (possibly unsuspected) relationships within the data

11.1 The Problems and the Data

The prediction of the course and outcome of a chemical reaction is one of the fundamental tasks in organic chemistry. Since chemical reactions are initiated by the breaking of one or more bonds in a molecule, a knowledge about reactive bonds, that is bonds that will easily break, is indispensable for the prediction of chemical reactions. This is the theme of the first study reported in Sections 11.1 – 11.6.

Chemists have derived their knowledge about chemical reactions largely from individual observations, have ordered these individual reactions, have generalized their observations by building models or by making predictions by analogy. In this inductive learning process, the classification of reactions into reaction types plays a major role.

With the availability of large reaction databases with some of them comprising millions of reactions, the automatic classification of reactions becomes of major interest. For, this will allow the mining of knowledge from reaction databases, knowledge that can be used for reaction prediction systems and for computer-assisted synthesis design. The classification of reactions is the theme of the second study reported in Sections 11.7 – 11.8.

Organic reactions are largely governed by polar processes, which break a bond in such a way as to generate a positive charge on one atom and a negative charge on the other. Such a polar bond breaking, also called *heterolysis*, can occur for each bond in two ways (Figure 11-1); both possibilities were investigated in the first study.

This example follows the work of V. Simon and coworkers of the Model Laboratory for Computer Chemistry at the Technical University of Munich, who have trained a neural network for predicting the polar breaking of bonds (Reference 11-6). Given any single bond in an aliphatic organic compound, such a neural network should be able to predict whether this bond will break easily, and how the charges will be shifted onto the atoms of the bond (Figure 11-1).

A dataset of 29 molecules is chosen so as to cover the diverse structural variations of aliphatic molecules; these molecules contain 385 bonds capable of 770 potential polar bond breaking modes. Considering only unique single bonds (e. g., only one C–H bond of a methyl group) leaves 373 chemically different polar breaking modes.

From among these 373, a series of 149 breaking modes are selected that can rather unequivocally be classified by chemists into 43 reactive and 106 non-reactive ones.

Figure 11-2 shows four of the 29 molecules and the 11 single functional group bonds they contain; The arrow in each bond points to the atom that obtains the negative charge. Plain bent arrows indicate reactive bonds, i.e., bonds that can be broken easily, while arrows with X's indicate non-reactive bonds. All other bonds are unclassified.

The breaking of a chemical bond is influenced by a variety of energetic, electronic or steric effects, for example: charge distribution, the inductive, resonance, and polarizability effects, bond dissociation energies, etc.

Organic chemists often discuss these effects only in a qualitative manner, but in recent years a number of empirical quantitative methods has been developed for such factors. We choose seven of these to describe a chemical bond in this study:

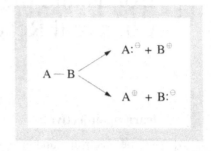

Figure 11-1: Two choices for the heterolysis of a chemical bond.

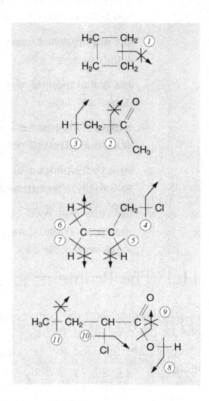

Figure 11-2: Four of the molecules and eleven of the bonds considered in this example. Plain bent arrows indicate reactive bonds whereas arrows that are crossed (X) show polar breaking modes that are difficult to achieve.

- the difference in total charge, Δq_{tot},
- the difference in π-charge, Δq_{π},
- the difference in σ-electronegativity, $\Delta\chi_{\sigma}$,
- a measure of bond polarity, Q_{σ},
- the amount of resonance stabilization, $R^{\pm}$, available for the charges generated upon heterolysis,
- the bond polarizability, α_b, and
- the bond dissociation energy, *BDE*.

Values of these variables are calculated and assigned to the bonds using a program package called PETRA (Parameter Estimation for the Treatment of Reactivity Applications; see Section 11.7, References 11-1 to 11-5).

Table 11-1 gives the values of these seven parameters for the eleven bonds shown in Figure 11-2.

bond	reactivity	Δq_{tot} [e]	Δq_{π} [e]	$\Delta\chi_{\sigma}$ [eV]	Q_{σ} [e]	$R^{\pm}$ [1/eV]	α_b [Å^3]	*BDE* [kJ/mol]
1	–	0.00	0.00	0.00	0.00	4.02	5.46	236
2	–	–0.16	–0.03	–2.08	0.17	0.00	4.89	240
3	+	0.04	0.00	–0.56	0.06	8.09	3.85	412
4	+	0.24	0.00	–1.96	0.33	5.16	5.83	338
5	–	–0.13	–0.01	0.61	–0.12	3.43	5.06	461
6	–	–0.15	0.01	0.37	–0.11	0.00	4.22	456
7	–	–0.15	0.01	0.37	–0.11	0.00	4.22	456
8	+	0.45	–0.10	–1.56	0.44	7.35	3.60	437
9	–	–0.54	0.08	–1.48	–0.22	0.00	5.35	445
10	+	0.30	0.00	–1.32	0.31	7.48	6.93	336
11	–	–0.04	0.00	–0.34	0.01	0.00	6.16	362

Table 11-1:   Values of the effects influencing reactivity for the chemical bonds shown in Figure 11-2.

Our task is now to relate these seven variables to the reactivity classification of a particular bond. As can be seen from Table 11-1, no single parameter suffices to separate reactive (+) and nonreactive (–) breaking modes. Classifying the reactivity of chemical bonds is clearly a multivariate problem.

11.2 Architecture of the Network for Back-Propagation Learning

We first approach the problem of classifying breaking modes as reactive or nonreactive by a two-layer neural network employing back-propagation learning. The number of input units is set to seven, the number of reactivity-controlling effects. These have different ranges, e.g., one from −0.2 to +0.2, another from 200 to 500; since the input units expect values between 0 and 1, each input value has to be scaled separately between its minimum and its maximum value. The output classification is coded as 0 for nonreactive breaking modes, or 1 for reactive ones.

A two-category (binary) classification can be achieved either by **two output neurons**, one for each class, or by **one output neuron** which is set to zero for one class, or to one for the other (Figure 11-3).

In the two-neuron case, the sum of the two output values is always 1.0, and the weights in the output layer have, in pairs, the same value with opposite signs (Figure 11-4).

We decide to work with a one-output neural network because an additional output does not provide any advantages.

Finally, we have to decide how many hidden neurons to use. We must always remember that having too many weights relative to the number of training data will probably lead to overtraining (Section 8.6.3); hence, the aim is always to work with as few neurons as possible.

A network with three hidden neurons turns out to be appropriate; our final architecture is (7 x 3 x 1), with 28 weights including the connections to the bias (Figure 11-5).

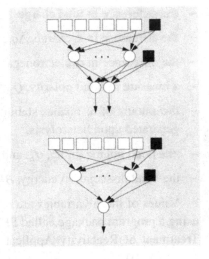

Figure 11-3: Two possible schemes for outputting a binary value.

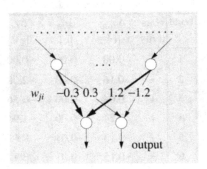

Figure 11-4: Two output neurons cause all pairs of weights connecting them to the level above to have the same value with opposite sign.

11.3 Using an Experimental Design Technique to Select the Training Set

> Two measures are used to determine the quality of a network. The *recall* gives the percentage of correctly recognized objects from the training set after training is over. The ability for *prediction* gives the percentage of correctly classified objects from a test set which was not considered during training.

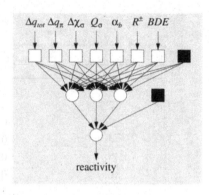

Figure 11-5: The network architecture for the prediction of the polar breaking of a bond.

Next, we therefore have to divide the dataset into a training and a test set. If no clear criterion exists, you can choose the training data randomly; that's what we did initially. However, we will show in this section that random selection is not a very good strategy.

The initial dataset contained 116 bond breaking modes; we divided it into 58 for training and 58 for testing. However, to cover the measurement space better and to allow a joint comparison of the different methods for selecting the training set, we found it necessary to **extend** the dataset to 149 modes: 64 for training and 85 for testing. Only the results obtained from this larger dataset are discussed in this chapter.

Table 11-2 shows the results of the initial trial; we randomly selected ten training sets (64 modes each) and trained the (7 x 3 x 1) neural network to recognize and predict reactivities of breaking modes with each of the ten sets separately.

The most important figure of merit for a multilayer neural network is the number of errors made in the test set. In predicted reactivity (yes or no), the number of wrong answers (out of 85) ranged from 3 to 12, with an average of 7.5.

training set	recall errors	prediction errors on a test set
1	1	7
2	0	3
3	0	12
4	0	10
5	0	6
6	0	7
7	1	5
8	0	9
9	0	7
10	0	5

Table 11-2:   Recall and prediction ability obtained on the neural network after training with 10 **randomly** selected datasets.

The selection of the training set is so important that it is practically a separate area of investigation, not only for neural networks, but for any approach on extracting knowledge from data, and even more so when the data come from (expensive and lengthy) experiments. Hence, a field of specialization called "experimental design" or "factorial design" has been developed.

Most importantly, the training set must cover the variable space in the most representative possible manner. A standard method is to select variables that are thought to influence the system under investigation, and divide the range of values of each variable into two or three fixed levels or intervals; for example, in a three-level design, we set up a low, a medium, and a high level. If exact values of the variables are difficult to obtain, the entire range of each variable is divided, into, let's say, three fixed intervals (low, medium and high), which can be called levels just as well. After the levels are determined, the data are selected so that the set contains values for all variables, representing all **combinations** of levels.

For example, if you have two variables for which three levels (intervals) of possible values are determined, the data should be selected so that each of the $3^2 = 9$ subspaces is filled with an experiment or a data point (Figure 11-6).

In our present example, **seven** variables require $3^7 = 2,187$ bond breaking modes for a complete three-level experimental design. This is a bit large, so we must look for ways to reduce the number of data

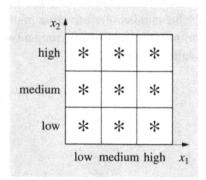

Figure 11-6: A complete three-level experimental design for a system with two variables (x_1 and x_2).

while still covering the information space as comprehensively as possible.

One way to do this is to reduce the number of variables to those representing the largest amount of **new** information. The correlation among the variables helps us make that choice: highly correlated variables carry similar information and are thus (more or less) redundant.

Based on an analysis of the correlation matrix shown in Table 11-3, four variables can be retained:

- the difference in σ-electronegativity, $\Delta\chi_\sigma$,

- the resonance stabilization of charges, $R^\pm$

- bond polarizability, α_b, and

- bond polarity, Q_σ,

	Δq_{tot} [e]	Δq_π [e]	$\Delta\chi_\sigma$ [eV]	Q_σ [e]	$R^\pm$ [1/eV]	α_b [Å^3]	*BDE* [kJ/mol]
Δq_{tot}	1.00	−0.31	−0.11	0.84	0.12	−0.03	0.02
Δq_π		1.00	0.19	−0.41	−0.14	0.03	0.01
$\Delta\chi_\sigma$			1.00	−0.74	0.07	0.00	−0.01
Q_σ				1.00	0.06	−0.02	0.04
$R^\pm$					1.00	0.30	−0.28
α_b						1.00	−0.72
BDE							1.00

Table 11-3:   Correlation matrix of seven variables for the breakability of bonds (based on 373 data).

With **four** variables, a three-level experimental design requires 3^4 = 81 data on bond breaking modes.

Now, if we sort our 116 bonds into these 81 boxes, how many boxes stay empty? The answer is 41. If we add the remainder of the original 382, 20 more boxes are filled (with as-yet-unclassified bonds), leaving 21 of the subspaces completely empty. That is, none of the 373 breaking modes has a combination of the four variable values that would allow it to fall into any of these 21 subspaces.

We mentioned above that the 116-bond set was extended to 149; this was done to put data points into the 20 subspaces that can be filled with these data; as regards the 21 that can **not** be filled, we must assume that these subspaces do not have any chemical significance.

For the training set, we chose **one** breaking mode **from each** of the 60 occupied subspaces. Afterwards, we noted that four subspaces contained both reactive and nonreactive modes. Apparently, these subspaces are borderline cases, indicating regions where reactivity changes. In order to account for that, both a reactive and a nonreactive mode were selected from these four subspaces, increasing the training set to 64.

This, then, is the "best" training set for the back-propagation (7 x 3 x 1) neural network given the original set of 29 molecules.

Now, as we did in the random selection, we need ten different datasets for comparison, all selected according to the same criteria, ensuring that all of the 64 bond breakings cover all 60 subspaces. The results of ten identical learning procedures in the same (7 x 3 x 1) network show much better results than those obtained from the random selection; see Table 11-4 and Table 11-2.

no.	random selection		experimental design	
	recall	prediction	recall	prediction
1	1	7	0	1
2	0	3	0	3
3	0	12	0	4
4	0	10	0	3
5	0	6	0	4
6	0	7	0	2
7	1	5	0	3
8	0	9	0	5
9	0	7	0	6
10	0	5	0	3

Table 11-4:   Errors in recall and prediction obtained after training with **random** training data, compared with results when the training data were determined by **experimental design**.

In results for the prediction set of 85 bond breaking modes, there are from one to six falsely classified modes, averaging 3.4. This is a remarkable improvement over the average 7.5 errors produced when a randomly selected dataset was used. This shows the importance and merit of experimental design techniques.

11.4 Application of the Kohonen Learning

The experimental design technique described above has two major disadvantages. First, we have to reduce the number of variables from seven to four; this is a rather tedious procedure, and necessarily leads to loss of information. Second, the choice of boundaries between the low, medium, and high intervals of the variable values is arbitrary and thus subject to user bias.

Is there another method that does not suffer from these problems? In this section we will show that a Kohonen neural network offers such an alternative.

In order to make the results comparable to those obtained in the previous Section, a (9 x 9) Kohonen network was chosen, containing 81 neurons that can be considered as equivalents of the "subspaces"; see Figure 11-7. When the dataset contains bonds with similar dependences on the seven controlling variables, they will map to the same neuron.

The network stabilizes after 30 training cycles, i.e., after all 373 polar bond breaking modes have been sent into the network 30 times. Six neurons are empty, 56 contain classified modes and 19 are occupied by nonclassified modes (Figure 11-8).

Remember, however, that a Kohonen network uses an unsupervised learning technique; since it does not use the class information when learning, it is remarkable that bonds of a particular classification end up in the same neuron. In all, 12 neurons carry only reactive bonds (and some unclassified ones), and 42 have only nonreactive (and some unclassified) bonds.

Only two neurons have conflicts, carrying both reactive and nonreactive bonds.

Also, the Kohonen map contains quite a lot of additional information that lends itself to chemical interpretation. This is further explained in Section 11.6.

Thus, the Kohonen network produces a basis for the selection of a training set. Again the (7 x 3 x 1) architecture is chosen and 10 different datasets are selected. This time only 56 bond breaking modes are necessary to cover the information space: one mode each from the neurons containing classified bonds.

One bond is chosen from each of the 56 neurons, plus two from those where conflicts occur. The only difference among the 10

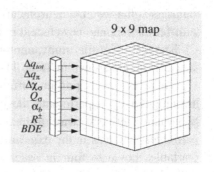

Figure 11-7: Kohonen network for mapping data on bond reactivity.

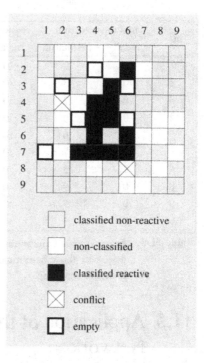

Figure 11-8: Kohonen map of the 373 polar bond breaking modes.

datasets is that when a neuron carries several modes of the same type, a different one may be selected in the next training set.

The results of this study, combined with the contents of Tables 11-2 and 11-4, are shown in Table 11-5.

There is one important difference between the selection of the training set by experimental design and by Kohonen mapping. The Kohonen network was trained by using **all seven** controlling variables, whereas in the experimental design study we had to cut the variables down to four in order to reduce the number of subspaces (which depends exponentially on the number of variables). Thus, one does not have to go through the tedious and time-consuming procedure for reducing the number of variables.

no.	random selection		experimental design		Kohonen network	
	recall	prediction	recall	prediction	recall	prediction
1	1	7	0	1	0	2
2	0	3	0	3	0	2
3	0	12	0	4	0	1
4	0	10	0	3	0	0
5	0	6	0	4	0	2
6	0	7	0	2	0	0
7	1	5	0	3	0	0
8	0	9	0	5	0	0
9	0	7	0	6	0	0
10	0	5	0	3	0	2

Table 11-5: Errors in recall and prediction obtained after training with random data, with those determined by experimental design, and by Kohonen mapping.

11.5 Application of the Trained Multilayer Network

The chemical importance of this example is that, by applying a **combination** of neural network techniques (a one layer (7 x 9 x 9) Kohonen network for selecting the training set, and (7 x 3 x 1) multilayer neural network with back-propagation learning), we end up with a trained network that is able to predict which single bonds will preferentially break in a polar manner and which will not **in any** aliphatic molecule. In addition, it even indicates in which direction the charges are shifted upon heterolysis.

Let's test this with a relatively complex molecule containing several functional groups. For all the bonds in this molecule, the seven controlling parameters were calculated for the two polar bond breaking modes. With 32 bonds in the molecule, this amounted to 448 (= 32 x 2 x 7) variables, which are input into the trained (7 x 3 x 1) network. From among the 64 bond breaking modes (two for each of the 32 bonds), only nine are found to be reactive; these are indicated in Figure 11-9. The corresponding unscaled values of the controlling parameters are shown in Table 11-6.

The network correctly predicts a high reactivity for the deprotonation of hydroxyl (9), of –NH, (7) and of methylene in α-position to the aldehyde group (3 and 4); the loss of a hydroxyl anion (8) is correctly predicted to be easy. It also predicts loss of the proton at the aldehyde group (1); this is usually not observed, since most bases are also strong nucleophiles, and would rather make a nucleophilic attack at the carbonyl. However, the predicted deprotonation does occur in formic esters, which contain the H–C=O group.

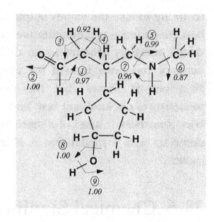

Figure 11-9: Predictions of bond breaking made by the (7 x 3 x 1) neural network trained as described above.

bond	Δq_{tot} [e]	Δq_π [e]	$\Delta \chi_\sigma$ [eV]	Q_σ [e]	$R^\pm$ [1/eV]	α_b [Å³]	BDE [kJ/mol]
1	−0.04	−0.03	−2.13	0.21	0.00	4.80	356
2	0.48	0.06	−3.03	0.60	3.69	4.81	425
3	0.03	0.00	−0.66	0.07	8.09	6.28	397
4	0.03	0.00	−0.66	0.07	8.09	6.28	397
5	0.32	0.00	−0.26	0.19	3.81	7.39	331
6	0.31	0.00	−0.39	0.20	0.00	5.97	343
7	0.44	0.00	0.31	0.24	0.00	5.37	382
8	0.45	0.00	−0.84	0.36	7.68	6.19	384
9	0.60	0.00	−0.88	0.42	0.00	3.93	437

Table 11-6: Values of the controlling parameters for the nine reactive bond breaking modes in the test molecule of Section 11.5. The modes are identified in Figure 11-9.

Although the network has been trained only for single bonds, it is able to further **generalize** and also assign the correct reactivity to the C=O double bond (2). The breaking of the two C–N bonds (5 and 6) is also considered feasible. These bonds are quite polar, but would need further activation in order to react.

The study of a series of other molecules has shown the overall correctness of most of the predictions on chemical reactivity. It is tempting to take the values output by the network as **probabilities** for bond breaking – as a quantitative measure of reactivity. However, this would certainly amount to an overinterpretation of the results; the network has been trained for classification and not for modeling. However, it shows that the border between classification and modeling is not hard and fast; had we used quantitative values for the bond reactivity instead of a mere binary classification, we could have come up with a model that is able to predict quantitative reactivity values.

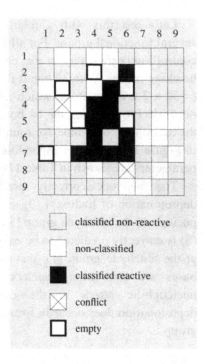

Figure 11-10: Kohonen map of the 373 polar bond breaking modes (cf. Figure 11-8).

11.6 Chemical Significance of the Kohonen Map

We mentioned in Section 11.4 that the Kohonen map for 373 bond breaking modes contains additional information capable of an interesting chemical interpretation.

Figure 11-10 is just a duplicate of Figure 11-8. Since not all the bonds were classified as reactive or nonreactive, unclassified bonds are spread throughout the network; they are indicated only in those neurons that contain neither reactive nor nonreactive modes, nor conflicts.

It can be seen that the reactive modes form a cluster in the center of the map (Figure 11-10). This is an indication that the self-organization that occurs during Kohonen learning perceives the similarity of **certain types** of modes and puts them into the same neurons. The Kohonen network even goes beyond that by recognizing the similarity of all reactive modes and putting them into neighboring neurons, thus forming the observed clustering of neurons with reactive modes.

There is one intruder: a bond classified unreactive by chemists occurs in the cluster of reactive modes (neuron at column 5, row 6 in Figure 11-10). This bond breaking mode is shown in Figure 11-11. The polar breaking of this bond is not observed in this molecule, and thus its classification as nonreactive is justified **for this molecule**.

However, in compounds having a bulkier group instead of C_2H_5, the breaking of this bond is observed. Thus, the bond breaking shown in Figure 11-11 can occur when it is not superseded by other types of reactions. In effect, this has to be considered as a **potentially** reactive bond, which makes reasonable its occurrence in the cluster of reactive bond breaking modes.

Neurons activated by nonreactive bond breaking modes and those activated by reactive ones touch each other in only a few places. This is a further indication of the success of Kohonen learning in differentiating reactive from nonreactive modes. The cluster of reactive bonds is surrounded by neurons mapping conflicts or nonclassified modes. The conflicts are a consequence of the transition from reactive to nonreactive; the nonclassified modes indicate how cautious chemists are about classifying bonds in doubtful cases.

After discussing the general features of the Kohonen map, let's take a closer look at the bond breaking modes ending up in individual neurons. In order to show some features in more detail, we have shifted the map of Figure 11-10 two columns left and one row down (which represents cutting the toroidal mapping surface at different places).

First, let's look at the reactive bonds (shaded cluster). All carbon-heteroatom breaking modes are at the right-hand side of this cluster, starting at the top with carbon-iodine and carbon-bromine bonds, then carbon-chlorine, carbon-oxygen and carbon-nitrogen bonds. This "top-to-bottom" sequence shows a clear tendency of decreasing polar reactivity.

The left-hand side of the shaded cluster shows modes that correspond to the dissociation of a proton. The more acidic O–H and N–H are more to the center of the cluster, and the less acidic C–H are towards the outskirts.

Second, the nonreactive C–H- and C–C-bonds are distributed over a wide area of the Kohonen map, because in the test set of molecules they have such differing first- and second-neighbors. A discussion of these small variations goes beyond the scope of this book (cf. Reference 11-6).

One more major feature of this map should be pointed out: the reactive modes are in the **lower left-hand corner**, while in the vicinity of the **upper right-hand corner** are those cases where a polar bond is broken **against** its inherent polarity (and thus are particularly unlikely to occur).

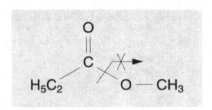

Figure 11-11: A bond considered unreactive by chemists that ends up in the cluster of reactive bond breaking modes.

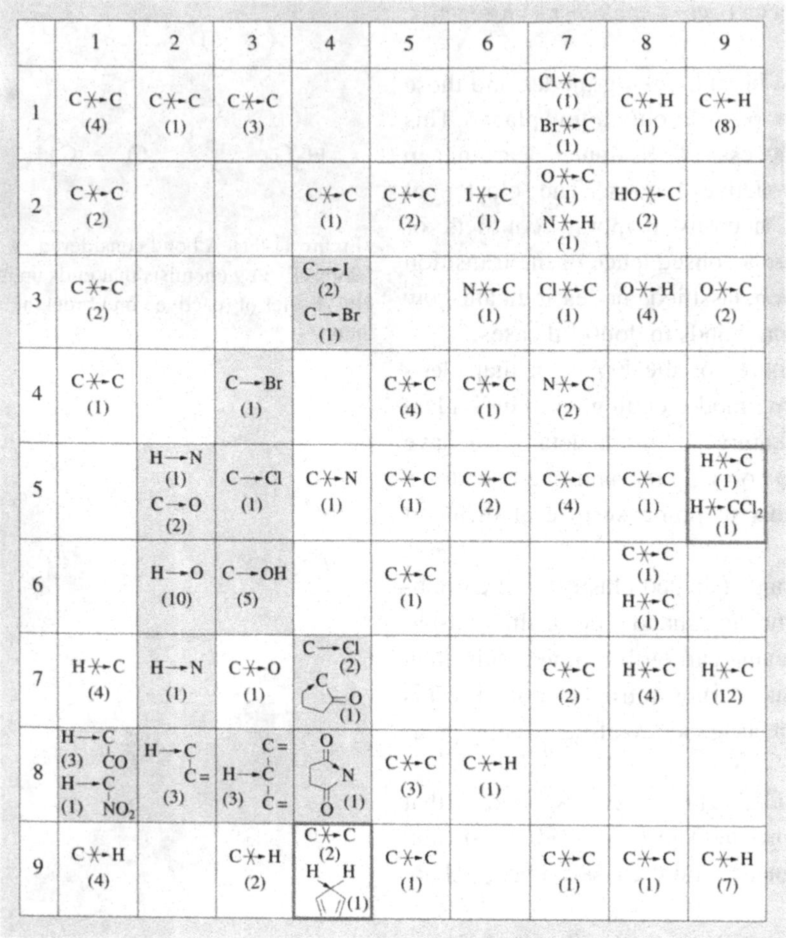

	1	2	3	4	5	6	7	8	9
1	C⚡C (4)	C⚡C (1)	C⚡C (3)				Cl⚡C (1) Br⚡C (1)	C⚡H (1)	C⚡H (8)
2	C⚡C (2)			C⚡C (1)	C⚡C (2)	I⚡C (1)	O⚡C (1) N⚡H (1)	HO⚡C (2)	
3	C⚡C (2)			C→I (2) C→Br (1)		N⚡C (1)	Cl⚡C (1)	O⚡H (4)	O⚡C (2)
4	C⚡C (1)		C→Br (1)		C⚡C (4)	C⚡C (1)	N⚡C (2)		
5		H→N (1) C→O (2)	C→Cl (1)	C⚡N (1)	C⚡C (1)	C⚡C (2)	C⚡C (4)	C⚡C (1)	H⚡C (1) H⚡CCl₂ (1)
6		H→O (10)	C→OH (5)		C⚡C (1)			C⚡C (1) H⚡C (1)	
7	H⚡C (4)	H→N (1)	C⚡O (1)	C→Cl (2) C=O (1)			C⚡C (2)	H⚡C (4)	H⚡C (12)
8	H→C (3) CO, H→C (1) NO₂	H—C C= (3)	C= H—C C= (3)	O⟩N⟨O (1)	C⚡C (3)	C⚡H (1)			
9	C⚡H (4)		C⚡H (2)	C⚡C (2) + H H ring (1)	C⚡C (1)		C⚡C (1)	C⚡C (1)	C⚡H (7)

Figure 11-12: Expanded and shifted version of Figure 11-10. Reactive bond types are drawn on a shaded background; the arrow indicates the polarization of the electron pair. The numbers in parentheses are the numbers of bonds mapped onto this particular neuron.

11.7 Classification of Reactions: The Data

The results reported in Section 11.6 show that there is a lot of chemical significance in the projection of bonds, as represented by physicochemical descriptors, into a Kohonen map. This observation can be taken one step further by considering all bonds that are broken in a chemical reaction, the reaction center, and projecting them into a two-dimensional Kohonen map. A two-dimensional arrangement of reactions has great advantages for the comparison of chemical reactions (Figure 11-13): the distance between two reactions in such a map can represent the degree of similarity; different directions in such a map can express different types of similarities.

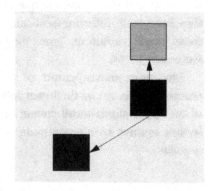

Figure 11-13: The representation of the degree and the type of similarity of reactions in a Kohonen map.

Clearly, it is rather easy to classify reactions that have different reaction centers, that have different types of atoms and bonds involved in the electron rearrangement during a reaction. But how about reactions that break and make the same type of bonds during a reaction but which chemists would classify into different reaction types because of the functional groups and substructures that are adjacent to the reaction center but are not directly involved in the electron relocation during a reaction?

In order to classify such reactions we have, as a typical example, retrieved reactions that involve the addition of a C–H bond to a C=C as indicated by the following scheme:

$$\ce{>C=C< + H-C<} \longrightarrow \ce{H-C-C-C<}$$

All those reactions, that contain this reaction center were retrieved from the 1992 edition of the ChemInform RX reaction database. Altogether, 120 reactions were obtained that a chemist would classify into different reaction types such as Michael additions, Friedel-Crafts alkylation of aromatic compounds by olefins, radical reactions, etc.

The next step is the choice how to represent these reactions. The most important driving forces of chemical reactions are electronic effects and, therefore, we performed calculations on charge distribution, inductive effects, as represented by orbital electronegativities, as well as effective polarizabilities by the empirical methods collected in the PETRA program package in order to account for the influence of functional groups onto the reaction site. Specifically, total charges, q_{tot}, σ- and π-electronegativities, χ_σ and χ_π, as well as effective polarizabilities, α_i, were calculated for all atoms of the reaction site. From all these variables chemists selected those that were deemed to be of importance for the reaction type under study, ending up with the seven physicochemical descriptors shown in Table 11-7.

11.8 Classification of Reactions: Results

The previous section has shown how the chemical reactions of the chosen data set are represented as points in a seven-dimensional space of electronic descriptors. A Kohonen network of planar topology with 12 x 12 neurons is chosen to project these 120 reactions from the seven-dimensional space into two dimensions.

electronic variable	C=C + H–C → H–C–C–C		
q_{tot}		*	*
χ_σ		*	*
χ_π		*	*
α_i			*

Table 11-7:   Physicochemical variables used to describe the reaction site. An asterisk indicates the atom for which the corresponding descriptor was used.

Figure 11-14: The Kohonen map for the 120 reactions containing the reaction center shown in Table 11-7 and described by seven physicochemical variables.

How was the size of the network chosen? For studies of this type, which is basically a similarity perception, it is a good starting point to choose approximately as many neurons (here 121) as there are objects (here 120) to investigate. Reducing the number of neurons forces more objects into the same neuron, reducing somewhat the resolution of the Kohonen network.

The resulting map with the individual reactions of the data set identified by their number is shown in Figure 11-14. How can this map be interpreted? In order to evaluate the chemical significance of this mapping, all reactions of the data set were intellectually classified into reaction types by chemists. Altogether, 14 reaction types were identified containing between 1 and 75 reactions as members. This assignment of reaction types was used for coloring the Kohonen map of Figure 11-14 to give Figure 11-15.

This Figure shows that reactions of the same type form coherent areas in this map. Thus, the Kohonen learning has identified on the basis of the seven physicochemical variables reaction types much in the same manner as they were attributed by a chemist.

The next question is now, how do we know where one reaction type changes into another one in this map? This question can be addressed by an analysis of the weight differences between weights of adjacent neurons. The weights do not have a uniform distribution throughout the Kohonen network. Rather, one can identify locations in the Kohonen network where the weight differences between adjacent weights are larger than in other places.

Figure 11-16 shows the weight differences that exceed the selected threshold of 1.5 as walls, with higher walls indicating larger weight differences. It can be discerned from Figures 11-15 and 11-16 that the barriers in the weight differences between adjacent neurons coincide

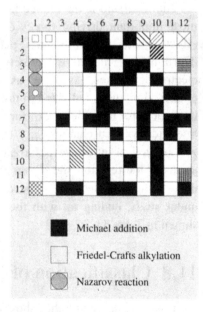

■ Michael addition

□ Friedel-Crafts alkylation

● Nazarov reaction

Figure 11-15: The Kohonen map of Figure 11-14, now with reactions identified by their respective reaction types. Three of the reaction types are identified by names.

with transitions from one reaction type to another. Thus, clusters of reactions can be found in the Kohonen maps of data sets of reactions that comprise reaction types.

A detailed discussion of the chemical inferences of these maps of chemical reactions and reaction types that truely form landscapes with mountains, passes, and valleys of weight differences goes beyond the scope of this book. The interested reader is suggested to contact the corresponding publications in journals.

Furthermore, methods have been developed that allow the automatic assignment of reaction types, thus superceding the time-consuming intellectual identification of reaction types.

Suffice to say that these landscapes of chemical reactions allow

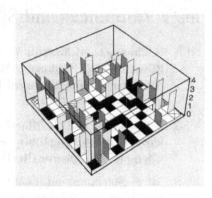

Figure 11-16: The Kohonen map of Figure 11-15 with the weight differences that exceed the threshold 1.5 shown as walls.

– the identification of reaction types,

– the clustering of reaction databases and of hits from reaction searches,

– the location of transitions between reaction types,

– the definition of the scope of a reaction type,

– the identification of special reactions,

– the extraction of knowledge from reaction databases for reaction prediction,

– the mining of reaction databases for synthesis design.

11.9 References and Suggested Readings

11-1. J. Gasteiger, M. Marsili, M. G. Hutchings, H. Saller, P. Löw, P. Röse and K. Rafeiner, "Models for the Representation of Knowledge about Chemical Reactions", *J. Chem. Inf. Comput. Sci.* **30** (1990) 467 – 476.

11-2. J. Gasteiger and M. Marsili, "Iterative Partial Equalization of Orbital Electronegativity – A Rapid Access to Atomic Charges", *Tetrahedron* **36** (1980) 3219 – 3228.

11-3. M. G. Hutchings and J. Gasteiger, "Residual Electronegativity – An Empirical Quantification of Polar Influences and its Application to the Proton Affinity of Amines", *Tetrahedron Lett.* **24** (1983) 2541 – 2544.

11-4. J. Gasteiger and M. G. Hutchings, "Quantification of Effective Polarisability. Applications to Studies of X-Ray Photoelectron Spectroscopy and Alkylamine Protonation", *J. Chem. Soc. Perkin 2* (1984) 559 – 564.

11-5. J. Gasteiger and H. Saller, "Berechnung der Ladungsverteilung in konjugierten Systemen durch eine Quantifizierung des Mesomeriekonzeptes", *Angew. Chem.* **97** (1985) 699 – 701; "Calculation of the Charge Distribution in Conjugated Systems by a Quantification of the Resonance Concept", *Angew. Chem. Int. Ed. Engl.* **24** (1985) 687 – 689.

11-6. V. Simon, J. Gasteiger and J. Zupan, "A Combined Application of Two Different Neural Network Types to the Prediction of Chemical Reactivity", *J. Am. Chem. Soc.* **115** (1993) 9148 – 9159.

11-7. L. Chen and J. Gasteiger, "Organische Reaktionen mit Hilfe neuronaler Netze klassifiziert: Michael-Additionen, Friedel-Crafts-Alkylierungen durch Alkene und verwandte Reaktionen", *Angew. Chem.* **108** (1996) 844 – 846; "Organic Reactions Classified by Neural Networks: Michael Additions, Friedel-Crafts Alkylations by Alkenes, and Related Reactions", *Angew. Chem. Int. Ed. Engl.* **35** (1996) 763 – 765.

11-8. L. Chen and J. Gasteiger, "Knowledge Discovery in Reaction Databases: Landscaping Organic Reactions by a Self-Organizing Neural Network", *J. Am. Chem. Soc.* **119** (1997) 4033 – 4042.

11-9. The ChemInform RX reaction database is produced by FIZ CHEMIE, Berlin, Germany and marketed by MDL Information Systems, Inc., San Leandro, CA, USA.

11-10. Information on the methods in the PETRA package can be found on: *http://www2.ccc.uni-erlangen.de/software/petra/*

11-11. A. Ultsch, G. Guimaraes, D. Korus and H. Li, "Proc. Transputer Anwendertreffen/World Transputer Congress TAT/WTC 93", Aachen, Springer Verlag, New York, USA, 1993, pp. 194 – 203.

11-12. H. Satoh, O. Sacher, T. Nakata, L. Chen, J. Gasteiger and K. Funatsu, "Classification of Organic Reactions: Similarity of Reactions Based on Changes in the Electronic Features of Oxygen Atoms at the Reaction Sites", *J. Chem. Inf. Comput. Sci.* **38** (1998) 210 – 219.

12 HPLC Optimization of Wine Analysis

---

learning objectives:

- an outline of the HPLC method of analysis

- a classical method for modeling (predicting the effectiveness of) the mobile phase of an HPLC column

- modeling the mobile phase with a neural network

- comparison of two different neural nets for this modeling task

---

12.1 The Problem of Modeling

The identification of individual components in a complex mixture is very common in analytical chemistry; but before the components can be identified and quantified, they first have to be separated. Chromatographic methods presently play a dominant role as separation techniques. High Performance Liquid Chromatography (HPLC) is a widely used method in which the mixture is distributed between a stationary solid or liquid phase on a solid support, and a mobile liquid phase applied at high pressure. The various components are distributed between the two phases to different extents and thus are separated.

Modeling is quite commonly applied in chemistry to a number of areas or problems, particularly to all kinds of optimization from complex procedures composed of many activities, to recipes for given products. To optimize a process or a recipe, or to find the best-fitting function to a number of experimental points, a model has to be found first; after this, the optimization procedure is performed using the response surface of the model as the basis for finding the best solution.

> Without a model, optimization does not lead to a general solution of the problem.

A model tries to describe in mathematical terms the response of a system to appropriate stimuli. Classical modeling is based on a relatively small number of variables – usually fewer than ten – and one response. If **more** responses from the **same set** of variables are required, then a **separate model** must normally be built **for each response** when classical methods are used (see Section 9.3). However, neural networks are able to do both one-response and multiresponse modeling.

As a first example of modeling, we will demonstrate both methods: classical modeling and modeling by neural networks. By comparing all steps in both procedures from the selection of data to the checking of the results, we can appreciate the differences and the advantages and disadvantages of both.

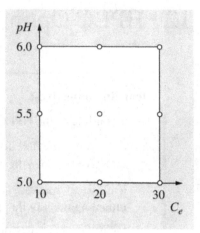

Figure 12-1: A full two-variable three-level experimental (factorial) design scheme. Circles represent properties of nine different liquid phases leading to nine chromatograms.

12.2 Modeling the Mobile Phase for HPLC by a Standard Method

In Chapter 10, we built a system that can identify the geographical origins of more than 500 Italian olive oils, based on analyses for eight fatty acids. (Such extensive analytical work is not unusual where products for human consumption must be checked.)

In this chapter we will follow an example worked out by the group of Professor Rius of the Analytical Laboratory of the Chemistry Department of the University Rovira i Virgili of Tarragona (see Section 12.5). They were faced with setting up an analytical procedure for routine analysis of wine samples. Since hundreds of identical analyses have to be carried out each day, it is important to automate the procedure and reduce the time needed for each analysis.

The chemical analysis was supposed to be carried out by HPLC. The chief figure of merit is the *performance factor* of an analysis, which includes two things: first, how well the components are separated on the column, and how long it takes for the entire analysis to be carried out.

It is evident that the performance factor is a compromise between two features: separability and time. Both features are influenced by

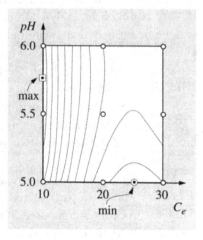

Figure 12-2: Model surface of the performance factor *PF* obtained by a standard modeling technique.

the selection of the mobile phase: if it passes through the column too quickly, then some components may not be well separated; if it passes through too slowly, then the time per analysis may be too great.

The first goal in setting up the analytical procedure is to make a model for the mobile phase, specifically for the performance factor as a function of its properties. Then we can select the most appropriate mobile phase for a given analysis.

In this example, only two properties of the mobile phase will be considered: the concentration of ethanol, C_e, and the acidity of the mobile phase (its pH). Experience has shown that C_e should be between 10 and 30%, while the pH should be between 5.0 and 6.0.

In order to build a model, we have to have some data, whether we will be using a classical method or a neural network. To obtain the data, a classical two-variable three-level experimental design is used, as shown in Figure 12-1. Nine pairs of variables (C_e, pH) are selected, and *HPLC* chromatograms are made for each; then a performance factor *PF* is assigned to each pair (Table 12-1).

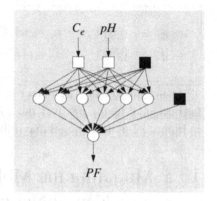

Figure 12-3: The (2 x 6 x 1) neural network used in modeling the HPLC mobile phase.

no.	C_e [%]	pH	PF
1	10	5.0	6.08
2	20	5.0	2.42
3	30	5.0	2.10
4	10	5.5	7.31
5	20	5.5	3.00
6	30	5.5	3.13
7	10	6.0	7.06
8	20	6.0	3.72
9	30	6.0	3.37

Table 12-1: The data for building a model of the HPLC process.

Standard modeling assumes a quadratic polynomial for the performance factor:

$$PF = ax^2 + by^2 + cxy + dx + ey + f \qquad (12.1)$$

where x stands for the concentration of the ethanol, C_e, and y stands for the pH. The parameters a, b, c, d, e, and f are obtained by any standard modeling or optimization technique. Setting up the model by SIMPLEX optimization leads to the following set of coefficients:

$$a = 0.018 \quad b = -1.42 \quad c = 0.0145$$
$$d = -0.995 \quad e = 16.16 \quad f = -31.86 \quad (12.2)$$

Using the above parameters in (12.1) gives an expression for the performance factor *PF* over the entire measurement space, as shown in Figure 12-2. This model maximizes *PF* at $C_e = 10.2\%$ and $pH = 5.7$.

12.3 Modeling the Mobile Phase for HPLC by a Neural Network

In the next step, we will use the same data (Table 12-1) for building a neural net with one hidden layer employing back-propagation learning strategy. (Back-propagation is chosen for learning because building a model **always** involves supervised learning.)

Evidently, the network should have two neurons for input and one for output. The entire training set involves only 27 different numbers, i.e., nine input vectors each having two input values, and one target. The number of weights trained by the back-propagation strategy should, at least in principle, **not exceed** this number.

A few small neural networks with three, four, five, and six neurons in the hidden layer were constructed and the model trained with the nine input vectors and their targets; the result were compared with the classical model. It turns out that the best model is the (2 x 6 x 1) network, with $(2 + 1) \times 6 + (6 + 1) \times 1 = 25$ weights (Figure 12-3).

The comparison of the neural network model with the model obtained by optimization is shown in Figure 12-4. The differences are small indeed; the optimum in the neural network model is at $C_e = 10\%$ and $pH = 5.6$. The overall agreement between their responses is very good.

> Note that no hypothesis about the model function was necessary for constructing the neural network model.

In the classical model, we had to guess a model function. However, this does have an advantage: the parameters of the final

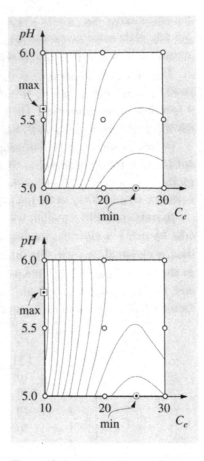

Figure 12-4: Above: the model obtained by the neural network using back-propagation learning; below: the model obtained by the classical modeling technique.

model indicate how the response depends on the particular variables. For example, the parameters (12.2) in combination with Equation (12.1) indicate that concentration of ethanol and the *pH* act in the quadratic and linear terms oppositely to each other, which apparently causes the coefficient c of the mixed term to be relatively small. Such evidence of how individual variables influence the measured quantity is very important; unfortunately, this information is missing from the neural network model.

12.4 Comparison of Networks with Identical Architectures

First, let us take a closer look at the weights of the neural network. We did two training runs of the same network with the same set of nine input vectors and targets, but with different initial weights. This leads to two completely different networks (as judged by the final weights of each network, Table 12-2). However, **the maps produced** by these different networks **are almost identical** as can be seen in Figure 12-5.

This means that the (2 x 6 x 1) network is too large for the problem, or to be more precise, the network is too large for the amount of data we have at disposal to train it. This fact becomes evident if one counts the number of weights in the network on Figure 12-3. This (2 x 6 x 1) network has, including the biases, in the hidden layer, i.e., in the layer between the input and the hidden nodes 3 x 6 = 18 weights and in the output layer 3 x 1 = 3 weights. Together there are 21 weights in the network. For nine input vectors (Table 12.1) these are far too many, and it should not come as a surprise that virtually any new random set of weights at the beginning of learning will yield a different set of weights after learning ended (the weights stabilized). This result is as anticipated because the problem of modeling the mobile phase for HPLC is a relatively simple case and will therefore not have many local minima or maxima but will have a response surface that is relatively smooth. The response surfaces yielded by both completely different networks are almost identical (Figure 12-5).

The smallest error-backpropagation network with one hidden layer that one can create for nine input objects represented by 2-dimensional input vectors has 5 nodes in the layout (2 x 2 x 1) as shown in Figure 12-6. Adding the bias weights this (2 x 2 x 1) network

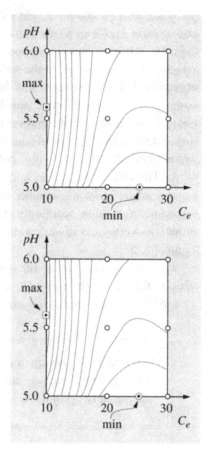

Figure 12-5: Identical maps obtained by two (2 x 6 x 1) networks having completely different weights.

has 9 weights. Someone might expect that the small network would now always converge to the same set of weights. We have tried more than 50 different randomization and found eight different sets of weights. For the novice to the field the result might come somewhat unexpected. There are eight different sets of weights, however, all are composed of the same set of absolute figures! Due to the symmetrical nature of the normalized input vectors these eight different weight sets turn out to be two groups of four sets with identical weights, the only difference being that the input nodes are changing the places (Table 12.3). The changes of places and signs of weights in these "different" sets are made in such a manner that the resulting outputs are always the same. Although seemingly different, the eight neural network models nevertheless always yield surface maps identical to those in Figure 12-5.

The reader can check the results exposed in this Chapter by visiting the site on the web with the address

http://www2.ccc.uni-erlangen.de/ANN-book/

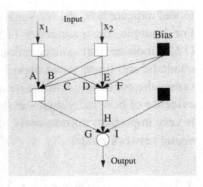

Figure 12-6: The smallest possible two layer error-back-propagation network for the problem of nine 2-dimensional input vectors.

> In order to obtain numbers comparable to the experimental values, the inputs and outputs of a neural network must be properly scaled.

		network 1 input			network 2 input		
w_{ji}^1		$i=1$	2	bias	$i=1$	2	bias
	1	−5.49	0.30	−0.15	−4.37	0.97	−0.78
	2	0.36	3.89	−0.93	1.17	−3.21	−0.81
$j=$	3	0.85	0.67	−0.19	−1.25	0.11	−1.15
	4	−2.85	−0.61	−0.68	6.58	−0.60	−0.41
	5	5.44	−0.32	0.15	1.45	1.64	−1.87
	6	−0.95	−1.16	−1.59	−1.45	0.46	−0.16
w_{ji}^2		$j=1$			$j=1$		
	1	4.51			3.26		
	2	2.88			−1.85		
	3	0.39			0.84		
$i=$	4	1.76			−4.60		
	5	−4.47			1.84		
	6	0.03			0.60		
	bias	−0.03			1.82		

Table 12-2: Weights w_{ji}^l of two (2 x 6 x 1) networks giving very similar maps.

network	weights on the first hidden neuron w_{i1}^1 $i=1$		2	bias	weights on the second hidden neuron w_{i2}^1 $i=1$		2	bias	weights on the output neuron w_{i1}^2 $i=1$		2	bias
1	A	B	−C		D	E	−F		−G	H	I	
2	A	B	−C		−D	−E	F		−G	−H	I	
3	−A	−B	C		D	E	−F		G	H	−I	
4	−A	−B	C		−D	−E	F		G	−H	I	
5 ≈ 1	D	E	−F		A	B	−C		H	−G	I	
6 ≈ 2	−D	−E	F		A	B	−C		−H	−G	I	
7 ≈ 3	D	E	−F		−A	−B	C		H	G	−I	
8 ≈ 4	−D	−E	F		−A	−B	C		−H	G	−I	

Table 12-3: Weights, w_{ji}^l, of eight "different" (2 x 2 x 1) neural network models which all yield the same response surface shown in Figure 12-5.

12.5 References and Suggested Readings

12-1. J. Zupan and F. X. Rius, "XYZ, A Program for Modeling, Validation, Prediction and Graphic Display of Data", *Anal. Chim. Acta* **239** (1990) 311 – 315.

12-2. M. Tusar, X. F. Rius and J. Zupan, "Neural Networks in Chemistry", *Mitteilungsblatt der GDCh-Fachgruppe "Chemie–Information–Computer" (CIC)* **19** (1991) 72 – 84.

12-3. D. L. Massart, B. G. M. Vandeginste, S. N. Deming, Y. Michotte and L. Kaufmann, *Chemometrics: A Textbook*, Elsevier, Amsterdam, NL, 1988.

12-4. S. N. Deming and S. L. Morgan, *Experimental Design: A Chemometrics Approach*, Elsevier, Amsterdam, NL, 1987.

12-5. K. W. C. Burton and G. Nickless, "Optimization via Simplex", *Chemom. Intell. Lab. Syst.* **1** (1987) 135 – 149.

12-6. S. N. Deming and L. R. Parker, "A Review of Simplex Optimization in Analytical Chemistry", *Crit. Rev. Anal. Chem.* (Sept. 1978) 187 – 202.

12-7. M. Tusar, J. Zupan and J. Gasteiger, "Neural Networks and Modelling in Chemistry", *J. Chim. Phys.* **89** (1992) 1517 – 1529.

13 Quantitative Structure-Activity Relationships

learning objectives:

- the basis for Quantitative Structure-Activity Relationships (QSAR)

- identification of factors controlling anticarcinogenic activity of carboquinones related to the identities of substituents on the basic skeleton

- use of a neural net to associate biological activity with a given profile (set of values for the controlling factors)

- comparison of neural network approach with classical statistical approach

- representation of structures of different size with the same number of descriptors

- selection of appropriate descriptors for representing the input objects

- using a combination of unsupervised and supervised neural networks to study a data set

- application of a genetic algorithm for the reduction in the number of variables

13.1 The Problem

"QSAR" stands for Quantitative Structure-Activity Relationships, that is, quantitative relationships between a chemical structure and its physical, chemical, or biological activity. The search for such relationships is one of the most important applications of modeling techniques.

Correlating the chemical structures of drugs with their pharmacological activities is of particular interest. Because of the large development costs of new drugs, a reliable quantitative prediction of activity **before** the compound is made is of great interest to synthesis laboratories.

Because neural networks can be developed into complex models they have gained large prominence in QSAR research. Particularly in pharmaceutical research and development many investigations on the relationships between structure and biological activity have been made. We will here investigate three different data sets, in the first and the third one, all compounds have the same skeleton and only the substituents on this skeleton have been varied. Such data sets are typical for many investigations that deal with the optimization of a pharmaceutical lead structure. The second data set encompasses a variety of structures having different skeletons. Such data sets are met in the search for a lead structure.

As a typical example, we will first discuss the work of Aoyama, Suzuki and Ichikawa (References 13-1 and 13-2). They chose a dataset that had previously been studied by Yoshimoto and coworkers (Reference 13-3) using traditional modeling techniques, (such as multilinear regression analysis (MLRA)), in order to compare those results with the results of a neural network. Aoyama and coauthors did their best to keep all variables, along with the selection of the training and test sets, as similar to the classical studies as possible.

The second data set comprises structures having different skeletons and different numbers of atoms in the molecules. Here, one has to face the problem of representing such a data set by a uniform set of descriptors, having the same number of variables for molecules of different size.

Neural networks, as any learning method, be it statistical or pattern recognition methods or neural networks, need the objects of a study to be represented by the same number of descriptors (variables).

The third data set is a collection of 55 biologically active flavonoid compounds, inhibitors of the enzyme tyrosine kinase. They all have the same skeleton on which different substituents are attached at different positions. Due to the initial 180-dimensional spectrum-like representation (for details on this representation see Chapter 21, Section 21.4) of their 3D structures the task to be solved in the example is the reduction of the 180 variable set into a smaller and more easily manageable set of variables which still contains the most relevant information about the biological activity of the flavonoids in question.

13.2 Dataset I

The dataset in this study involves modifications of the basic carboquinone skeleton shown in Figure 13-1. Many carboquinones exhibit varying degrees of *anticarcinogenic* activity.

This quantitative structure-activity relationship study was designed to predict the minimum dose of a drug required to produce a 40 percent extension of the lives of the test animals, BDF_1 mice that had been inoculated with lymphoid leukemia L-1210 cells.

This *minimum effective dose* depends on the concentration, C, of the substance necessary to give the desired effect, and is given as $log(1/C)$. The more effective the drug is, the smaller will be the concentration necessary. (Since the required concentrations of different drugs vary over several orders of magnitude, it is more convenient to use the logarithm, $log(1/C)$, as a measure of the effective dose.)

As expected, the anticarcinogenic activity depends on the identities of the substituents R^1 and R^2. In the standard multilinear analyses, this substituent is described by physicochemical variables that describe the combined influence of the substituents R^1 and R^2:

- the molar refractivity $\mathcal{MR}_{1,2}$

- the substituents' contribution to the hydrophobicity $\pi_{1,2}$

- the sum of the substituent constants for the field effect $\mathcal{F}$

- the sum of the substituent constants for the resonance effect $\mathcal{R}$

along with two local variables, describing only the influence of one substituent R^1:

- the molar refractivity $\mathcal{MR}_1$

- the contribution to the hydrophobicity π_1.

The assignment of substituents as R^1 and R^2 is based on their molar refractivities: $\mathcal{MR}_1 \leq \mathcal{MR}_2$.

The study used eleven different substituents R^1 (consisting primarily of short alkyl groups like methyl, ethyl, and propyl) and about 30 different substituents R^2 (mostly substituents having longer chains and bearing additional functionalities like $-CH_2CH_2OCH_3$ and $-CH(OCH_3)CH_2OCONH_2$). Two of the compounds are shown in Figure 13-2.

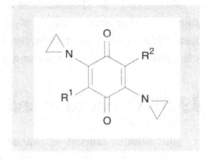

Figure 13-1: The characteristic structure of carboquinones, a class of compounds with anticarcinogenic activity. The substituents on the skeleton can be quite varied (see Figure 13-2).

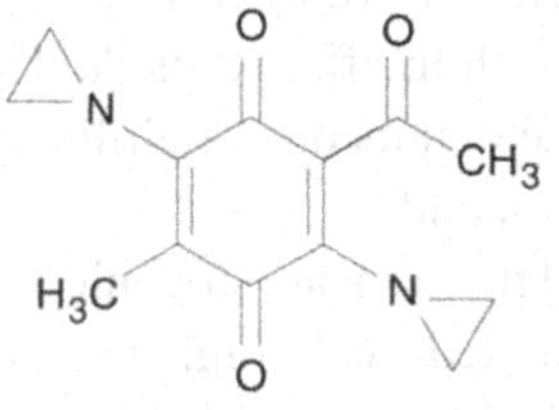

a) **1**

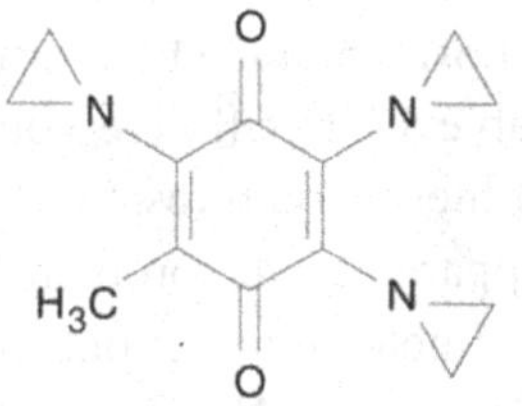

b) **38**

Figure 13-2: The most (a) and the least (b) anticarcinogenic compound among the carboquinones.

Altogether, 35 different carboquinones were selected; for all of them, six variables, $MR_{1,2}$, $\pi_{1,2}$, MR_1, π_1, $\mathcal{F}$, and $\mathcal{R}$ were either measured (MR) or taken from the literature (π, $\mathcal{F}$, $\mathcal{R}$), and the minimum effective dose was measured and given as $\log(1/C)$.

13.3 Architecture and Learning Procedure

Our inputs consist of variables (MR_1, $MR_{1,2}$, π_1, $\pi_{1,2}$, $\mathcal{F}$, $\mathcal{R}$) describing the structure, and our target data are values of $\log(1/C)$; thus, a supervised learning method should be used. In this example, we will try to find a model that can predict the minimum effective dose $\log(1/C)$ for each set of the six input variables, $MR_{1,2}$, $\pi_{1,2}$, MR_1, π_1, $\mathcal{F}$, and $\mathcal{R}$ for any carboquinone derivative. Hence, our network requires six input units and one output neuron.

As in most applications, one hidden layer turns out to be sufficient; after some trial and error, twelve neurons were placed into the hidden layer (Figure 13-3).

The ($6 \times 12 \times 1$) neural network, with one hidden and one output layer, was trained with 35 carboquinones by the back-propagation algorithm; afterwards the $\log(1/C)$ output values were compared with those obtained by multilinear regression analysis on the same set of 35 compounds.

The anticarcinogenic activity of 17 of the carboquinones is predicted with higher accuracy than in the multilinear regression analysis study, for six compounds the results are of about equal quality, and for 12 structures they are worse. Overall, the results of the neural network (NN) are significantly (but not dramatically) better than those obtained by multilinear regression analysis (MLRA).

Table 13-1 will give you an impression of the data and results for six of these structures.

Apparently, the problem under investigation is adequately handled by a linear model, but the neural network does lead to slight improvements. **Nonlinear** QSAR problems will show much larger improvements when modeled by neural networks.

13.4 Prospects of the Method

In investigations of the biological activity of a series of compounds, several different biological activities are often monitored.

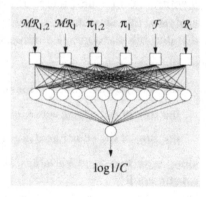

Figure 13-3: Network architecture for studying the anticarcinogenic activity of carboquinones.

| no. | substitutents | | variables | | | | | | log(1/C) | | |
---	R^1	R^2	$MR_{1,2}$	$\pi_{1,2}$	MR_1	π_2	$\mathcal{F}$	$\mathcal{R}$	exp.	MLRA	NN
1	CH$_3$	COCH$_3$	1.69	−0.05	−0.55	0.57	0.28	0.07	3.94	4.12	4.39
3	CH$_3$	(CH$_2$)$_3$C$_6$H$_5$	4.50	3.66	3.16	0.57	−0.08	−0.26	3.93	4.23	4.18
6	CH$_3$	CH$_2$C$_6$H$_5$	3.57	2.51	2.01	0.57	−0.12	−0.14	4.74	4.77	4.67
10	C$_2$H$_5$	C$_2$H$_5$	2.06	2.00	1.00	1.03	−0.08	−0.26	4.94	5.01	4.99
32	C$_2$H$_5$	(CH$_2$)$_2$CONH$_2$	3.09	0.95	−0.05	1.03	−0.08	−0.26	5.98	5.55	5.59
38	CH$_3$	N(CH$_2$)$_2$	2.13	0.68	0.57	0.18	0.06	−1.05	6.54	6.30	6.31

Table 13-1:   Input and output variables of six compounds used in the QSAR study of anticarcinogenicity of carboquinones by Aoyama and coworkers.

For example, when investigating the anesthetic activity of a compound, one will also monitor its toxicity; or, in investigations of the carcinogenicity of a compound, two different types of carcinogenicity tests might be performed.

The two – or more – different biological activities quite often depend on the same types of structural variables, e. g., the value of the coefficient, log$P$, indicating the distribution of the compound between aqueous and lipid phases. In such cases, as we have mentioned in previous examples (see, for example Section 9.3) standard techniques develop **separate** modeling equations for the **two** biological activities, expressed as logarithms of the inverses of some threshold concentrations C_1 and C_2:

$$\log\frac{1}{C_1} = c_{01} + ... + c_{i1}\log P + ... \tag{13.1}$$

$$\log\frac{1}{C_2} = c_{02} + ... + c_{i2}\log P + ... \tag{13.2}$$

With neural networks, however, it becomes feasible to model **both** biological activities **simultaneously** in one network. Then, one output neuron will be used to output the first activity (expressed as log(1/C_1)), whereas a second output neuron is used to indicate the second activity (log(1/C_2)).

Figure 13-4 shows the part of a two-layer neural network that expresses the influence of log$P$ on two biological activities. Thus, while MLRA requires two coefficients to express the influence of log$P$ on the two activities, a neural network with one hidden layer containing, say, three neurons provides **nine** weights for expressing

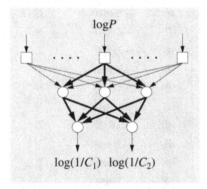

Figure 13-4: Propagation of the influence of one input variable (log$P$) on two different biological activities by a two-layer neural network with three hidden neurons.

these influences. With n neurons in the hidden layer, $3n$ weights are available to convey the influence of one input variable onto two activities. This indicates quite clearly the higher flexibility of a neural network compared to a statistical analysis.

13.5 Dataset II

The second data set comprises 31 steroids having different binding affinity to the *corticosteroid binding globulin* (CBG) receptor. Table 13-2 gives the full list of compounds with their binding affinity data and a classification into high, intermediate, or low affinity.

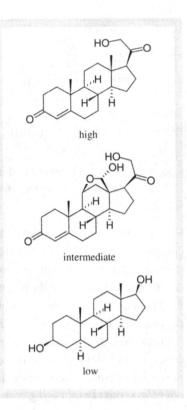

Figure 13-5: One steroid each with high, intermediate, or low binding affinity to the *corticosteroid binding globulin* (CBG) receptor.

compd	CBG affinity (pK)	activity class[a]	compd	CBG affinity (pK)	activity class[a]
1	−6.279	2	17	−5.225	3
2	−5.000	3	18	-5.000	3
3	−5.000	3	19	-7.380	1
4	−5.763	3	20	-7.740	1
5	−5.613	3	21	-6.724	2
6	−7.881	1	22	-7.512	1
7	−7.881	1	23	-7.553	1
8	−6.892	2	24	-6.779	2
9	−5.000	3	25	-7.200	1
10	−7.653	1	26	-6.144	2
11	−7.881	1	27	-6.247	2
12	−5.919	2	28	-7.120	2
13	−5.000	3	29	-6.817	2
14	−5.000	3	30	-7.688	1
15	−5.000	3	31	-5.797	2
16	−5.225	3			

Table 13-2: *Corticosteroid binding globulin* (CBG) affinity data.
[a] 1, high; 2, intermediate; 3, low; this classification was obtained by dividing the data set into three classes of comparable size.

Figure 13-5 shows one structure each with high, intermediate, or low binding affinity to the CBG receptor, respectively. The full data set, with the structures encoded as connection tables, is contained on the web site for this book
(*http://www2.ccc.uni-erlangen.de/ANN-book/*).
See the Appendix for further information.

This data set was chosen because it had been selected for the introduction of the widely used CoMFA method and has also been studied by a variety of other methods. Although all compounds of this data set are steroids they, nevertheless, comprise different skeletons having A- and B-rings with or without double bonds or, in some cases, aromatic A-rings. Furthermore, the substituents at various positions differ quite extensively, and the number of atoms in the set of compounds also varies.

13.6 Structure Representation by Autocorrelation of the Molecular Electrostatic Potential

Neural networks, as many learning methods, need the objects of study to be represented by the same number of input descriptors. Thus, with a data set as just described, one is faced with the task of transforming the structure into a preset number of descriptors. In this book, we will present a variety of methods for such a mathematical transformation of chemical structure information into a set of descriptors (cf. Chapter 21). The choice on the structure encoding scheme should somehow take into consideration the factors that are thought to be involved in the property investigated.

The electrostatic potential on the surface of a molecule (cf. Figures 19-2, 19-13, and 19-14) is one of the most important factors for binding a ligand to its receptor. The question is then, how can this property be encoded into a fixed number of descriptors? This task was achieved by autocorrelation, as indicated in Equation (13.3)

$$A(d) = \frac{1}{m}\sum_{i,j} p(i)\, p(j) \qquad \text{with } d_l < d_{ij} < d_u \qquad (13.3)$$

First, the molecular electrostatic potential (MEP) is calculated for a set of points evenly distributed over the van der Waals surface of the molecule with a selected density. Then, the products of this property, p, (the MEP), at points i and j is calculated whereby the distance d_{ij} between these two points must be between a lower, d_l, and upper, d_u, bound (say, between 3Å and 4Å). All these products are collected into a single value of $A(d)$; in this case, $A(3)$. This value is normalized by the total number, m, of distances in this interval (Figure 13-6). With a series of distance intervals with different upper and lower bounds, a

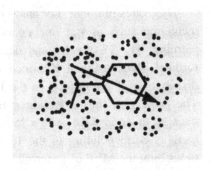

Figure 13-6: Autocorrelation of a property on a molecular surface (see Equation 13.3).

vector of autocorrelation coefficients is obtained. In our case, we have collected the products between 1Å and 2Å, all the way to 12Å and 13Å, thus providing an autocorrelation vector of length 12. The values of $A(d)$ are displayed at the center of each interval. Figure 13-7 illustrates this process of calculating an autocorrelation vector of the MEP using the steroid *corticosterone* as an example. In effect, the molecular electrostatic potential at the van der Waals surface was encoded into an autocorrelation vector of 12 values for each of the 31 steroids. Clearly, this encoding scheme is a drastic reduction of the information on the MEP. However, the goal, to encode a molecule into a preset number of descriptors (here 12), irrespective of the size of the molecule, was achieved.

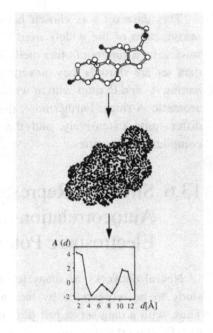

Figure 13-7: Calculation of the autocorrelation vector of the molecular electrostatic potential for *corticosterone*.

13.7 Verification of Structure Representation by Unsupervised Learning

The final goal of this study was to find a quantitative relationship between a structure encoding and the binding affinities to the CBG receptor by a back-propagation neural network. The back-propagation algorithm is such a powerful modeling technique that it will establish apparent relationships, albeit of low predictive power, even between input and output data that have only a small correlation. We have, therefore, found it highly recommendable to first establish whether the variables describing the objects are, in fact, significant for the property under investigation. In our case, the question is: Is there a relationship between the encoding of the MEP of a steroid by a 12-dimensional autocorrelation vector and the activity of binding to the CBG receptor?

This question can be answered by an unsupervised learning technique such as the one contained in Kohonen neural network learning. The 12-dimensional descriptor space was projected into a toroidal plane using a Kohonen network in order to visualize the distribution of the objects in the high-dimensional descriptor space. The projection into a Kohonen map was performed by training a network that consisted of 15 x 15 neurons, with each neuron having 12 weights corresponding to the 12-dimensional autocorrelation vector describing the MEP of a steroid. After projection of the data set of 31 steroids into this two-dimensional Kohonen map, the projection was visualized by marking those neurons having obtained a steroid with high, intermediate, or low binding affinity with a filled square, an asterisk, or with a cross, respectively. (Recall, that this activity level was not used in the training of the Kohonen network).

The Kohonen network used had the topology of a torus, i.e., neurons at the left and the right side of the network, and neurons at the upper and the lower part of the networks are directly connected (cf. Figures 6-6 and 6-7). Therefore, four identical copies of the resulting Kohonen map were arranged like tiles (cf. Figure 6-14) in Figure 13-8 in order to better present the clusters formed by the steroids. The compounds of high, intermediate, and low activity form three clearly perceivable clusters in the Kohonen map as indicated in Figure 13-8. Only one compound, a steroid of intermediate activity, is not grouped together with compounds of the same activity class, but is surrounded by highly active compounds, instead.

The ability of the Kohonen network to here distinguish between compounds belonging to different activity classes shows, that the autocorrelation vector fulfills one of the prerequisites for a successful quantitative analysis: compounds that are similar to each other in the descriptor space exhibit similar biological activity. The visualization proved that the compounds group together in the descriptor space corresponding to their biological activity. Therefore, we were encouraged to quantitatively model the binding affinity with a feed-forward neural network trained by back-propagation as the next step.

We have also investigated the 12-dimensional descriptor space by another unsupervised learning method, a principal component analysis (PCA). Figure 13-9 shows the clustering of the steroids in a plot of the first against the second principal component. The compounds are by far not as well separated as in the Kohonen map of Figure 13-8. The principal component analysis performs a rotation of the coordinate axes of a high-dimensional space, trying to put as much variance as possible into the first few component. In our case, we are apparently left with more than two components and, thus, a plot of only two components cannot quite separate the compounds into their activity classes. The learning in a Kohonen network, on the other hand, knows from the very beginning that it has to end up with two dimensions and therefore places as much information as possible into these two dimensions.

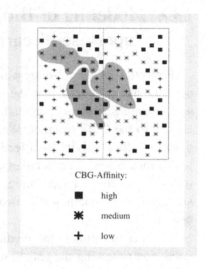

Figure 13-8: Arrangement of four identical Kohonen maps obtained from the 12-dimensional MEP autocorrelation space showing the separation of steroids of high (squares), intermediate (asterisks), and low activity (crosses).

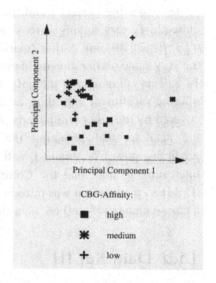

Figure 13-9: Plot of the first two components of a principal component analysis of the 12-dimensional descriptor space.

13.8 Modeling of Biological Activity by Supervised Learning

Projection of the 12-dimensional descriptor space by a Kohonen network had indicated that the encoding by autocorrelation of the molecular electrostatic potential has a relationship to the binding affinity to the CBG receptor. We, therefore, generated a quantitative model of CBG activity by a feed-forward neural network trained by back-propagation. The architecture of the network used was as follows: twelve input units corresponding to the twelve autocorrelation coefficients, two hidden neurons, and an output neuron (Figure 13-10).

In order to estimate the predictive power of the approach, cross-validation following the leave-one-out scheme was performed. In 31 independent experiments, the network was trained with the data of 30 steroids. After training, the network was used to predict the activity of the 31st compound. This procedure was repeated 31 times, until the biological activity of each compound had been predicted by a neural network that had not included this compound in the training set. Figure 13-11 shows the results in the form of a plot of the predicted affinity values against the experimental ones. Although the predicted values show the correct trend, the quality of the predictions is not quite satisfactory, having a cross-validated correlation coefficient r^2 of 0.63. Especially one outlier (marked by a circle) can be identified – the very same outlier already identified in the Kohonen map of Figure 13-8. This compound is the only one in the entire data set bearing a fluorine substituent and, thus, apparently outside the structure space covered by the other compounds. After omitting this compound from the data set and repeating the cross-validation, a much better predictive power is obtained, with a cross-validated r^2 of 0.84. It is interesting to note that the Comparative Molecular Field Analysis (CoMFA) method that was introduced with this data set achieved only a cross-validated r^2 of 0.66, on a subset of 21 steroids.

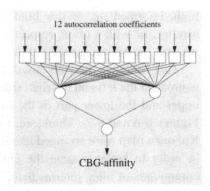

Figure 13-10: Architecture of the feed-forward network used for modeling CBG binding affinity from the 12 autocorrelation coefficients.

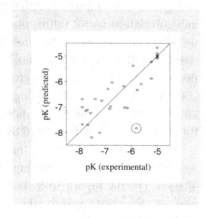

Figure 13-11: Crossvalidation of CBG activity of 31 steroids modeled by back-propagation.

13.9 Data Set III

The third set of data consists of 55 *flavonoid* derivatives which are low molecular weight substances found in most parts of all plants. Due to their broad variety of biological activities they are often called "*bioflavonoids*". In this particular case, we shall explore flavonoid substances which are inhibitors of the *protein tyrosine kinase* (PTK) and are, therefore, important factors in cellular signal transduction. The skeleton of the flavonoids is given in Figure 13-13. The biological activity, *a*, in this study is the logarithm of the inverse experimental biological activity IC50, i.e., the molar concentration necessary for 50% of maximal inhibition of PTK in comparison with the experiment without the flavonoid. Table 13-3 gives the full list of 55 compounds and their corresponding activities. The compounds are described with their substituents and the positions at which they are bonded to the flavonoid skeleton. The data are taken from three papers by Cushman et al (References to Chapter 13).

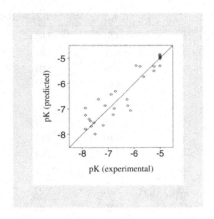

Figure 13-12: Crossvalidation of CBG activity of 30 steroids modeled by back-propagation.

13.10 Structure Representation by *Spectrum– Like* Uniform Representation

It has been explained many times that modeling either with artificial neural networks or any other method requires uniform representation of input data. Because different molecules are assembled from different number of atoms a direct 3D description by coordinate triplets $(x,y,z)_j$ of each of the constituent atom j does not fulfill the requirement of uniformness. One of the possible alternatives which is used in QSAR modeling quite often is a description of a molecule by several topological and electronic descriptors. In the present case a different approach featuring the so called *spectrum-like* structure representation will be used. In this Section only a brief explanation of the coding principles are given while in Section 21.4 a more detailed description how to calculate such a representation for any molecule is outlined.

The main idea of the *spectrum-like* representation is to mimic a "light source" placed somewhere close to the molecule which casts "shadows" of atoms onto the surface of an imaginary sphere drawn around the light source (Figure 13-14). The positions and intensities of the shadows of the atoms on the surface of the sphere depend on the relative positions of the atoms and the light source. The complete shadow of all atoms on an arbitrary equator of the imaginary sphere resembles a "spectrum" (Figure 13-15), hence, the name of the representation.

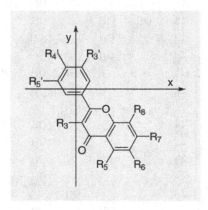

Figure 13-13: The flavonoid skeleton.

	R3	R5	R6	R7	R8	R3'	R4'	R5'	activity *a*	class
1	OH	OH		OH		OH	OH		4.88	6
2	OH			OH		OH	OH		4.86	6
3		OH		OH			OH		4.83	6
4		OH					OH		4.80	6
5			OH				OH		4.80	6
6		OH		OH					4.71	6
7		OH		OH		OH	OH		4.46	5
8				OH			OH		4.41	5
9			OH			OCH_3	OCH_3	OCH_3	4.22	5
10	OH	OH		OH		OCH_3	OH	OCH_3	4.16	4
11	OH	OH		OH		OH		OH	4.00	4
12			OH				OH		3.93	4
13				OH	OH	OCH_3	OH	OCH_3	3.92	4
14			OH				OR		3.92	4
15			OH			OCH_3	OH	OCH_3	3.89	4
16				OH			OH		3.78	3
17				OH	OH	OH			3.75	3
18	OH	OH		OH					3.53	3
19		OH	OCH_3				OH		3.55	3
20		OH					OH		3.50	3
21				OH	OH				3.50	3
22				OH					3.47	3
23			OH			OCH_3	OR	OCH_3	3.43	3
24				OH	OH	OCH_3	OCH_3	OCH_3	3.40	3
25				OH			OR		3.01	2
26				OH		OCH_3	OH	OCH_3	2.90	1
27				OH		OCH_3	OR	OCH_3	2.82	1
28			OH				NH_2		5.92	9
29		OH		OH			NH_2		5.13	7
30						OCH_3	OH	OCH_3	4.57	5
31				OH			NH_2		3.86	4
32							NH_2		3.68	3
33	COOMe						OH		3.36	2
34							OH		3.30	2
35	COOMe						NH_2		3.09	2
36	COOH			OCH_3			OH		2.99	1
37	COOH						OH		2.80	1

Table 13-3:   Flavonoid compounds with activities *a*. The substituents are marked
        according to the assignment of atoms in Figure 13-13. The activity
        values are divided into nine classes. The substituent OR stands for
        $OSi(Me)_2C(Me)_3$.

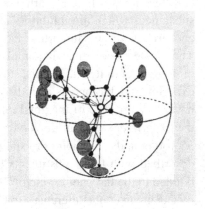

Figure 13-14: Shadows of atoms on a spherical surface with an arbitrary radius.

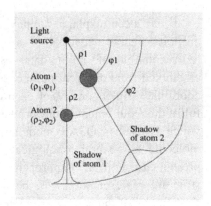

Figure 13-15: "Spectrum" of two atoms' shadows on the equator.

	R3	R5	R6	R7	R8	R3'	R4'	R5'	activity a	class
38		NH_2	OH	NH_2			NH_2		4.74	6
39		NH_2	OH	NH_2		NH_2			4.34	5
40			OCH_3		NH_2	NH_2			4.25	5
41			NH_2				NH_2		3.99	4
42			NH_2		NH_2		NH_2		3.97	4
43			OH		NH_2		NH_2		3.93	4
44					NH_2		NH_2		3.91	4
45			NH_2	OH			NH_2		3.85	4
46			NH_2			NH_2			3.70	3
47		OH	NH_2				NH_2		3.65	3
48		OH			NH_2		NH_2		3.49	3
49				OH	NH_2		NH_2		3.48	3
50			OCH_3		NH_2		NH_2		3.42	3
51			NH_2	OH		NH_2			3.30	2
52			NH_2	OH	NH_2		NH_2		3.12	2
53				OH					2.81	1
54		OCH_3			NH_2		NH_2		2.79	1
55				OH	NO_2		NO_2		2.73	1

Table 13-3: Flavonoid compounds with activities a. The substituents are marked
(continued) according to the assignment of atoms in Figure 13-13. The activity
values are divided into nine classes. The substituent OR stands for
$OSi(Me)_2C(Me)_3$.

In order to compare *spectrum-like* representations of several
compounds their structures must be aligned in the same manner. For
example, to compare the flavonoids all their skeletons must be
oriented in the same direction and superimposed onto each other. In
other words the coordinate origin of the "light source" must be placed
at exactly the same relative position of the internal coordinate system
of the skeletons. In the present example the coordinate origin (or the
"light source") for all representations is placed at the point (–1.7Å,
3.9Å, –0.6Å relative to atom no. 2 in the benzopyran ring system)
within the benzene ring as shown in Figure 13-13.

The intensity s_{ij} of the shadow of an atom j on the equator at point
i is described by the Lorentzian function (Equation 13.4). The
Lorentzian function is chosen because of its simplicity. It could be any
other appropriate function. In order to acquire information of the
entire 3D structure, the "shadows" of the atoms are projected onto
three mutually perpendicular circles. Figure 13-15 shows how for

each atom j its shadow's intensity s_{ij} depends on the angle φ_i on one circle:

$$s_{ij} = \frac{\rho_j}{(\varphi_j - \varphi_i)^2 + \sigma_j^2} \qquad (13.4)$$

The position of the atom j is determined by the polar coordinates (ρ_j, φ_j) in the internal coordinate system of the "light source". The peak width parameter σ_j of the Lorentzian function (Equation 13.4) is used to describe any individual property of the atom j (atomic or ionic radius, atomic number, ionization energy, electron affinity, charge, etc.). Throughout this example for all atoms in all molecules the parameters σ_j are set to $\sigma_j = 1 + charge\ on\ atom\ j$. If the charge on atom j is negative σ_j is less than *1*, otherwise it is larger than *1*.

If more atoms j are taken into account the intensities s_{ij} at all positions φ_i are clearly additive. The cumulative formula for any variable s_i of the additive *spectrum-like* representation of the whole structure consisting of n atoms can be written as:

$$s_i = \sum_{j=1}^{n} \frac{\rho_j}{(\varphi_j - \varphi_i)^2 + \sigma_j^2} \qquad (13.5)$$

with φ_i running from φ_1 to φ_{360}

The number of variables s_i in each representation depends on the number of angles φ_i which divides the equator around the molecule – the finer the division, the more precise the description. If the resolution of 1° radial degree is chosen for the projection on each equator, one spectrum has 360 intensities. Hence, a complete *spectrum-like* representation of each molecule (projections of its structure into three perpendicular equators) has 1080 intensity values.

As can be easily understood, such representations are too large for most applications. In the present example the interval of division on each circle is 6°. Therefore, the spectrum on one circle is composed of 60 intensities, what means that the entire *spectrum-like* representation of any flavonoid has 180 intensities. Although much smaller, even this number is too high for handling 55 compounds and must therefore be reduced.

13.11 Selection of the Most Important Variables Using a Genetic Algorithm

A genetic algorithm is one of the most effective optimization methods for problems involving large number of variables. Its idea is to mimic the optimization by natural selection of living organisms in real life. The three main factors governing the natural selection are:

- **survival of the fittest**,

- changing the individual genetic material by **cross-over** of chromosomes, and

- changing of individual genetic material by **mutation** of genes.

All three mentioned factors are implemented in the computer simulated optimization called genetic algorithm, or GA for short. Before explaining these three factors in detail a few more parallelisms between living organisms and the objects to be optimized must be explained.

It is assumed that properties of each living being in nature are determined by genes "stored" in chromosomes. The presence or absence of genes that might be beneficial to or dangerous to the survival of an individual influences the chance on whether the subject will live long and have many offsprings or whether it will die having few or no offsprings at all.

Let us consider the case that in a world of fixed and limited resources there is a pool of living organisms that have only one chromosome with exactly 180 genes. This means that there are only 180 properties which can be important for their lives. For increasing the chance of survival some properties are good, some bad, and some irrelevant. Hence, the best suited individual for the given world would be the one whose chromosome would have only the genes assuring the good properties and none of those causing the bad ones. The crucial question in the optimization is how to find out which genes are responsible for the good and which for the bad properties. Unfortunately, we do not know this; all we know is only the individual's behavior in the defined world – its overall performance. If the individual has many good genes it will survive and have a lot of offsprings. The number of offsprings the individual "produces" is therefore influenced by the selection criterion which ranks each individual. For objects to be optimized this criterion is called a **fitness**

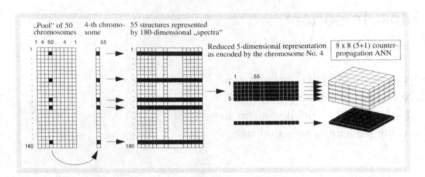

Figure 13-16: Pool of 50 chromosomes (left part). The 4-th chromosome is taken out and 55 flavonoid structures are represented by 5 intensity representation. With the set of 55 five-dimensional spectra an (8x8)-counter-propagation network model is generated and the RMS on the output layer evaluated (right part).

function. Definition of a good fitness function is always the crucial task in any optimization process.

One more parallelism between the natural selection of living organisms and the problem of selection of the most relevant variables in terms of a GA is that reduced *spectrum-like* representations can be described by one chromosome – like the above mentioned living organism. Each reduced representation features different set of intensities (genes) picked out from the 180 possible ones. One reduced representation might have only three intensities, another one fifty, still another one fourteen, and so on. How to decide which is the best?

The best, the fittest, or the optimized, is the reduced *spectrum-like* representation that would yield the best model. So, all we have to do is to represent all 55 objects with one of the possible reduced representations, make a model, test it; then represent the 55 objects again with another reduced representation, make a new model, test it and compare both results. By consecutive testing of various reduced representation with the generated models one can gradually find better and better representations.

The unpleasant part of such testing, however, is that a chromosome having 180 genes (bits) offers 2^{180} ($\approx 10^{54}$) possible reduced structure representations! Because there is no real chance to check even a minute part of this huge number of possibilities, we will try to use a genetic algorithm which promises to find, if not the very best, at least a very good one.

First, we set up a pool of 50 chromosomes each with 180 bits randomly turned to either ones or zeros. These strings of bits are "genetic codes" for 50 different reduced representations (Figure 13-16, left part). By considering the first "generation" of 50 reduced representations (out of the 10^{54} possible ones) we shall explain how the testing is carried out. Once more, all 55 *flavonoids* are encoded 50 times, each time with a different reduced representation as suggested by the corresponding 180-bit-string – if the bit is turned to one, the intensity is taken into the account and otherwise not.

With each reduced representation a small 8 x 8 x (5 + 1) counter-propagation network (Figure 13-16, right part) is trained. The number of weights in the Kohonen layer in each of the 50 networks depends on the number of bits turned to one in the chromosome suggesting the tested reduced representation. In a small network of only 64 neurons on the average about 20-30% of neurons is expected to be excited by two or more different objects. The fitness function, *ff*, and, thus, a measure of quality of the tested representation is the RMS value (cf. Equation 7.6) or the square root of the sum of squared differences between the experimental activities of compounds exciting the same neuron and the response given by this neuron:

$$ff_k = \sqrt{\frac{\sum_{i=1}^{n_e}\sum_{j=1}^{n_i}\left(a_i - w_j^{out}\right)^2}{n_e \cdot n_i}} \tag{13.6}$$

for *k = 1...50* reduced representations

The first summation runs over all n_e neurons that are excited at least twice, while the second summation runs over all n_i experimental biological activities a_i of compounds that have excited the *j*-th neuron having in the output layer the weight (output activity) w_j^{out}. The smaller the fitness function (Equation 13.6), the better the reduced representation. Once all 50 different counter-propagation models are made, their rank list can easily be made by sorting the corresponding outcomes of the fitness functions { ff_k }.

The second step of a GA is natural selection. This step involves the selection of the best chromosomes for mating, allowing them to have offsprings, and omitting ("killing") the others with low values of fitness function. From the ranked list of fitness functions { ff_k } sixteen best ones, approximately one third of all, i.e., 50/3 ≈ 16, are selected and the rest is ignored. By allowing to mate each of these sixteen chromosomes to three randomly selected partners 48 chromosomes of the new generation are produced. Additionally, the very best chromosome mates once more (four times altogether) and its clone (the identical chromosome) is added to the new generation. In this way 50 new chromosomes (reduced representations) are obtained.

In the GA, the offsprings of the mating process are generated by the **cross-over** procedure. For each mating pair, a random gene position is determined and from that position the cross-over (twisting) procedure is applied. The resulting two new chromosomes are obtained by exchange of the twisted parts (Figure 13-17). Each of the two offsprings has one part (lower or upper one) of the gene sequence from one parent and the other part from the second parent.

In a relatively small population of only 50 chromosomes it may well happen that all genes at a given position have the same value ("0" or "1") in all of them. Such gene position can never be changed by the cross-over procedure only. Therefore, a process called **mutation** is applied. Mutation means random switching of a small percentage of bits to its opposite value. If chosen for a mutation, the gene with value "0" turns to "1" and *vice versa*. In order to not disturb the improvements due to the cross-over breeding process too much, the mutation procedure ought to be applied with care. The probability at which a gene is subject to mutation should be low (usually below 1%).

Once all three steps of a GA – natural selection, cross-over and mutation – have been applied, new testing or fitness function evaluation of all 50 chromosomes of the new generation can start again. Due to the fact that the clone of the best chromosome from the previous generation is always present in the next one, constantly increasing values of the fitness function are assured. The number of generations required to obtain an optimum varies from case to case. In the present example 500 generations each consisting of 50 chromosomes have been tested (Figure 13-18). In the entire GA process, 25,000 times all 55 structures were encoded into different reduced representation and 25,000 counter-propagation networks were built and tested. The improvement of the RMS value as the fitness function in the GA process is shown in Figure 13-18.

After 500 generations the best fitness function as defined by Equation (13.15) on the 8 x 8 x (5 + 1) counter-propagation network was RMS = 0.167. This value was obtained using a reduced representation of 18 variables only.

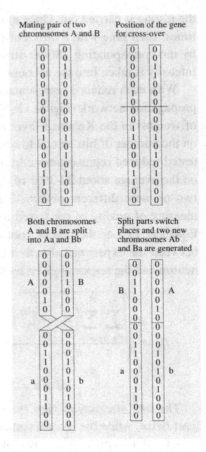

Figure 13-17: Cross-over procedure for making two new offsprings Ab and Ba from two mating chromosomes Aa and Bb.

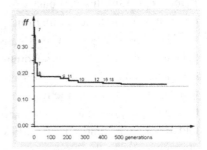

Figure 13-18: GA optimization evolution through 500 generations. The lengths of the reduced representations has increased from the starting 7 to the final 18 intensities.

13.12 Cross-validation of the Counter-Propagation Model Obtained by the Optimal Reduced Representation

The first task in the modeling process was to obtain the short uniform representation consisting of 18 variables for all 55 *flavonoids*. Out of the 180 intensities of the complete *spectrum-like* representation, the following 18 intensities were selected: 30th (180°), 32nd (192°), 33rd (198°), 44th (264°), and 46th (276°) from the first part (x,y-projection), 83rd (138°), 93rd (198°), 98th (228°), 108th (288°), 111th (306°), and 117th (342°) from the second part (x,z-projection, intensities from 61 to 120) and finally, 127th (42°), 129th (54°), 133rd (78°), 143rd (138°), 145th, (150°), 146th (156°), and 166th (276°) of the y,z-projection intensities from 121 to 180 (Figure 13-19). In parentheses, the corresponding angle in radial degrees is given. Because these selected variables represent defined space windows of 6° radial degrees it is easy to conclude that the presence or absence of substituent atoms in these eighteen directions comprises the most influential factor in the biological activity of flavonoids.

It was mentioned in Section 13.11 that the RMS value of the recognition results on the 8 x 8 counter-propagation neural network model is applied as fitness function. At this place two more questions have to be elaborated in more detail. The first one is why the counter-propagation network was chosen for modeling and not the error back-propagation, and, second, why the choice was made on a 8 x 8 network and not something else, let us say a 7 x 7 or 10 x 10 network?

Both answers are relatively simple. Because in the Kohonen layer of the counter-propagation network the formation of clusters on the basis of input representations is achieved, one can evaluate the formation of the optimized reduced representation that discriminates between the compounds in question better than with the error back-propagation which delivers only the model and no internal information about the representations. The other argument for the preference of counter-propagation over the error back-propagation model is the number of training epochs necessary for the networks to converge. The convergence rate of the former is two orders of magnitude better than that of the latter one.

The choice of the 8 x 8 layout of the counter-propagation network (Figure 13-20) is based on the following reasoning. In a 49 neurons

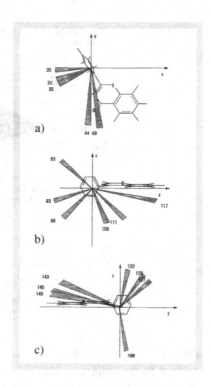

a)

b)

c)

Figure 13-19: Each of the selected 18 variables define a 6° degree wide window in the space where the most relevant substituents are lying. There are 5 directions in the (x-y) plane a), 6 in the (x-z) plane b), and 7 directions in the (y-z) plane c).

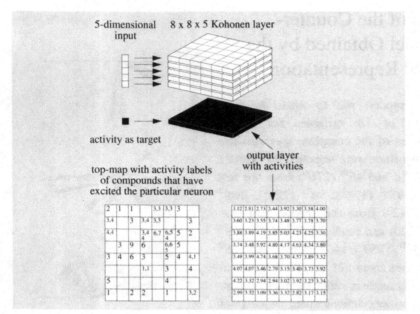

Figure 13-20: Counter-propagation network used in the GA procedure and for the final model (above). The distribution of 55 hits on the output map (lower part left). Double and triple hits are marked by the numbers 2 or 3, respectively. The final output layer with resulting activities (lower part right).

network (i.e., 7 x 7 layout) there is not even a theoretical chance that each of the 55 objects would excite its own neuron. On the other hand in a large network of, let us say, 400 = 20 x 20 neurons, each object would almost inevitably excite its own neuron, hence no good quantitative measure of how structurally similar *vs.* dissimilar and at the same time biologically more-active *vs.* less-active compounds would be clustered. Therefore, the first larger network of 64 neurons (8x8 layout) that theoretically allows exciting of 55 different neurons is a reasonable choice because it enables the theoretical possibility of the formation of a network which would have an RMS value equal to zero (each object would excite its own neuron). However, in the optimization process, while testing different representations, the RMS values will undoubtedly vary quite significantly, thus allowing to make quantitative comparisons of the representations. Even more, the final output map of the counter-propagation network will show the distribution of all objects depending on the optimized representation.

Figure 13-20 (lower part) shows the final output map of the 8 x 8 counter-propagation network of 55 flavonoids. Nine neurons were excited twice and three of them three times. This counter-propagation model generated with all 55 flavonoids represented by 18 variables in 900 epochs has a correlation factor of estimates a_i^{est} *vs.* experimental biological activities a_i of 0.95.

In order to check the reliability of both the model and the selected reduced representation, a cross-validation test is performed. The cross-validation leave-one-out test is one of the possibilities how to simulate real-life conditions in shortage of a test set. The leave-one-out test (known also as jack-knifetest) requires to make 55 models by the same modeling procedure with the same representation, but each time with one object omitted from the modeling procedure. The test of the prediction of this model obtained on 54 objects only is executed by input of the left-out object into it as an "unknown". The correlation coefficient r between 55 predictions $a_i^{est\ by\ CV}$ in the cross-validation procedure and the actual experimental activities a_i gives a fair estimate how the actual model obtained by the same procedure on 55 objects will perform when encountered with really unknown objects of the same type. The cross-validation correlation factor r obtained in our study was 0.86. The table of actual biological activities *vs.* the predictions obtained by the cross-validation predictions is given in Table 13-4.

Sample No.	a_i	$a_i^{est\ CV}$	Sample No.	a_i	$a_i^{est\ CV}$	Sample No.	a_i	$a_i^{est\ CV}$
1	4.88	4.74	21	3.50	3.56	41	3.99	4.02
2	4.86	4.73	22	3.47	3.49	42	3.97	3.57
3	4.83	4.98	23	3.43	3.22	43	3.93	3.89
4	4.80	4.46	24	3.40	3.87	44	3.91	3.81
5	4.80	4.47	25	3.01	3.22	45	3.85	3.90
6	4.71	4.12	26	2.90	3.53	46	3.70	3.66
7	4.46	4.66	27	2.82	3.37	47	3.65	3.78
8	4.41	4.61	28	5.92	4.88	48	3.49	3.74
9	4.22	3.75	29	5.13	4.97	49	3.48	3.77
10	4.16	3.53	30	4.57	4.06	50	3.42	3.70
11	4.00	3.98	31	3.86	3.82	51	3.30	3.59
12	3.93	3.71	32	3.68	3.64	52	3.12	3.14
13	3.92	4.25	33	3.36	3.23	53	2.81	3.21
14	3.92	3.50	34	3.30	4.05	54	2.79	2.80
15	3.89	4.23	35	3.09	3.23	55	2.73	3.10
16	3.78	3.83	36	2.99	3.61			
17	3.75	3.63	37	2.80	3.32			
18	3.53	4.12	38	4.74	4.10			
19	3.55	3.64	39	4.34	4.02			
20	3.50	3.63	40	4.25	3.98			

Table 13-4: Comparison of the experimental biological activities a_i and activities obtained by the cross-validation process $a_i^{est\ CV}$. The correlation factor r between these two series is *0.86*. The estimated $a_i^{est\ model}$ as yielded by the complete model on 55 objects are even better giving *r=0.95*.

13.13 References and Suggested Readings

13-1. T. Aoyama, Y. Suzuki and H. Ichikawa, "Neural Networks Applied to Quantitative Structure-Activity Relationships", *J. Med. Chem.* **33** (1990) 905 – 908.

13-2. T. Aoyama, Y. Suzuki and H. Ichikawa, "Neural Networks Applied to Quantitative Structure-Activity Relationship (QSAR) Analysis", *J. Med. Chem.* **33** (1990) 2583 – 2590.

13-3. M. Yoshimoto, H. Miyazawa, H. Nakao, K. Shinkai and M. Arakawa, "Quantitative Structure-Activity Relationships in 2,5-Bis(1-aziridinyl)-p-benzoquinone Derivates against Leukemia L-1210", *J. Med. Chem.* **22** (1979) 491 – 496.

13-4. R. D. Cramer, III, D. E. Patterson, and J. D. Bunce, "Comparative Molecular Field Analysis (CoMFA). 1. Effect of Shape on Binding of Steroids to Carrier Proteins", *J. Am. Chem. Soc.* **110** (1988) 5959 – 5967.

13-5. G. Moreau, and P. Broto, "Autocorrelation of Molecular Structures: Application to SAR Studies", *Nouv. J. Chim.* **4** (1980) 757 – 764.

13-6. M. Wagener, J. Sadowski and J. Gasteiger, "Autocorrelation of Molecular Surface Properties for Modeling Corticosteroid Binding Globulin and Cytosolic Ah Receptor Activity by Neural Networks", *J. Am. Chem. Soc.* **117** (1995) 7769 – 7775.

13-7. E. A. Coats, "The CoMFA Steroids as a Benchmark Dataset for Development of 3D QSAR Methods" in *3D QSAR in Drug Design*, Vol. 3, Eds.: H. Kubinyi, G. Folkers and Y. C. Martin, Kluwer / ESCOM, Dordrecht, NL, 1998, pp. 199 – 214.

13-8. J. Zupan and M. Novic, "General Type of a Uniform and Reversible Representation of Chemical Structures", *Anal. Chim. Acta* **348** (1997) 409 – 418.

13-9. R. T. Burke, "Protein - Tyrosine Kinase Inhibitors", *Drugs of the Future* **17** (1992) 119 – 131.

13-10. M. Cushman, D. Nagarathnam and L. R. Geahlen, "Synthesis and Evaluation of Hydroxylated Flavones and Related Compounds as Potential Inhibitors of the Protein-Tyrosine Kinase p56", *J. Nat. Products* **54** (1991) 1345 – 1352.

13-11. M. Cushman, D. Nagarathnam, L. D. Burg and L. R. Geahlen, "Synthesis and Protein-Tyrosine Kinase Inhibitory Activities of Flavonoid Analogues", *J. Med. Chem.* **34** (1991) 798 – 806.

13-12. M. Cushman, H. Zhu, L. R. Geahlen and J. A. Kraker, "Synthesis and Biochemical Evaluation of a Series of Aminoflavones as Potential Inhibitors of Protein-Tyrosine Kinases p56, EGFr, p60", *J. Med. Chem.* **37** (1994) 3353 – 3362.

13-13. M. Novic, Z. Nikolovska-Coleska and T. Solmajer, "Quantitative Structure-Activity Relationship of Flavonoid p56lck Protein Tyrosine Kinase Inhibitors. A Neural Network Approach", *J. Chem. Inf. Comput. Sci.* **37** (1997) 990 – 998.

13-14. B. Hibbert, "Genetic Algorithm in Chemistry", *Chemom. Intell. Lab. Syst.* **19** (1993) 277 – 293.

13-15. D. E. Goldberg, *Genetic Algorithms in Search Optimization and Machine Learning*, Addison-Wesley, New York, USA, 1989.

13-16. G. Jones, "Genetic and Evolutionary Algorithms", in *Encyclopedia of Computational Chemistry*, Eds.: P. v. R. Schleyer, N. L. Allinger, T. Clark, J. Gasteiger, P. A. Kollman, H. F. Schaefer III and P. R. Schreiner, Wiley, Chichester, UK, 1998, pp. 1127 – 1136.

13-17. V. Venkatasubramanian and A. Sundaram, "Genetic Algorithms: Introduction and Applications", in *Encyclopedia of Computational Chemistry*, Eds.: P. v. R. Schleyer, N. L. Allinger, T. Clark, J. Gasteiger, P. A. Kollman, H. F. Schaefer III and P. Schreiner, Wiley, Chichester, UK, 1998, pp. 1115 – 1127.

13-18. J. Devillers (Ed.), *Neural Networks in QSAR and Drug Design*, Academic Press, London, UK, 1996.

13-19. J. Devillers (Ed.), *Genetic Algorithms in Molecular Modeling*, Academic Press, London, UK, 1996.

14 The Electrophilic Aromatic Substitution Reaction

learning objectives:

- structural factors influencing electrophilic aromatic substitution

- representation of molecular structures by a connection table

- three forms of structure representation:

  1) connection table

  2) specifying formal charges for all atoms of the ring

  3) use of electronic and steric parameters **specific to a reaction site**

14.1 The Problem

The substitution of a hydrogen atom of a monosubstituted benzene derivative by another group (e.g., nitro, halogen, acyl, alkyl) is a remarkable reaction on many grounds. First, it is of great industrial importance, many basic chemicals being produced by this reaction. Second, nearly all these reactions occur by the same fundamental mechanism: the attack of an electrophilic group, an agent with electron demand, on the benzene derivative.

This electrophilic substitution of a proton by another group can occur, in principle, in three different positions: *ortho* (*o*), *meta* (*m*), and *para* (*p*) (Figure 14-1).

The rationalization for the relative reactivities of various monosubstituted benzene derivatives and for the distribution of *ortho*-*meta*-, and *para*-substituted products is a paradigm of the methods of physical organic chemistry. In undergraduate organic chemistry courses, it is a standard case for explaining the influence of various

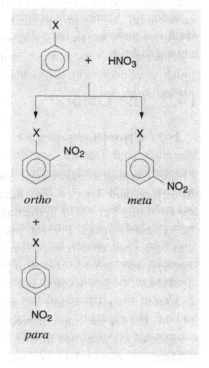

Figure 14-1: Product distribution in the electrophilic aromatic substitution reaction.

electronic effects, particularly the inductive and resonance effects, on the reactivity and selectivity of a chemical reaction. In spite of this apparently well-settled matter, there are still a lot of unanswered questions. The ratio of *ortho* to *para* product is hard to predict, although it is thought to be largely influenced by steric effects. Reaction conditions, particularly the solvent, can have a drastic influence on product distribution; this is hardly understood at all.

Furthermore, taking what we know about isomer distribution in the reaction of monosubstituted benzene derivatives, and using it to predict product ratios obtained from polysubstituted benzene derivatives is not very successful (Figure 14-2).

Nevertheless, at an elementary level, electrophilic aromatic substitution (EAS) shows some distinct characteristics. Most substituents on the aromatic ring can be classified into two categories: groups that donate electron density, either by an inductive or a mesomeric effect (Figure 14-3) primarily give *ortho* and *para* substitution, while groups that are mesomeric electron acceptors (Figure 14-4) react by direct substitution at the *meta* position.

Since the factors that determine the *ortho/para* product ratio are much less understood, the yields of these two products are quite often lumped together.

14.2 The Data

For the present example, we will use the work and data of Elrod, Maggiora, and Trenary (see Reference 14-1), who investigated the distribution of products in the nitration of a series of monosubstituted benzene derivatives. For reasons just mentioned, the yields of *ortho* and *para* product were combined; thus, they worked with the ratio of *meta* product to *ortho* plus *para* product.

Table 14-1 shows ten out of the 50 substituents used in this example, arranged in order of increasing percentage of *meta* product obtained in the electrophilic reaction.

One of the purposes of this example is to discuss the problem of finding an appropriate coding scheme for the structures of the compounds or substituents to be input into the neural networks. Two coding schemes are investigated.

Any coding scheme for this problem must, of course, be capable of representing these substituent effects. The first approach takes the *partial atomic charges* on the six carbon atoms of the ring (as

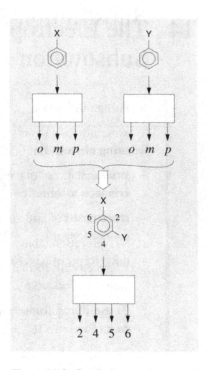

Figure 14-2: Predicting product distribution in the electrophilic aromatic substitution of disubstituted benzene derivatives, based on a knowledge of product distribution in the EAS of the two corresponding monosubstituted benzene derivatives.

no.	substituent	yield of *meta* product
1	$-NH_2$	0
2	$-NHCOCH_3$	2
3	$-CH_3$	4
5	$-CH_2NH_2$	10
5	$-CH_2COOH$	22
6	$-SiMe_3$	40
7	$-CCl_3$	64
8	$-CONH_2$	70
9	$-COOH$	80
10	$-SO_2CH_3$	100

Table 14-1: Ten substituents, and the corresponding yield of the *meta* product.

calculated by a semiempirical quantum mechanical method, MOPAC – see Reference 14-3) as a representation of the electronic effects influencing product distribution (Figure 14-5).

In the other scheme, the structure of a substituent is represented by a (5 x 5)-connection table, in which each row represents one atom. The five entries arc the atomic number, the ID number of this atom, the ID number of the atom it is bonded to, the bond order of this bond and the formal charge on this atom (Figure 14-6).

For each nonhydrogen atom of the substituent, we need a new row, starting with the atom directly bonded to the ring. If the substituent has fewer than five nonhydrogen atoms, the remaining rows are filled with zeros; if the substituent contains more than five atoms, the representation is cut after five.

Next, we need a canonicalization for the connection table, i.e., a set of rules valid in **all** cases to produce a **unique numbering** scheme for the atoms. Particularly in larger connection tables the atoms in a given substituent can be numbered in different ways, so that different inputs will be produced in the neural network, and consequently many different outputs will result.

14.3 The Network

With two essentially different representations for monosubstituted benzene derivatives, we will need two different architectures for the neural network. The first representation leads to an architecture with

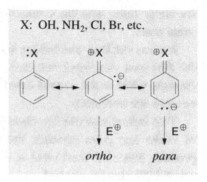

Figure 14-3: An atom X with a free electron pair can donate electron density to the *ortho* (o) and *para* (p) positions by a so-called *plus mesomeric effect*, and thus direct the attack of the electrophile primarily to the *ortho* and *para* positions.

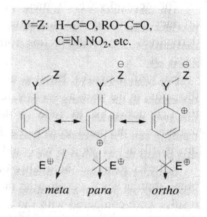

Figure 14-4: A group Y=Z with a multiple bond can reduce the electron density at the *ortho* (*o*) and *para* (p) positions by the so-called *minus mesomeric effect*. This makes the attack of the electrophile at these positions particularly difficult, leading to preferred *meta* substitution.

six input units, and the connection table representation requires 25 input units.

It turns out that one hidden layer is enough for both networks; in the first case (six input neurons), 10 neurons in the hidden layer are enough, while in the second case (25 input neurons), only five hidden neurons are necessary.

Two output neurons are chosen, one for the combined percentage of *ortho* and *para* product, the other for the percentage of *meta* product. Since the total yield is 100%, it is enough to report only the percentage of *meta* product.

With two neurons in the output layer, the complete architecture of the first network is (6 x 10 x 2), with (6 + 1) x 10 + (10 + 1) x 2 = 92 weights. The second network, (25 x 5 x 2), has (25 + 1) x 5 + (5 + 1) x 2 = 142 weights. The architectures of both networks are shown in Figure 14-7.

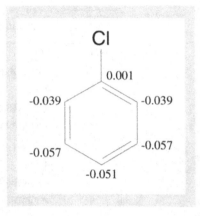

Figure 14-5: Partial charges on the six atoms of the benzenoid ring of *chlorobenzene* used as input representation.

14.4 Learning and Results

The networks were trained by error back-propagation using the product ratios of the nitration of 37 monosubstituted benzene derivatives. The product ratios from 13 other compounds were used as a test set.

For both networks, Elrod et al. used 100,000 epochs (!) to reduce the errors in the training set to values as small as 0.3% in the better of the two networks. In spite of this excellent recall ability for the training set, the predictions had a much higher average error (12.1%); this is still not as good as we could desire.

In order to judge the quality of the results, the authors used two other approaches for estimating the amount of *meta* product: first, the results were compared with those obtained from CAMEO, an expert system for predicting the products of reactions (see Reference 14-4). Second, the 13 examples of the test set were given to three organic chemists, who were asked to predict the percentage of *meta* product. The predictions of these three chemists for all 13 compounds were averaged and compared with the results obtained by the two neural networks and by the CAMEO expert system. All this is summarized in Table 14-2.

Both neural networks gave better results than the expert system CAMEO. The average error of the three chemists is lower than that obtained from CAMEO and the neural network based on charge

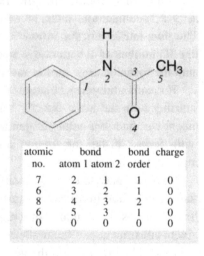

atomic no.	bond atom 1	atom 2	bond order	charge
7	2	1	1	0
6	3	2	1	0
8	4	3	2	0
6	5	3	1	0
0	0	0	0	0

Figure 14-6: The substituent *acetanilide* and its connection table representation.

distribution. However, **the three chemists were outperformed by the network based on the connection table representation**.

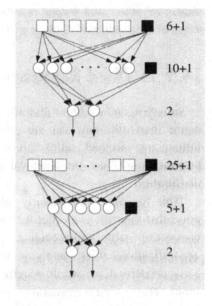

Figure 14-7: Two different neural network architectures for the two different structure-coding schemes.

system	training set	test set
neural network (6 x 10 x 2)	5.2	19.8
neural network (25 x 5 x 2)	0.3	12.1
CAMEO (expert system)	18.0	22.6
human experts		14.7

Table 14-2: Errors in recall and predictions for the amount of *meta* product by the two neural networks, by an expert system and by chemists (in percent).

Of the two neural networks, the one built on the connection table representation clearly shows the better results. This might be surprising, because this representation is much simpler to obtain than the one using partial atomic charges.

Why is the connection table network so good? A better question might be, why the other one is worse. In fact, it is not too surprising that the network based on the partial charges did not perform very well. The ground state charge distribution is only **one** of the various electronic factors influencing product distribution in electrophilic aromatic substitution, and, if considered **alone**, is clearly insufficient for representing the results of the nitrations.

Before continuing, it is prudent to admit that this study (like all studies) has limitations:

– The product ratio in the nitration of benzene derivatives depends strongly on reaction conditions, particularly on the concentration of sulfuric acid; this is not accounted for in the present study.

– The amounts of *ortho* and *para* product are combined, which prevents us from studying the important problem of the *ortho* effect. Furthermore, a separate treatment of the *ortho* and *para* distribution is a prerequisite for any attempt at predicting product distributions in di- and polysubstituted benzene derivatives.

This, of course, does not detract from the importance of the work by Elrod, Maggiora, and Trenary; but it is important to stress that the choice of input and output representations strongly determines the scope of an application and its prediction ability.

14.5 A Third Representation of Data

Satisfying as we might find the results obtained by the (25 x 5 x 2) neural network, they can not give an **explanation** of the effects influencing product ratios, since the connection table coding is arbitrary, and hides the chemical effects responsible for the product distribution.

This becomes particularly clear in the nitration of di- or polysubstituted benzene derivatives (Figure 14-8). Then, the above connection table representation is of no help at all in making generalizations. First, applying this representation to disubstituted benzene derivatives would require two (5 x 5)-connection tables; 50 input units would generate quite a different neural network architecture. Even worse, the two connection tables would be identical for the *ortho-*, *meta-*, or *para*-disubstituted benzene derivatives: they are insufficient to distinguish among these three different starting materials.

Clearly, we need a better coding scheme for di- or polysubstituted benzene derivatives. We will explore another representation that addresses the problem at the point where the reaction occurs, i.e., at the *ortho*, *meta*, or *para* position on the benzene ring.

The connection table representation is fine for predicting the yield ratio of *meta* to (*ortho + para*) product. However, if:

– the influence of sulfuric acid is to be accounted for

– the amounts of *ortho* and *para* product are to be distinguished

– the electronic effects governing product distribution in EAS reactions are to be deciphered, and

– the predictions on product ratios in the nitration of di- and polysubstituted benzene derivatives have to be made,

we need more general representations on the input side for the starting materials and reaction conditions, and on the output side for the product distribution.

For example, the input vector should be coded using a **reaction site-specific** representation; this means that the representation for a given substituent should be **different** for each output ring position, so

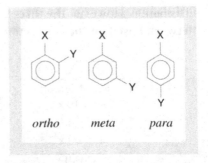

Figure 14-8: *Ortho-*, *meta-*, and *para*-disubstituted benzene derivatives.

that different variables may be input for different positions (while there is always only one answer: percent of the product corresponding to substitution at the position for which the variables are input).

In principle, there are five hydrogens on a monosubstituted benzene derivative that can be substituted by an electrophilic agent (e.g., nitro). However, because the molecule has a plane of symmetry, there are two equivalent *ortho* and two equivalent *meta* positions, so that only three different positions, *ortho*, *meta* and *para* have to be considered, but because of the symmetry, *ortho* and *meta* must be weighted twice as much as *para*.

Furthermore, we need one (or more) additional input units to account for one (or more) variables on reaction conditions; in this case the concentration of sulfuric acid should be included.

The problem of individually representing the three substitution positions of a monosubstituted benzene derivative was addressed by A. Fröhlich and coworkers of the Model Laboratory for Computer Chemistry at the Technical University of Munich (see Reference 14-5). They used **one** steric and **four** electronic variables.

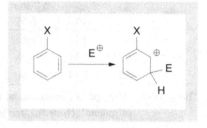

Figure 14-9: Intermediate state leading to the *meta* product in an electrophilic aromatic substitution ($E^\oplus$ is the electrophile).

Figure 14-9 shows the intermediate formed in an electrophilic aromatic substitution; we can use electronic variables for the carbon atom where the electrophile E is bonded as controlling parameters: the σ-electronegativity, χ_σ; the π-electronegativity, χ_π; and both the average inductive stabilization, $\chi_\sigma^{av}(o,p)$, and the resonance stabilization, R^+, of the positive charge generated at this carbon atom (Figure 14-9). These parameters can be calculated by empirical methods.

Two additional input units are provided: one for an estimate of the amount of steric hindrance, *Ster*, at the reaction position obtained from the van der Waals radii of the atoms, and one for the concentration of sulfuric acid, $[H_2SO_4]$ (Figure 14-10).

The values of the input parameters for each of the three sites of phenol are given in Table 14-3.

With a site-specific representation of the starting material, we need only one output neuron, the amount of product at the site being considered (*meta* position, in Figure 14-10). Seven hidden neurons complete the architecture of the neural network in this study (Figure 14-10).

Note that the neural network is trained with data for each individual position separately. This means that if we input the six parameters χ_σ, χ_π, $\chi_\sigma^{av}(o,p)$, R^+, *Ster*, and $[H_2SO_4]$ for the **para**

site	χ_σ [eV]	χ_π [eV]	$\chi_\sigma^{av}(o,p)$ [eV]	R^+ [1/eV]	Ster [$Å^3$]	$[H_2SO_4]$ [%]	yield [%]
ortho	8.45	5.41	8.66	21.1	5.85	74.1	18.0
meta	8.25	5.34	8.38	6.0	5.40	74.1	0.5
para	8.23	5.34	8.66	21.1	5.04	74.1	63.0

Table 14-3:   Site-specific input and output parameters for the three different sites of phenol. (Note that the total yield is 18 x 2 + 0.5 x 2 + 63 = 100)

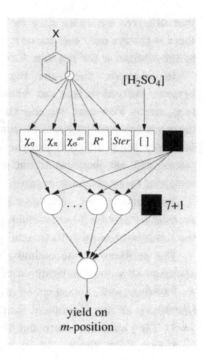

Figure 14-10: Architecture for prediction of the yield of product at a specific site (*meta*, in this case).

position, then as a target for training the network, the percentage of ***para*** product should be given. However, if data for the ***ortho*** or ***meta*** positions are used, the network outputs only half the expected yield, since in actuality, each ring has two such positions ("statistical factor", or "symmetry factor").

Thus, one single neural network is trained to predict the yield of substitution products at **each individual** position.

This network was trained by back-propagation of errors, using as a training set the product distributions in the nitration of 23 monosubstituted benzene derivatives at various concentrations of sulfuric acid. Altogether, these 23 compounds provided 159 data on the yields of substitution products at different positions, for different concentrations of sulfuric acid. The data are easily learned, with an average error of 6% on recall.

Three disubstituted benzene derivatives were then used for testing the predictive performance of the neural network. Since each disubstituted benzene derivative contains four sites for potential substitution, predictions for each of the twelve sites for nitration were individually made by inputting the electronic and steric variables of each site together with a preset concentration of sulfuric acid (Figure 14-11). The average error in the prediction of the yield for substitution at the various positions amounts to 10%.

Figure 14-12 summarizes the three different approaches for setting up a multilayer neural network capable of predicting the regioselectivity in electrophilic aromatic substitution, and for coding a monosubstitued benzene derivative as input.

Figure 14-12a) shows the representation used by Elrod, Maggiora and Trenary, in which the charges at the six positions of a monosubstituted benzene derivative are input in order to predict the yields of *meta* and (*ortho* + *para*) products by using **two** output neurons. In Figure 14-12b), the structure of the **substituent** is coded by a (5 x 5) connection table; again, the *meta* and (*ortho* + *para*) product yields are obtained on **two** outputs.

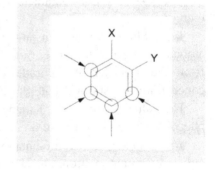

Figure 14-11: The four sites for potential substitution in an *ortho*-disubstituted benzene derivative.

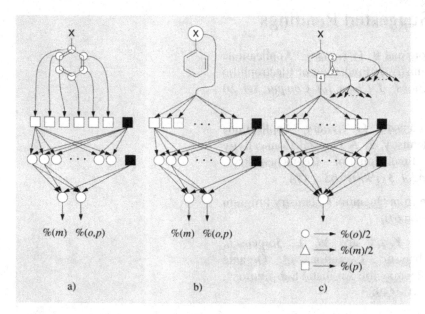

Figure 14-12: Three different architectures and input representations for learning the regioselectivity of the electrophilic aromatic substitution reaction: a) global charge vector, b) connection table of the substituent, c) local electronic, steric, and reaction condition representation.

Figure 14-12c) diagrams the site-specific neural network that takes as input one steric and four electronic variables for one ring site, as well as the concentration of sulfuric acid. **One** output neuron suffices to predict the amount of product at that site.

14.6 Concluding Remarks

Go back and read the quotation of Bernhard Widrow at the very end of Chapter 9. The important message to carry away from this study is that your method of representing information determines the scope of the predictions that can be made. A **global** representation of the substituent in a benzene derivative by a connection table only allows us to make global predictions, e.g., the amount of *meta* product vs. the sum of *ortho* and *para* product. A *local* representation of the influence of a substituent on each individual ring position allows us to make **local** predictions: the amount of product at each individual position. Furthermore, only a local representation of a mono-substituted benzene derivative can be generalized to make predictions about product ratios in the nitration of di- and polysubstituted derivatives.

14.7　References and Suggested Readings

14-1.　D. W. Elrod, G. M. Maggiora and R. G.Trenary, "Applications of Neural Networks in Chemistry. *1*. Prediction of Electrophilic Aromatic Substitution Reactions", *J. Chem. Inf. Comput. Sci.* **30** (1990) 477 – 484.

14-2.　D. W. Elrod, G. M. Maggiora and R. G. Trenary, "Applications of Neural Networks in Chemistry. *2*. A General Connectivity Representation for the Prediction of Regiochemistry", *Tetrahedron Comput. Methodol.* **3** (1990) 163 – 174.

14-3.　MOPAC program, available from Quantum Chemistry Program Exchange (QCPE 455, Version 6.0).

14-4.　M. G. Bures, B. L. Roos-Kozel and W. L. Jorgensen, "Computer-Assisted Mechanistic Evolution of Organic Reactions (CAMEO). *10*. Electrophilic Aromatic Substitution", *J. Org. Chem.* **50** (1985) 4490 – 4498.

15 Modeling and Optimizing a Recipe for a Paint Coating

learning objectives:

- factors involved in adjusting an industrial recipe (ingredients plus processing conditions) while maintaining consistent results

- identification of controlling variables as **little correlated** with each other as possible

- use of experiments to select training data

- comparison of 3-D and 2-D (sectioned) display of the results: **partial models**

15.1 The Problem

This example is a typical industrial application of modeling. The chemical, pharmaceutical, food, and many other industries rely heavily on recipes for their products. A recipe usually consists of two types of quantities: the first comprises the amounts of components that should be put together or processed; the second type consists of process parameters, like temperature of processing, pH of solvent, time of mixing, etc., necessary to make the ingredients into a product.

In order to maintain preset levels of quality and/or price, the recipe must be followed exactly; but components may change in price, or suppliers may change **their** products, and all such difficulties require an adjustment of the recipe.

If, for example, the pH of the process is critical, then the recipe will have to be adjusted if some component from a new supplier has a higher or lower acidity; and adjusting the quantity of one component probably calls for adjustments of all other components to maintain consistent properties and quality of the product.

Recipes are as delicate as complex circuits; adjusting one is often considered an "art", since an analytical dependence of product properties on compound properties (and on component **interactions**) is unknown.

> Remember that there are **multiple** product properties that must be kept within prescribed tolerances.

We want to find a non-analytical means of adjusting the conditions and ingredient quantities to bring product properties back within specs; and if this is not possible, the method should tell us **that**, too.

15.2 The Data

The example that will be followed here was worked out by Tusar and coworkers at the National Institute of Chemistry in Ljubljana, Slovenia, for a chemical factory which produces different kinds of paints, paint coatings, and related products. The goal was to obtain full knowledge about the dependencies among the various properties (coordinates in measurement space) of a product that we will call simply "paint coating".

During intensive studies of this product, it was found that there are three highly significant and **non-correlated** (or only slightly correlated) variables: concentration of the polymer component, C_p, concentration of the catalyst, C_c, and the temperature T used for heating the product. Six properties have to be controlled and adjusted to the norms for this product:

- hardness, H

- elasticity, E

- adhesiveness, A

- resistance to *methyl-isobutyl-ketone*, *MIBK*

- stroke resistance, *SR*

- contra-stroke resistance, *CSR*

Using standard modeling techniques, six models are set up:

$$H = f_1 (C_p, C_c, T, a_h, b_h, c_h, ...)$$

$$E = f_2 (C_p, C_c, T, a_e, b_e, c_e, ...)$$

$$A = f_3 (C_p, C_c, T, a_a, b_a, c_a, ...)$$

$$MIBK = f_4(C_p, C_c, T, a_{mibk}, b_{mibk}, c_{mibk}, ...)$$

$$SR = f_5(C_p, C_c, T, a_{sr}, b_{sr}, c_{sr}, ...)$$

$$CSR = f_6(C_p, C_c, T, a_{csr}, b_{csr}, c_{csr}, ...) \qquad (15.1)$$

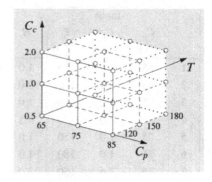

Figure 15-1: Full three-variable, three-level experimental (factorial) design for determining the dataset for training and testing the model.

This means that we need to invent (or guess!) functions f_1, f_2, ..., f_6 that give values of these six properties in terms of the three controlling variables. Furthermore all six sets of adjustable coefficients a_k, b_k, c_k, ... should be determined in such a way that the set of experimental values $\{H_s, E_s, A_s, MIBK_s, SR_s, CSR_s\}$ will fit the set of calculated ones at given values of $\{C_{sp}, C_{sc}, T_s\}$ as well as possible.

As we already know, supervised neural network learning does not require any prior assumptions or hypotheses about the function types, numbers of parameters, etc. All that is needed is a well selected architecture and the input data and targets to which the model should be adapted.

The data for the model were selected using a full three-variable three-level experimental design requiring 27 measurement points (Figure 15-1).

Accordingly, 27 cover paints were made in the test laboratory and for each paint all 6 properties were measured. Altogether a matrix of 27 x (3 + 6) = 243 values was obtained: 81 were used as input vectors (27 x 3) and 162 as targets (27 x 6). A few of the experiments used in developing the model are given in Table 15-1; output values in this Table are given on a quality scale, on which 1.0 represents superior quality, while values represented by 0.0 actually mean that the measured property is so bad that it may not even be measurable.

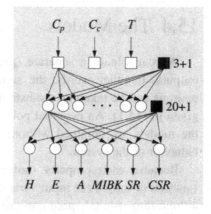

Figure 15-2: (3 + 1) x (20 + 1) x 6 neural network for modeling properties of paint coatings.

15.3 The Network and Training

As usual in our examples, a one-hidden-layer neural network was chosen, having three input units, six output neurons, 20 neurons in the hidden layer, and 189 weights (Figure 15-2).

The 81 input values were normalized between 0 and 1, while the 162 output values were additionally scaled into the interval 0.2 – 0.8. This further scaling is often recommended for output neurons having a nonlinear squashing function (Equation (2.39)), but trained to yield nearly linear outputs (Figure 8-15). After about ten thousand epochs, the network became stabilized.

	input			output					
	C_p [%]	C_c [%]	T [°C]	H	E	A	$MIBK$	SR	CSR
1	65	1.0	150	1.0	0.3	0.0	1.0	0.0	0.0
2	65	2.4	150	0.0	0.7	0.8	1.0	0.7	1.0
3	75	0.2	150	1.0	0.9	1.0	1.0	0.0	0.0
4	75	1.0	120	0.5	1.0	0.9	0.9	0.6	0.6
5	75	1.0	150	1.0	0.9	1.0	1.0	0.7	0.8
6	75	1.0	180	1.0	0.9	1.0	1.0	0.0	0.0
7	75	2.4	150	1.0	0.8	0.3	0.8	1.0	0.5
8	85	0.2	150	0.0	1.0	1.0	0.0	1.0	1.0
9	85	2.4	120	0.0	1.0	1.0	0.0	1.0	1.0

Table 15-1:   Some of the experiments used for building the model. The output properties are normalized between zero and one: 1.0 signifies an excellent value of the property, while 0.0 represents a bad (or unmeasurable) value.

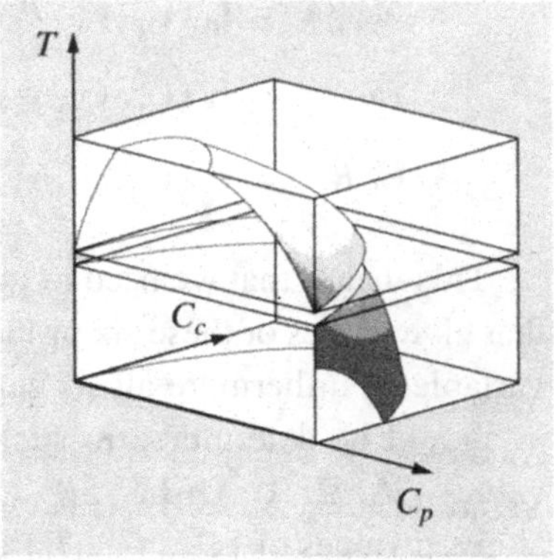

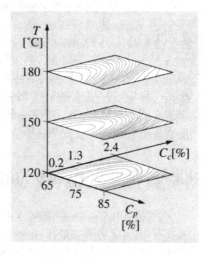

Figure 15-3: Three-dimensional display of surface presenting possible values of the predicted property *H*. A cut is made at a specific temperature and the two "boxes" have slightly been moved apart for better visualization.

15.4 The Models

We can obtain a separate (partial) model from each of the six outputs. In other words, the signal at each one of the six output neurons can be taken as a substitute for one of the explicit equations given in (15.1). An essential point of this example is that, as always, the models are obtained without any *a priori* knowledge about the behavior of the system.

Because each property (output variable) is a relatively simple function of only three input variables, it can be visualized as a plot in a three-dimensional space comprising variables C_p, C_c and T (Figure 15-3). Unfortunately, when looking at this figure, it becomes clear that only one variable (one surface) can be shown on one such picture.

However, even this three-dimensional display does not offer more than a qualitative description of the property behavior. In order to get usable quantitative output, we must produce two-dimensional maps (each at a constant value of the third parameter). Such cross-sectional planes are shown in Figure 15-4; once the model is obtained, these may be drawn at any value of the third variable.

The 18 maps in Figure 15-5 show how each of the controlled properties *H*, *E*, *A*, *MIBK*, *SR*, or *CSR* behaves. The three maps under each output are only three of many possible constant-*T* cross-sections.

Figure 15-4: Three two-dimensional cross-sections describe the property *H* better than the three-dimensional surface in Figure 15-3.

As discussed above, an analogous map at any temperature between 150°C and 180°C can be obtained at any of the six outputs. These 18 maps illustrate the richness of information that can be obtained from the trained network.

This can easily be programmed on a personal computer, a simple user interface added and the whole package handed over to engineers for use. Since the calculation of all six properties from any triplet of variables C_p, C_c and T can be made almost instantaneously, the entire measurement space can be thoroughly inspected.

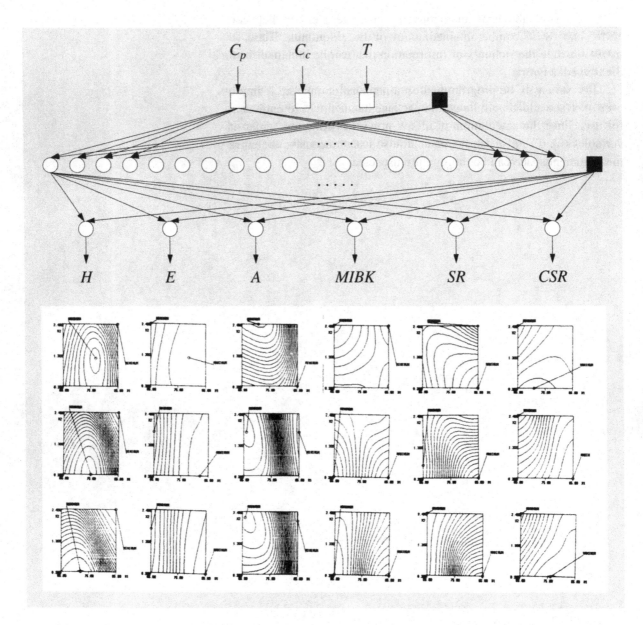

Figure 15-5: Two-dimensional sections through the 3-D input space. Each column of three maps shows the dependance of the indicated output property (*H*, *E*, etc.) upon C_p (abscissa) and C_c (ordinate); each map is at a different constant value of *T*.

15.5 References and Suggested Readings

15-1. L. Tusar, M. Tusar and N. Leskovsek, "A Comparitative Study of Polynominal and Neural Network Modelling for Optimization of Clear Coat Formulations", *Surf. Coat. Intern.* **78** (1995) 427 – 434.

15-2. N. Leskovsek, L. Tusar and M. Tusar, "Empirical Modeling of Rheological and Mechanical Properties of Paint", *Rheology* **5** (1995) 140 – 145.

15-3. M. Tusar, L. Tusar, N. Barle, N. Leskovsek and M. Kumaven, "A Study of the Influence on the Hiding Power of the Composition of a Paint and its Film Thickness", *Surf. Coat. Intern.* **78** (1995) 473 – 479.

15-4. A. P. de Weijer, L. Buydens, G. Kateman and H. M. Heuvel, "Neural Networks Used as a Soft Modeling Technique for Quantitative Description of the Relation Between Physical Structure and Mechanical Properties of Poly(ethylene terephthalate) Yarns", *Chem. Intell. Lab. Systems* **16** (1992) 77 – 86.

15-5. A. P. de Weijer, C. B. Lucasius, L. Buydens, G. Kateman and H. M. Heuvel "Using Genetic Algorithms for an Artificial Neural Network Model Inversion", *Chem. Intell. Lab. Systems* **20** (1993) 45 – 55.

16 Fault Detection and Process Control

learning objectives:

- complex processes: priorities and problems involved in identifying malfunctions, and in controlling conditions to achieve constant results

- comparison of classical methods and neural networks

- one danger of careless choice of data representation: introducing spurious correlations into the dataset; one way to avoid this

- special requirements of time-dependent data, and one technique (the "moving window") for dealing with them

- how the counter-propagation network behaves as a self-organizing lookup table

- example of fault detection in a catalytic conversion reactor

- comparison of results from back-propagation and counter-propagation networks

- relationship between **forward** model (input → results) and **inverse** model (results → input)

16.1 The Problems

Two basic problems have to be faced in running chemical processes; first, the recognition and detection of faults, and second, control of the process itself. Fault detection is regarded mainly as qualitative, since it is necessary only to identify the defective part(s) in the system. On the other hand, controlling the process is quantitative, since the values of process variables must be continuously monitored and maintained within prescribed limits. This information is, of course, a prerequisite for detecting (and identifying) system faults.

Immediate recognition of faults prevents loss of time, of production quotas, or even of the equipment. Therefore, a permanent on-line diagnosis based on the process variables is needed in addition to equipment testing and preventive maintenance, operator training, etc.

Control of the process requires feedback calculated on the basis of a model believed to govern the entire process. The better the model, the more reliable are the calculated responses and the smoother is the process. The model should be set up either on the basis of the known chemical and technological subprocesses, or learned on the basis of past monitoring.

There may be many sources of difficulty in achieving fault prediction and/or process control based on continuous monitoring of the process variables:

— reliable models that yield accurate behavior of the process under all circumstances are often not known

— the dependence of the behavior of the process on the controlling variables is in most cases nonlinear

— the data monitored as a basis for the feedback can be noisy or uncertain

— there exists no one-to-one correspondence between the set of observed symptoms and the correct diagnosis.

Since it is important to know as much as possible about the behavior of the process, especially under unusual conditions, there has been considerable research into the theory of process control. Different approaches to the problem have been tried:

— statistical analyses

— linear models using classical techniques

— pattern recognition

— knowledge (rule) based expert systems

— neural networks

The quality and accuracy of predictions based on rule-based expert systems strongly depend on the extent and quality of the rule database, which must be painstakingly collected from human experts, a time-consuming and expensive process. Therefore, neural network techniques in general and the back-propagation algorithm in particular have stirred enormous interest among chemical engineers.

Some of the more important advantages of neural networks over classical approaches are believed to be that:

- neural networks can learn from examples, making analytical models unnecessary;

- the handling of nonlinear models is easy for neural networks, because of their inherently nonlinear response;

- neural networks can perform association, i.e., they can handle incomplete or slightly corrupted data, which is important for good modeling outside the trained regions,

- neural networks can handle continuous variables as well as discrete ones (in rule-based expert systems, discrete variables are usually used;

- neural networks can model inverse functions, an important feature for generating feedback control systems.

Hoskins and Himmelblau (Reference 16-1) have discussed the applications of neural networks in various areas of chemical process engineering, such as fault detection, diagnosis, process control, process design, and process simulation. Sometimes it is difficult to clearly separate one task from another, but we will try to summarize them in this chapter.

16.2 The Data

A chemical process, whether on plant or pilot scale, is monitored by measuring a number of variables, x_i, at different points of the process, and supervised by feedback signals to various controllers. The task of a *controller* is to keep the measured variables in a given state (on/off), or at a defined level of a given variable (the *set-point*), or within a defined interval. These variables may include flow-rate, temperature, concentration, pressure, liquid level, etc. The controllers supervise heaters, pumps, valves, stirring devices, etc.

The assembly of all measured variables and states at a given time t, x_{it}, is called the process vector $\boldsymbol{P}_t$:

$$\boldsymbol{P}_t = (x_{1t}, x_{2t}, ..., x_{it}, ..., x_{mt}) \qquad (16.1)$$

When teaching the neural network to recognize and to predict faults in a chemical process, the training input vectors $\boldsymbol{X}$ are sets of measured variables known to result in a "smooth" or "no-fault" process, along with sets that describe faulty conditions. For each of the latter, a vector describing the faulty state(s) must be given as a target vector $\boldsymbol{Y}$ that pinpoints the faults caused by that particular $\boldsymbol{X}$.

Quite often, the causes are binary or discrete states, such as presence or absence of fluid flow (pump working/not working), etc. The cause of a malfunction can be, of course, a combination of faulty states.

The best way to handle multiple discrete states of a variable is to transform it into a *distributed representation*, i.e., to transform it into as many binary variables as there are different states for the original discrete variable.

For example, suppose that a variable x_v called "state of the valve" can have one of three possible states: "left pipeline", "shut", and "right pipeline" (Figure 16-1). The process vector $\boldsymbol{X}$ should have **three** new sub-variables to describe these three states as 0 or 1:

$$\boldsymbol{X} = \left(x^{left}, x^{shut}, x^{right}, ... \right) \qquad (16.2)$$

Then for given values of the other variables, $\boldsymbol{X}$ may be:

$$\boldsymbol{X}_i = (1, 0, 0, ...) \quad \text{valve open to the left pipeline}$$

$$\boldsymbol{X}_j = (0, 1, 0, ...) \quad \text{valve shut} \qquad (16.3)$$

$$\boldsymbol{X}_k = (0, 0, 1, ...) \quad \text{valve open to the right pipeline}$$

Obviously, any other combination, such as $\boldsymbol{X} = (1, 1, 0, ...)$ or $(0, 1, 1, ...)$ corresponds to impossible states. It might seem a waste to use three variables and then throw away six of their nine possible states; perhaps a new variable having three values, +1, 0 and −1, would be more efficient.

This representation can of course be used, but it has some unintended side effects. The sequence of values (+1, 0, −1) implies that the states are **not equivalent**: the state "shut" (0) is closer to both "open" states than these two states are to each other. Such *unjustified*

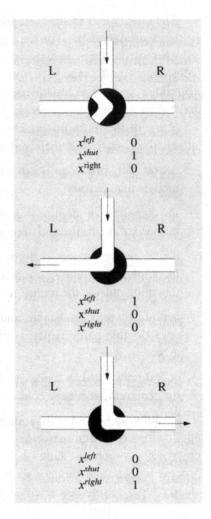

Figure 16-1: Description of three states of a valve and the corresponding value of the process vector.

correlation among the equivalent states, if included in the learning procedure, may have an undesired influence on the results. The more equivalent discrete states a variable has, the worse is the situation when the condensed distribution is used. A very similar case of unwanted correlation among the states of one variable is discussed in Chapter 17, where coding of proteins by amino acids is described. The situation is solved exactly as described here: each multistate variable is replaced by as many binary inputs as the variable has different states.

In order to predict how the process variable x_i will vary with time, **at least two of its values** in each input vector of the training set must correspond to **consecutive times**, e.g., x_{it} and x_{it+1}. (In terms of the moving window technique described in Section 9.5, the past horizon of the training vector must be at least two events long.) The same variable x_i should also appear in the target vector Y with the value corresponding to the time $t + 2$. As usual, the future horizon of the training vector, or target, can be only one event long.

However, the future horizon during **prediction** may be longer than one, if predictions are to be made several time units ahead. However, it is advisable to keep the horizons only a few time intervals long, because there is always noise in the measured data. **Usually, the past horizon is at least two times longer than the future horizon.**

In order to give you a feeling for how the training vectors are composed, a sequence of training vectors with a past horizon of length 3 and a future horizon of length 1 is given in Table 16-1. Each process vector $P_t = (x_{1t}, x_{2t}, ..., x_{it}, ..., x_{mt})$ (Equation (16.1)) has the same form. All inputs from $t = 0$ to $t = r - 4$ form **one epoch**.

In many cases, especially if the model will be used for quantitative control, it is not necessary that **all** process vectors P_t in the training vector contain **all** process variables $x_1, x_2, ..., x_i, ..., x_m$; only those variables that must be **controlled** and those that can be **manipulated** by the operators are mandatory. Although it may be desirable to include other variables that influence the process, one must be careful not to inject too much redundancy into the input data.

If, for example, the controlled variable is xc and the variable that can be manipulated by the operator is xm, the training vector should look something like this (Figure 16.2):

$$X_t = (..., xc_t, xm_t, ..., xc_{t+1}, xm_{t+1}, ..., xc_{t+2}, xm_{t+2}, ...) \quad (16.4)$$

time	input			target
t	X_t			Y_t
0	$(P_0,$	$P_1,$	$P_2)$	(P_3)
1	$(P_1,$	$P_2,$	$P_3)$	(P_4)
2	$(P_2,$	$P_3,$	$P_4)$	(P_5)
	. . . .			
$t-2$	$(P_{t-2},$	$P_{t-1},$	$P_t)$	(P_{t+1})
$t-1$	$(P_{t-1},$	$P_t,$	$P_{t+1})$	(P_{t+2})
t	$(P_t,$	$P_{t+1},$	$P_{t+2})$	(P_{t+3})
$t+1$	$(P_{t+1},$	$P_{t+2},$	$P_{t+3})$	(P_{t+4})
	. . . .			
$r-5$	$(P_{r-5},$	$P_{r-4},$	$P_{r-3})$	(P_{r-2})
$r-4$	$(P_{r-4},$	$P_{r-3},$	$P_{r-2})$	(P_{r-1})
	— new epoch —			
$r-3$	$(P_0,$	$P_1,$	$P_2)$	(P_3)
$r-2$	$(P_1,$	$P_2,$	$P_3)$	(P_4)
$r-1$	$(P_2,$	$P_3,$	$P_4)$	(P_5)
r	$(P_3,$	$P_4,$	$P_5)$	(P_6)
$r+1$	$(P_4,$	$P_5,$	$P_6)$	(P_7)
	. . . .			

Table 16-1:   Data sequence in a moving window learning scheme; the historical database contains r process vectors, P_0 to P_{r-1}.

while as target, only the manipulated variable *xm* can be given:

$$Y_t = (xm_{t+3}) \qquad (16.5)$$

The "future" values of the variables should be obtained from either a historical database, or calculated by a theoretical model.

16.3 The Methods

Until now, all applications of neural networks in chemical engineering have used the back-propagation algorithm, because fault detection, modeling of processes, and feedback control applications require supervised learning.

The size of a network is mainly influenced by the numbers of input and output variables. All those back-propagation neural networks described in the chemical engineering literature are small compared to those used in some other applications, such as spectrum-structure correlations, or the determination of the secondary structures of

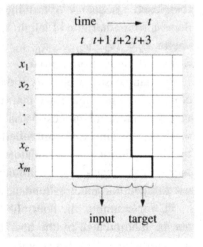

Figure 16-2: The shape of the moving window can have various forms; here, a window is shown with only one variable from the future horizon.

proteins. The number of input and output parameters in chemical processes are a few tens at most, so there is no need for more neurons in the hidden and output layers.

However, the learning times can be quite long in spite of the small sizes of the networks, especially when models are sought, because a large number of different input vectors must cover the problem space quite meticulously if the model is to be reliable.

We believe that the counter-propagation architecture (see Chapter 7) offers attractive opportunities in process control applications of neural networks. Two layers of neurons are combined in a counter-propagation network: a Kohonen layer influenced by the input vectors, and an output layer influenced by the targets. The counter-propagation network acts like a self-organizing lookup table (Section 7.2): all required answers, i.e., the responses for different variables, are pre-calculated and stored in as many lookup tables as there are output variables.

These lookup tables are of the same size and aligned one upon the other; hence, although the complete n-variable output vector Y (y_1, y_2, ..., y_n) is stored in n different lookup tables, all answers y_1, y_2, ..., y_n are stored in a hypercolumn that links **all** n lookup tables at the corresponding row and column (Figure 16-3).

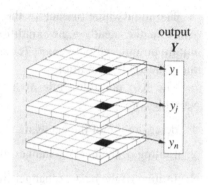

Figure 16-3: All components of the output vector, i.e., the multivariable responses, are stored in all n lookup tables at corresponding locations.

16.4 Predictions of Faults

This example follows the work by K. Watanabe, and coworkers (Reference 16-3), who investigated the catalytic conversion of *heptane* into *toluene* occurring in a reactor:

$$C_7H_{16} \longrightarrow C_7H_8 + 4\,H_2 \qquad (16.6)$$

Three variables were measured: the outlet concentration of the product *toluene*, c_p, the heater outlet temperature T_h, and the output signal s_h of the controller that regulates the temperature T_h in the reactor; five different faults can be deduced from these three measured variables. The simplified process scheme is shown in Figure 16-4.

A constant flow of *heptane* is maintained by pump no. 1. The temperature in the reactor is sustained by a heater operated by no-fault controller and the pump (no. 2) that cycles steam between the reactor and the heater. *Toluene* leaves the reactor with an outlet concentration of c_p and with the temperature T_o; depending on T_o, the controller adjusts the signal s_h that goes to the heater.

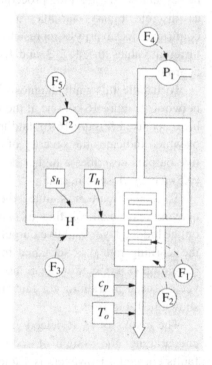

Figure 16-4: The catalytic reactor for the conversion of *heptane* into *toluene*.

The input value of each of the three variables is normalized to its value in the steady state condition. Therefore, all input values are within an interval between 0.75 and 1.33, with 1.00 corresponding to the : "no-fault" condition.

The five possible faults to be determined from the above three variables are:

1: deterioration of the catalyst

2: fouling of the heat exchanger in the reactor

3: fouling of the heat exchanger in the heater

4: malfunction of pump no. 1

5: malfunction of pump no. 2

Faults 4 and 5 also include the clogging of the pipes.

These faults correspond to the output values y_1 to y_5. They are represented in the learning procedure in two ways; first, they are used as discrete binary variables (yes/no); second, when a fault is confirmed, its severity is represented as a discrete variable having five different values: 0.5, 1, 2, 3, and 4, which requires five output (yes/no) neurons.

To handle this fault diagnosis scheme, they set up six different networks (Figure 16-5): one at the top level, determining which faults have occurred (Figure 16-6), and five networks one level below, each of which indicates the severity of each of the five faults. Each of the five outputs describes a higher degree of deterioration; the first level (0.5) means close to normal.

To make matters simple, the architectures of all six neural networks are identical: they all have three input nodes, four neurons on the hidden layer, and five output neurons. In the training procedure, the learning rate η is set equal to 0.1, and the momentum μ to 0.9 (Equation (8.1) given in Sections 8.3 and 8.7). The input and output vectors used in training for fault recognition (top level network) are shown in Table 16-2.

The results of fault detection in the catalytic process were rather encouraging: the system of six networks is able to detect the five faults correctly. However, as Table 16-3 shows, the determination of the **exact level** of fault is not as clear-cut. Nevertheless, if we take the largest value of the output as the degree of fault, the correct answer is always obtained.

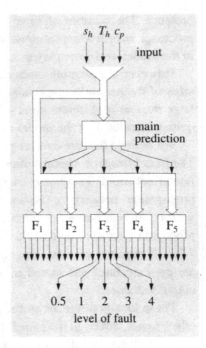

Figure 16-5: Six single-hidden-layer neural networks for the prediction of five four-level faults from three input variables.

fault	input X			output Y				
	x_1	x_2	x_3	y_1	y_2	y_3	y_4	y_5
	s_h [mV]	T_h [K]	c_p [gmol/m^3]					
1	219	885	498	1	0	0	0	0
2	240	906	524	0	1	0	0	0
3	248	889	524	0	0	1	0	0
4	212	878	550	0	0	0	1	0
5	201	889	524	0	0	0	0	1
normal	223	889	524	0	0	0	0	0

Table 16-2:   Data for training the first level recognition of faults: yes/no (Watanabe et al., AIChE Journal 1989).

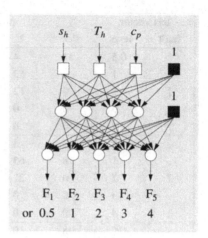

Figure 16-6: One-hidden layer (3 x 4 x 5) neural network for the prediction of faults. The same (3 x 4 x 5) architecture is used for five second-level networks, each predicting the severity of the corresponding fault.

Where totals are less than 100%, it is because the **first** network misidentified the fault. The results are obtained from 1000 test vectors. (Each row of Table 16-3 represents a separate test input. When a fault of type 1 at level 0.5 was input, the system identified it correctly 84% of the time; 4% of the time, it identified the fault as type 1, level 1; and 12% (100–88) of the time, it misclassified the fault type altogether.)

16.5 Modeling and Controlling a Continuously Stirred Tank Reactor (CSTR)

As described earlier, two things are required in order to control a given process reliably. First, we need a model M able to predict the critical parameters of the process for a few time intervals in the future. Second, it must be possible to determine adjustments to the correction variable(s), i.e., the variable(s) that can be manipulated, from the predicted data accurately enough so that the system will return to normal if these adjustments are applied. In other words, *a model* M^{-1} *inverse to the initial model* M *must be obtained.*

This section discusses the control of a nonisothermal continuously stirred tank reactor (CSTR). We will first follow it as it was originally described by a group from the Department of Chemical Engineering at the University of Pennsylvania (see Reference 16-11).

270   *Fault Detection and Process Control*

| test vector | | predicted level of fault in % | | | | | |
fault	level	0.5	1	2	3	4	total
1	0.5	84	4	0	0	0	88
	1	9	63	13	0	0	85
	2	0	12	75	13	0	100
	3	0	0	3	73	3	79
	4	0	0	0	2	85	87
2	0.5	83	3	0	0	0	86
	1	8	63	17	0	0	88
	2	0	22	63	10	1	96
	3	0	0	3	49	18	70
	4	0	0	0	16	70	86
3	0.5	98	0	0	0	0	98
	1	2	77	18	0	0	97
	2	0	22	67	11	0	100
	3	0	0	1	59	15	75
	4	0	0	0	13	70	83
4	0.5	95	0	0	0	0	95
	1	2	74	8	0	0	84
	2	0	10	81	3	0	94
	3	0	0	3	86	0	89
	4	0	0	0	2	92	92
5	0.5	91	0	0	0	0	91
	1	0	98	0	0	0	98
	2	0	1	80	7	0	88
	3	0	0	2	82	0	84
	4	0	0	0	0	100	100

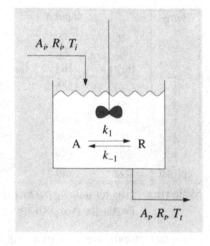

Figure 16-7: Nonisothermal continuously stirred reactor. A_i and R_i are the inlet concentrations of the reactants, and T_i is the corresponding temperature. A_t, R_t, and T_t are the same variables at time t.

Table 16-3:   Ability of the system of five networks to predict the type and level of fault.

These authors used the back-propagation algorithm; we will show that the **counter-propagation** approach can yield comparably good results, and some additional information as well.

A first-order reversible reaction:

$$A \rightleftharpoons R$$

occurs in the reactor, which is shown schematically in Figure 16-7.

A long, consecutive history of process data must be available in real-world applications to set up a model M and its inverse model M^{-1}. In the present case, a long sequence of data triplets measured at equal time intervals t is required: the concentration of the starting material, A_t, the concentration of the product, R_t, and the reaction temperature, T_t. These measurements should be influenced by as many different circumstances as can possibly occur during the process, for example, sudden fluctuations in the concentrations A or R, and/or changes of temperature T.

For simple processes, a reliable theoretical model can be found. In this example, the chemical process can be described by the following equations:

$$\frac{dA_t}{dt} = \frac{A_i - A_t}{\tau} - k_1 A_t + k_{-1} R_t$$

$$\frac{dR_t}{dt} = \frac{R_i - R_t}{\tau} - k_1 A_t + k_{-1} R_t$$

$$\frac{dT_t}{dt} = \frac{T_i - T_t}{\tau} - H(k_1 A_t + k_{-1} R_t) \qquad (16.7)$$

$$k_1 = Be^{-\frac{C}{T}}$$

$$k_{-1} = De^{-\frac{E}{T}}$$

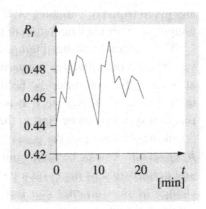

Figure 16-8: Changes of R_t if the input temperature T^m randomly varies by about 3%.

The inlet values A_i and R_i, and the parameters in the model Equations (16.7) have the values and units given in Table 16-4.

A_i	=	1 mol/l	R_i	=	0 mol/l
B	=	5×10^3 sec^{-1}	C	=	5033 K
D	=	1×10^6 sec^{-1}	E	=	7550 K
τ	=	60 sec	H	=	5 lK/mol
T_i	=	410 K			

Table 16-4: Values and units of the parameters used in the system of Equations (16.7).

The database of state variables A_t, R_t and T_t can be calculated at any time t by inserting the above parameters into Equations (16.7) and integrating them.

The process can be manipulated mainly by the temperature at which the reaction is run. Therefore, a historical database covering many different cases can be obtained from Equations (16.7) by taking any given triplet of variables A_t, R_t and T_t and randomly changing the temperature for the calculation of the new triplet A_{t+1}, R_{t+1} and T_{t+1}.

This randomly changed temperature is considered as an additional variable T^m (*manipulated temperature*); T^m specifies the temperature at which the controller is holding the system, while T_t is the temperature of the system if it is left to itself. (Of course, very few reaction systems can be "left to themselves"!)

In the present case, the goal of the process is to maintain the yield R_t constant. As can be seen (and calculated from Equations (16.7)) any perturbation of the system variables will change R_t; uncontrolled change of the variables can lead to serious problems (Figure 16-8 shows the fluctuation of R_t if the temperature T^m randomly varies by about ±3%).

Neural networks will be used twice in the construction of a controlling device for the CSTR: first, to develop a model M that will quantitatively predict the correct value of the yield R_{t+1} for any triplet of data A_t, R_t and T_t and the manipulated temperature T^m; and second, to develop the inverse model M^{-1} capable of predicting the needed adjustment of T^m from four variables – the triplet A_t, R_t and T_t **and** the yield R_{t+1}. That is, we adjust T^m to compensate **for all** changes away from optimum conditions.

Because we are concerned with neural networks and not process control as such, we will be satisfied with the scheme shown in Figure 16-9, in spite of its rather schematic form. The scope of this book does not permit too many details; if you are interested in this area, see the literature cited in Section 16.6 (Bhat, Ungar, Himmelblau, Bulsari, etc.).

As shown in Figure 16-9, there are four general steps in this algorithm, as follows. The state variables A_t, R_t and T_t of the process P are measured at regular time intervals dt (a). The model M checks the outcomes of the process P by regular forward prediction (b). If the controlled variable R_t changes for some reason, corrective action must be taken.

Therefore, the inverse model M^{-1} calculates the corrections T^m needed (c). This correction is input to the process P (d) and the actual consequences of the correction are monitored by comparison of the process data with the set-point data and/or with those predicted by the model.

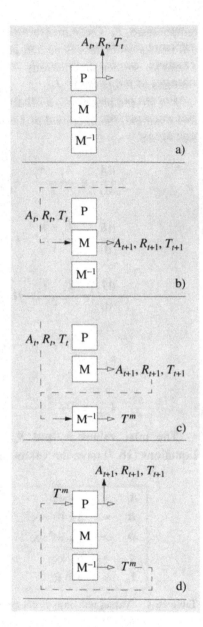

Figure 16-9: Control of the process P with the model M and the controller C (C = M^{-1}, the inverse model) shown in four phases.

Back-propagation approach. Here, we will follow the work of Watanabe et al. (see Reference 16-3). In back-propagation, two networks must be used: first for learning the model, M, and second, for learning its inverse model, M^{-1}. In the Watanabe approach, the first network has an architecture of (4 x 8 x 3) and the second one (4 x 8 x 1). Both networks are shown in Figures 16-10 and 16-11. Setting up the model M is called *forward learning*, while setting up the inverse model M^{-1} is called *inverse learning*.

The most important requirement of the model M is that, at time t, it predicts the future yield R_{t+1} (at time $t + 1$) as precisely as possible. Forward learning thus requires input of A_t, R_t, T_t and T^m at time t, and targets of A_{t+1}, R_{t+1}, and T_{t+1} at time $t + 1$.

The inverse learning, however, requires as input the state variables A_t, R_t and T_t **and** the variables of the future process vector A_{t+1}, R_{t+1}, T_{t+1}, while the target is the manipulated temperature T^m.

Of course, an input for the inverse model composed of two complete consecutive triplets of state variables, (A_t, R_t and T_t) and (A_{t+1}, R_{t+1}, T_{t+1}) requires a larger network with more weights to adjust, and consequently longer learning times. To keep the network for M^{-1} as small as possible, only the most important variable, i.e., the controlled variable, R_{t+1} from the future horizon, is retained for learning M^{-1}.

Recall the discussion of the moving window concept (Sections 9.5 and 16.2). The calculation of the model M and its inverse M^{-1} is a typical application of the moving window learning approach, although the past and future horizons are only one time step long.

A database of about 400 triplets taken at 30 second intervals was calculated using Equations (16.7) and the parameter values given in Table 16-4. By varying the manipulated temperature T^m randomly within 12 degrees around 410 K, a comparatively good distribution of all possibilities is obtained. The first 200 triplets were taken as the learning set and the rest as the test set.

Both networks were trained by standard back-propagation equations (Section 8.7), with all inputs and outputs scaled between 0 and 1. As shown in Figure 16-12, the neural network predicts the yield R_{t+1} within a few percent of the calculated values.

Theoretically, we should be able to perform a consistency check: the inverse model M^{-1} should yield for **any** input X an output Y, which, if input to the forward model M, will produce an output Y' **exactly equal** to the input X originally supplied to the inverse model

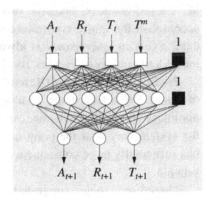

Figure 16-10: The neural network for learning the model M.

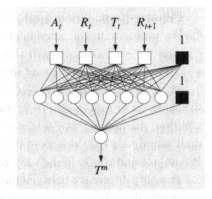

Figure 16-11: The neural network for learning the inverse prediction model, M^{-1}.

M^{-1}. Unfortunately, because **two** networks have to be trained to accomplish the entire controlling scheme from **two different** sets of data, a certain discrepancy is **always** present between them (see the inputs and outputs of Figures 16-11 and 16-12).

Psichogios and Ungar (References 16-11) reported a discrepancy of about 5% between the results obtained from M and M^{-1}. In a nonlinear controlling scheme, especially if the control needs to keep the system very near the point of maximum yield, even a small error has sufficiently large consequences to invalidate the above controlling scheme.

Therefore, they suggested a slightly altered (but more sophisticated) controlling technique employing a delay device on the output of the model M. With this, they were able to use the forward and inverse models developed by the neural networks without any problems. For more details, consult Reference 16-11.

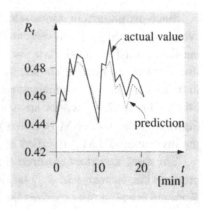

Figure 16-12: Differences between the calculated yield R_t and the one predicted by the neural network model M.

Counter-propagation approach. Here, we will try to solve the CSTR problem using a counter-propagation neural network (see Chapter 7). This example will show that counter-propagation can be as good at modeling as the back-propagation model.

The training and test data sets used in this example are obtained by the same set of Equations (16.7), using the same initial conditions. In addition, the number and selection of input and output variables and their scaling are the same as in the training and test sets employed by Psichogios and Ungar in the back-propagation model.

The only difference between the data (training and test) used in the two methods is that the datasets used in counter-propagation are more than twice as large.

Counter-propagation learning can be regarded as a *nonlinear smoothing* procedure; therefore, larger amounts of input data are necessary for a given accuracy. Fortunately, in counter-propagation learning, the total training time does not rise rapidly as a function of dataset size. Since the number of epochs needed for training is similar to those needed in Kohonen learning, the number of epochs needed for stabilizing the counter-propagation network is orders of magnitude smaller than for back-propagation learning.

The main goal of this example is to show that a counter-propagation network trained as the forward model M can be used as the inverse model, M^{-1}, too.

The counter-propagation network selected for this example is made up of 5000 neurons spread out in two (50 x 50) layers. The size of the network (2500 neurons) offers enough room for 1000 training vectors, (A_t, R_t, T_t, T^m), without too many conflicts. (Because some of the training vectors are quite similar to each other, conflicts cannot be completely avoided.)

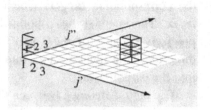

Figure 16-13: A three-dimensional lookup table. By selecting two coordinates, j' and j'', in the network's map, a "box" containing values of three different variables is always obtained.

> Remember, the trained counter-propagation network is a lookup table in which all the **multiresponse** answers are already calculated, stored in boxes, and waiting to be retrieved.

Hence, a (50 x 50) network can handle 2500 **multivariate** answers. The real problem actually is, to select the proper output neuron for the needed answer. Figure 16-13 shows a three-dimensional lookup table.

The neurons in the first (Kohonen) layer have four weights that are linked to the input signals, A_t, R_t, T_t, T^m. The neurons in the second (output) layer have three weights from which the output values, A_{t+1}, R_{t+1}, and T_{t+1}, are recalled. The described counter-propagation network, which has (50 x 50 x 4) + (50 x 50 x 3) = 35,000 weights, is schematically shown in Figure 16-14.

Recall that in the counter-propagation network, the output is obtained differently from other networks (see Chapter 7 for general and Section 7.4 for particular explanations about output in this type of network). The difference is that the output neurons in the counter-propagation network **do not** calculate the answer from the weights using the equations given in Chapter 2 ((2.9) and (2.39)); instead, the adapted weights **are** the answers already. The output weights in the counter-propagation network, w_{ji}^{out}, are labeled as c_{ji} (Section 7.7), in order to distinguish them from the weights w_{ji} in the Kohonen layer.

A counter-propagation network uses supervised learning, which is essentially a competitive or Kohonen learning with an additional adaptation of weights, c_{ji}, in the output layer to make them close to the targets Y_s. As explained above, the targets in our example are the three process variables, A_{t+1}, R_{t+1}, and T_{t+1} ($t + 1$ is the next time interval).

Learning begins in the Kohonen layer and consists of the self-organization of the multivariate inputs exactly in the same way as described in Chapter 6 for Kohonen learning.

Simultaneously with the self-organization going on in the Kohonen layer, a similar self-organization is being carried out among

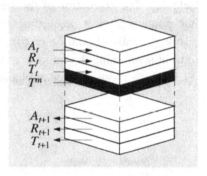

Figure 16-14: The architecture of the counter-propagation network used for learning the forward model M.

the corresponding **targets** which are **input** into the lower (output) layer. The method has obtained the name "counter-propagation" precisely because of these two "inputs" coming from two opposite directions: the input vectors into the Kohonen layer and the targets into the output layer.

The only difference between the two self-organization procedures is that the winning neuron is determined only once – in the Kohonen layer. Once the position (j', j'') of the winning neuron c ("central") in the Kohonen layer is determined:

$$out_c \leftarrow \min \left\{ \sum_{i=1}^{m} (x_i - w_{ji})^2 \right\} \qquad (16.8)$$

the neuron from the output layer at the **same position** (j', j'') is selected. The correction of the weights w_{ji} in the Kohonen layer uses the input vector X according to the Kohonen strategy (Equation (16.9); see also Equation (7.6)). The correction of weights c_{ji} in the output layer using the target Y is made according to Equation (16.10), (7.7). For a more detailed explanation, see Chapter 7.

$$w_{ji}^{(new)} = w_{ji}^{(old)} + \eta(t)\, a\, (d_c - d_j)\left(x_j - w_{ji}^{(old)} \right) \qquad (16.9)$$

$$c_{ji}^{(new)} = c_{ji}^{(old)} + \eta(t)\, a\, (d_c - d_j)\left(y_j - w_{ji}^{(old)} \right) \qquad (16.10)$$

The learning rate $\eta(t)$, which plays a more important role in Kohonen learning than in back-propagation learning, is calculated using Equation (6.5) (Chapter 6):

$$\eta(t) = (a_{max} - a_{min}) \frac{t_{max} - t}{t_{max} - 1} + a_{min} \qquad (16.11)$$

The parameters a_{max} and a_{min} determine the maximum and minimum corrections of the central neuron (a $(d_c - d_j) = 0$) at the beginning of training ($t = 1$, $\eta(t) = a_{max}$), and at the end of training ($t = t_{max}$, $\eta(t) = a_{min}$). The values of a_{max} and a_{min} must be defined within the interval 1.0 and 0.0.

The fact that the counter-propagation network is "merely" a lookup table, able to give only a limited number of different answers, may appear at first glance to be a serious drawback. Real models giving continuous answers to different sets of input variables might seem to be much better. However, more important than the **number** of possible answers is the **size of the error** of these answers. If the manipulated variable can take any value within a 5% interval around the reference point, let us say ±20 K around 400 K, this means that the 2500 answers can cover this interval in temperature steps as small as 0.016 K.

But in real applications, one temperature value can be associated with different combinations of values of other variables, which means that several boxes in the lookup table will have the same value of the temperature – which means that the temperature steps could **not** be as small as 0.016 K. Nevertheless, if the number of answers or the precision associated with this approach is not sufficient for a given application, the network can easily be enlarged.

In any case, the problem of evaluating the correct answer is reduced to finding the box with the most appropriate answer in the lookup table.

Because all the answers must be calculated in advance, rigorous experimental design with a uniform distribution of all possible cases is extremely important. This requirement was the basis for the initial selection of the network size. Figure 16-15 shows the distribution on the 50 x 50 map of the process vectors from the training set.

In the present example, 1000 input vectors $X = (A_t, R_t, T_t, T^m)$ with 1000 targets $Y = (A_{t+1}, R_{t+1}, T_{t+1})$ were used for the forward model. Altogether 20 epochs (i.e. 20,000 inputs) were needed for the network to stabilize when the value of the correction function $\eta(t)$ was:

$$\eta(t) = (0.5 - 0.01) \frac{20000 - t}{20000 - 1} + 0.05 \qquad (16.12)$$

Learning for 20 epochs means that in the total training period:

- process vectors are input 20,000 times;

- the central neuron, $c = (j', j'')$, is found each time:

  - all neurons in the corresponding neighborhood are corrected

  - the target process vector is input into the location (j', j'') in the output layer

  - the neurons in the corresponding neighborhood around the (j', j'') location in the output layer are corrected.

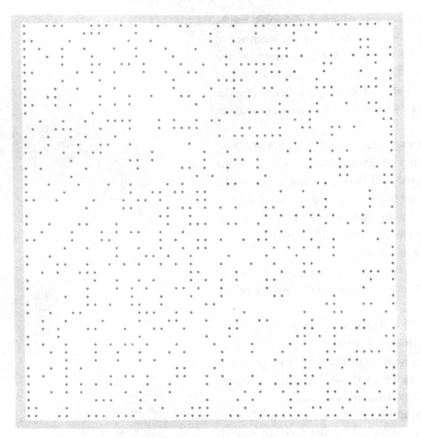

Figure 16-15: Distribution on a (50 x 50) map of the process vectors used in the training set. Because toroidal mapping was **not** used, slightly more objects (points) appear at the border lines of the map.

Once the counter-propagation network is trained, retrieval is straightforward: the neuron in the Kohonen network having weights most similar to the input variables is located; the weights of the neuron at the same position (j', j'') in the output layer contain the answer.

The trained counter-propagation network was tested with the thousand process vectors not used in the training; it has excellent prediction ability for all three process variables (Table 16-5).

variable	recall		prediction	
	σ	$\sigma/\sqrt{1000}$	σ	$\sigma/\sqrt{1000}$
A [%]	0.004	1.2 $\times 10^{-4}$	0.005	1.6 $\times 10^{-4}$
R [%]	0.004	1.2 $\times 10^{-4}$	0.005	1.6 $\times 10^{-4}$
T [K]	0.86	0.027	1.03	0.032

Table 16-5:   Recall and prediction ability of the (50 x 50 x 4) + (50 x 50 x 3) counter-propagation network.

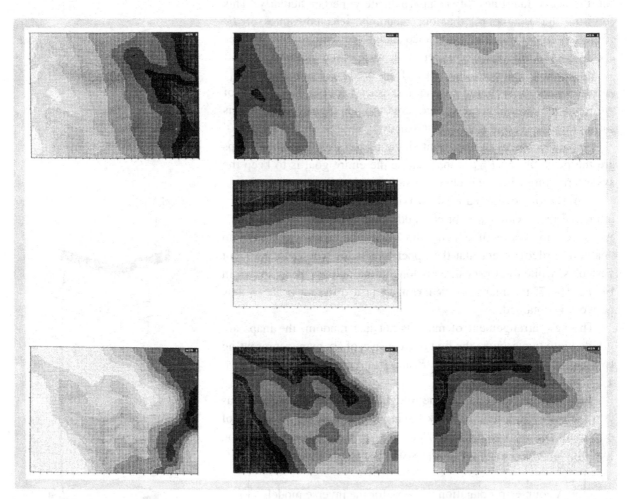

The standard error σ of 0.86 in the temperature prediction is equivalent to an error of less than 1 degree Kelvin. Errors of the means were calculated for 1000 tests; they have no particular significance.

Each level of weights, whether in the Kohonen or in the output layer, can be represented as a map (see Figures 7-16 and 7-17 of the tennis example in Section 7.5), and "contoured" by drawing lines (*iso-value lines*) connecting neurons having the same weight. The resulting seven maps (four input and three output maps) of weights trained to give the forward model M are shown in Figure 16-16.

These two-dimensional maps are very informative. For example, if we overlay the fourth input map (T^m), upon any of the other three input maps (A_t, R_t and T_t), it can be seen that the iso-value lines of T^m

Figure 16-16: The seven process variable maps (four input and three output) obtained by the counter-propagation neural network. The first three (input A_t, R_t, T_t), and the fourth (input manipulated temperature, T^m) were obtained in the Kohonen layer, while the three output maps (A_{t+1}, R_{t+1}, T_{t+1}) were obtained from the weights in the output layer. The darker parts of the map shows higher values.

cut the iso-value lines of T_t approximately perpendicularly. This confirms the assumption that our randomly selected values of T^m cover the entire variable space adequately: the selection of T^m does not depend on the choice of either of the three other variables.

In addition, the iso-value lines of the predicted temperature T_{t+1} (seventh map, third output map) almost exactly follow the iso-lines of the input T^m (manipulated temperature), though slightly shifted. This shows that the control is executed exactly.

Of course, the most important single features of **any** of the maps are the peaks in the R_{t+1} map, because the entire goal is to keep the system running with the highest possible yield of R.

The set of seven maps shown in Figure 16-16 offers a very attractive idea, which was briefly addressed at the end of Chapter 7. All seven maps are of exactly the same size, with apparently no "natural" order (except that the upper four maps belong to the input variables, while the lower three belong to the output). If, as shown in Figure 16-17, the maps are **rearranged** in a different order, a new network is obtained.

The new arrangement of maps is not just random; the maps are divided into two groups: the first one consists of six maps representing two consecutive process vectors, P_t and P_{t+1}, i.e., (A_t, R_t, T_t, A_{t+1}, R_{t+1}, T_{t+1}); the T^m map stands alone.

This rearranged stack of the weight maps can be regarded as corresponding to a new network having a Kohonen layer capable of accepting the six variables A_t, R_t, T_t, A_{t+1}, R_{t+1}, T_{t+1} as input, and yielding an output vector Y composed of the single variable T^m. Thus:

> A counter-propagation network for the inverse model M^{-1} can be obtained by **rearranging** the previous forward model network, M.

That is, we have just designed a new (50 x 50 x 6) + (50 x 50 x 1) counter-propagation network by rearranging the weight maps of another one – without the need for any training.

The rearranged network, M^{-1}, was tested twice: first, by predicting 1000 T^m values from the dataset on which the original network was based; second, by predicting 1000 T^m values from a new set of data obtained from a set of random process vectors. Both results are given in Table 16-6. Once again, remember that this network is a lookup table: the input variables are used to determine the neuron (j', j''), i.e.,

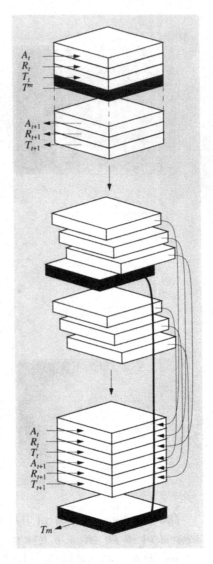

Figure 16-17: Rearrangement of seven weight maps of the forward

to determine the address of a neuron in the output layer, whose weights are the desired answer.

variable	1st set		2nd set	
	σ	$\sigma/\sqrt{1000}$	σ	$\sigma/\sqrt{1000}$
T [K]	3.4	0.11	3.8	0.12

Table 16-6:   Predictive ability of the inverse model obtained by rearranging the forward one.

The prediction ability of the inverse model is worse than that of the forward model; however, the standard errors are still below 4 K, which means errors less than 1%. Figure 16-18 shows a few predictions made by this inverse model.

It is surprising that such a simple lookup table, containing only 2500 answers and obtained as a byproduct of learning the forward model, can produce results that good. According to some authors (see Reference 16-15), even much more sophisticated and rigorous methods fail to produce satisfactory inverse models, because they rely on higher order derivatives, and are therefore highly sensitive to noise and numerical errors. Hence, this achievement is at least comparable to if not better than most of the other contemporary methods.

Of course, there are still a number of questions to be answered before this method of obtaining the inverse model can be generally accepted. For example: Is the obtained answer always the best possible one? Is it possible that there is an even better answer in another box that happens to show slightly worse agreement with the input variables? If this is the case, how can such local minima be found? How robust is this method?

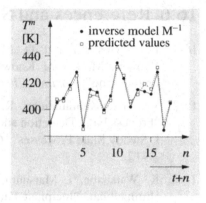

Figure 16-18: Comparison of test and predicted values obtained by the inverse counter-propagation model.

16.6 References and Suggested Readings

16-1. J. C. Hoskins and D. M. Himmelblau, "Artificial Neural Network Models of Knowledge Representation in Chemical Engineering", *Comput. Chem. Eng.* **12** (1988) 881 – 890.

16-2. V. Venkatasubramanian, R. Vaidyanathan and Y. Yamamoto, "Process Fault Detection and Diagnosis Using Neural Networks I. Steady State Processes", *Comput. Chem. Eng.* **14** (1990) 699 – 712.

16-3. K. Watanabe, I. Matsuura, M. Abe, M. Kubota and D. M. Himmelblau, "Incipient Fault Diagnosis of Chemical Process via Artificial Neural Networks", *AIChE Journal* **35** (1989) 1803 – 1812.

16-4. P. Bhagat, "An Introduction to Neural Nets", *Chem. Eng. Prog.* **86** (1990) 55 – 60.

16-5. V. Venkatsubramanian and K. Chan, "A Neural Network Methodology for Process Fault Diagnosis", *AIChE Journal* **35** (1989) 1993 – 2002.

16-6. L. H. Ungar, B. A. Powel and S. N. Kamens, Adaptive Networks for Fault Diagnosis and Process Control", *Comput. Chem. Eng.* **14** (1990) 561 – 572.

16-7. N. Bhat and T. J. McAvoy, "Use of Neural Nets for Dynamic Modeling and Control of Chemical Process Systems", *Comput. Chem. Eng.* **14** (1990) 573 – 583.

16-8. N. Bhat, P. A. Minderman, Jr., T. J. McAvoy and N. S. Wang, "Modeling Chemical Process Systems via Neural Computation", *IEEE Control Systems Magazine* (April 1990) 573 – 582.

16-9. J. Leonard and M. A. Kramer, "Improvement of the Back-Propagation Algorithm for Training Neural Networks", *Comput. Chem. Eng.* **14** (1990) 337 – 41.

16-10. A. Bulsari and H. Saxen, "Applicability of an Artificial Neural Network as a Simulator for a Chemical Process", *Proc. 5-th Intern. Symposium on Computer and Information Sciences*, Nevsehir, Turkey, (October 1990) 143 – 151.

16-11. D. C. Psichogios and L. H. Ungar, "Direct and Indirect Model Based Control Using Artificial Neural Networks", *Ind. Eng. Chem. Res.* **30** (1991) 2564 – 2573.

16-12. J. R. Lang, H. T. Mayfield, M. V. Henly and P. R. Kromann, "Pattern Recognition of Jet Fuel Chromatographic Data by Artificial Neural Networks with Back-Propagation of Error", *Anal. Chem.* **63** (1991) 1256 – 1261.

16-13. R. Hecht-Nielsen, "Counterpropagation Networks", *Appl. Optics* **26** (1987) 4979 – 84.

16-14. D. G. Stork, "Counterpropagation Networks: Adaptive Hierarchical Networks for Near Optimal Mappings", *Synapse Connection* **1** (1988) 9 – 17.

16-15. C. E. Economu and M. Morari, "Internal Model control. 5. Extension to Nonlinear Systems", *Ind. Eng. Chem. Process Des. Dev.* **25** (1986) 403 – 411.

16-16. J. Zupan and M. Novic, "Counterpropagation Learning Strategy in Neural Networks and Its Application in Chemistry", in *Further Advances in Chemical Information,* Ed.: H. Collier, Roy. Soc. of Chem., Cambridge, UK, 1994, pp. 92 – 108.

17 Secondary Structure of Proteins

learning objectives:

- fundamentals of description of protein structure in terms of amino acid subunits, and of protein secondary structure: α-helix, β-sheet and coil

- how to code the amino acid sequence for input into a network

- use of the moving window scan to process the protein chain as a series of overlapping neighborhoods

- comparison of results from network with results of a traditional structure prediction method

17.1 The Problem

Polypeptides and proteins are made up of elementary building blocks, the amino acids (a polypeptide is a short chain of amino acids; a protein is a long chain); apart from some special cases, only 20 of the many different amino acids occur in proteins. Figure 17-1 shows two of these amino acids, along with their abbreviations (a three- and a one-letter code).

These amino acids are arranged sequentially in a protein; the exact sequence is called the *primary structure*. Figure 17-2 shows the sequence of amino acids in a segment of a protein. (Amino acids in a protein are generically called "residues".)

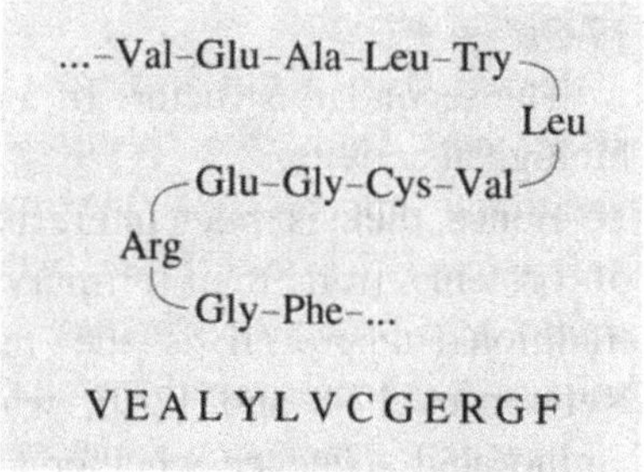

This linear sequence folds and turns into a unique **three-**dimensional structure, which contains global features that are referred to as the *secondary structure*. There are three types of secondary structures: α-helix, β-sheet, and random coil.

In an α-helix structure, the protein chain turns continuously in the same direction to form a "spiral"; in a β-sheet, two or more parts of the same chain are aligned parallel in space; the term "coil" collects all the other more or less irregular three-dimensional arrangements of

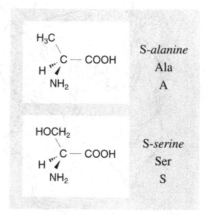

Figure 17-1: Two naturally occurring amino acids, their structures and their three- and one-letter abbreviations.

...–Val–Glu–Ala–Leu–Try

                      Leu

  Glu–Gly–Cys–Val

Arg

  Gly–Phe–...

V E A L Y L V C G E R G F

Figure 17-2: Part of the primary structure of the protein *insulin*.

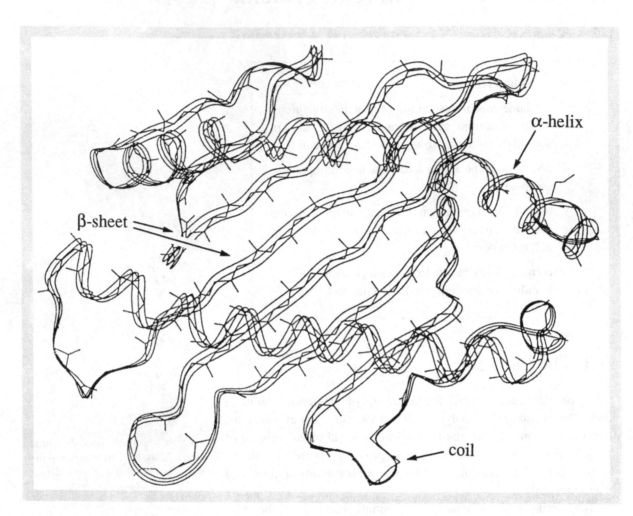

amino acids. All of these can be found in one protein. See Figure 17-3.

The secondary structure of a protein is of utmost importance to its biological activity.

Hence, there is much interest in predicting the secondary structures of proteins from their primary structures. The most widely used traditional approach is the method of Chou and Fasman (see References 17-8 and 17-9), which allows one to predict from the amino acid sequence whether a certain amino acid is part of an α-helix, a β-sheet, or a coil structure with about 50 – 53% correctness.

In recent years, numerous papers have been published on the use of neural networks to predict secondary structures of polypeptides from their amino acid sequences. The pioneers in this field were Qian

Figure 17-3: Residues 1 to 180 of human lymphocyte antigen A2. The three secondary structural features of a protein chain (α-helix, β-sheet, and coil) can be clearly seen. H_1, H_2, H_3 are three α-helices; the arrows indicate five "pleats" of a β-sheet; all the remaining parts of the molecule comprise a meandering "random coil". (Picture courtesy by Gerhard Müller and Horst Kessler, Org. Chem. Institut, TU München).

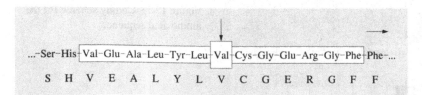

Figure 17-4: Segment (window) of the amino acid sequence thought to influence the secondary structure of the protein at a site centered on the middle amino acid (here, *valine*, Val).

and Sejnowski. Since quite a few other research groups adopted the essentials of their input representation, we will look at their work in some detail (see Reference 17-1).

The basic assumption in the work of Chou and Fasman and of Qian and Sejnowski is that the identities of an amino acid and its neighbors determine the secondary structure of that neighborhood. A sort of "window scan" over a whole polypeptide segment might in principle give the secondary structure of the whole chain (Figure 17-4).

17.2 Representation of Amino Acids as Input Data

In order to determine the dependence of secondary structure on the amino acid sequence, we must input the amino acid under consideration and a certain number (in this example six) of amino acids preceding and following it, a total of 13 amino acids. The sequence of 13 one-letter amino acid symbols, x_i^{orig} (original input variable), will be referred as to the **original input vector**, X^{orig}.

Each of the 20 naturally occurring amino acids is coded as a 21-bit string with **one** specific bit turned on and the others all zero; for example, *proline* is represented by a 1 in position 14 of this string. The 21st position is special, and will be explained in the next section. Because of its discrete character, each original variable, x_i^{orig} (in this case, each amino acid label) must be represented by 21 binary or bipolar (see Section 4.2, Equation (4.1)) variables. Such a coding scheme is called a *distributed representation* (Figure 17-5).

We discussed in Section 16.2 (Equation (16.3)) the (very good) reasons for substituting a discrete variable by a bit string containing as many bits as the variable has discrete values. Representing an amino acid by a sequential number running from 1 to 21 would imply that the numbering of amino acids is a quantitative measure for the similarity between them; that is, if two amino acids have numbers that

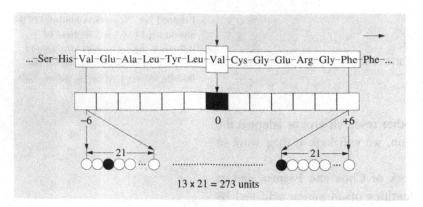

Figure 17-5: Coding scheme for the amino acid sequence.

differ by only one (e.g. 6 and 7) this would imply that these two amino acids are more closely related than, say, numbers 6 and 20.

However, what is important in determining secondary structure is not similarities in the structures of the amino acids, but similarities in their sequence order, that is, the relative position of a given amino acid from the center of the window plays the crucial role in the decision making, and not its structure.

(We are speaking here from an **information science** point of view, not a chemical one. Of course, the substituent groups ("R-groups") that distinguish one amino acid from another are primarily responsible for protein structure – but, as a first approximation, we **do not care** why the amino acids do what they do.)

Therefore, in this example the 13 original input variables x_i^{orig} are replaced by a 13 x 21 = 273-element **binary** input vector x_i. This means that the same number, 273, of units is required for input in the network, each input unit receiving one binary value, (0 or 1).

> The i-th original input variable, x_i^{orig}, tells us which of the twenty amino acids is presently at the $(i - 7)$-th position relative to the central amino acid in the 2 x 6 + 1 = 13-residue-long window.

The concept of a window bracketing a certain neighborhood, whether of a topological, sequential, or time-dependent nature, was used in the process control examples. See Sections 9.5 and 16.2.

Input of the entire sequence of amino acids of a protein (primary structure) is achieved by moving this window of 13 amino acids along the entire sequence in steps of one amino acid at a time. At each of

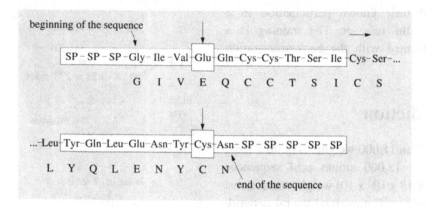

Figure 17-6: When the moving window is at the end of the chain, it has to be filled out with spacers.

these steps the corresponding window of amino acids is input into the network.

Special provision has to be made when the *moving window* is at the beginning or end of an amino acid chain; if it extends beyond the end, it will span fewer that the usual 13 residues (Figure 17-6).

So as not to complicate the algorithm, we will assume that there are always 13 residues in the window; at the beginning or end of the chain, any empty spaces will be filled with a special code called a *spacer*, coded using the 21st position of the bit vector.

(This coding scheme was inspired by Sejnowski's (very successful) work in training a neural network, NetTalk, to derive the pronunciation of a letter in an English word from the letters surrounding it.)

17.3 Architecture of the Network

In the network for predicting the secondary structure of proteins, three output neurons were used, one each for α-helix, β-sheet, and random coil. Qian and Sejnowski tried different numbers of hidden neurons (0 – 80) and decided that the optimum number is forty. Thus, the two-layer neural network has an architecture of (273 x 40 x 3), amounting to (273 + 1) x 40 + (40 + 1) x 3 = 11,083 weights, including those to the bias (Figure 17-7).

The training set consists of 106 proteins, having altogether 18,105 amino acid residues. Each of these is accompanied by a specification of the kind of secondary structure it is embedded in. (A given amino acid can be in different types of structures in different proteins, or even in different parts of the same protein.) Another 15 proteins with a total of 3,520 amino acids and their known participation in a secondary structure are taken as the test set. The training is a supervised learning process performed with the back-propagation algorithm.

17.4 Learning and Prediction

A network consisting of more than 11,000 weights is quite a large one. Ten epochs of training with 18,000 amino acid sequences requires about 2 billion ($11 \times 10^3 \times 18 \times 10^3 \times 10$) weight corrections. Obviously, this is a major undertaking. Commercial neural-network software is now available, and, for problems of this size, special accelerator boards for plugging into your PC. Some of these can accelerate the calculation by factors of 50 or more compared to a 33 MHz 486 computer.

The network gave 62.7% right answers on the test set. This is a remarkable improvement over the method of Chou and Fasman, which has a predictive ability of only 50 – 53%.

The publication of Qian and Sejnowski stirred quite some interest among protein-structure chemists. Since then (October, 1988), a number of papers on this subject have been published, and a number of improvements and suggestions have been made, from enlarging the window to learning a larger number of amino acid sequences.

In fact, the tide of scientific work has continued to rise since publication of the first edition of this book. The work up to 1996 has been summarized by B. Rost. An excellent overview of all aspects of protein structure prediction has appeared by the same author in the Encyclopedia of Computational Chemistry.

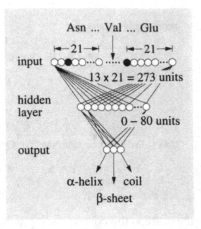

Figure 17-7: Architecture of the neural network for deriving the secondary structures of proteins from their amino acid sequences.

17.5 References and Suggested Readings

17-1. N. Qian and T. J. Sejnowski, "Predicting the Secondary Structure of Globular Proteins Using Neural Network Models", *J. Mol. Biol.* **202** (1988) 865 – 884.

17-2. L. H. Holley and M. Karplus, "Protein Secondary Structure Prediction with a Neural Network", *Proc. Natl. Acad. Sci. USA* **86** (1989) 152 – 156.

17-3. D. G. Kneller, F. E. Cohen and R. Langridge, "Improvements in Protein Secondary Structure Prediction by An Enhanced Neural Network", *J. Mol. Biol.* **214** (1990) 171 – 182.

17-4. H. Bohr, J. Bohr, S. Brunak, R. M. J. Cotterill, B. Lautrup, L. Norskov, O. H. Olsen and S. B. Petersen, "Protein Secondary Structure and Homology by Neural Networks. The α-Helices in Rhodopsin", *FEBS Lett.* **24** (1988) 223 – 228.

17-5. H. Bohr, J. Bohr, S. Brunak, R. M. J. Cotterill, F. Fredholm, B. Lautrup, O. H. Olsen and S. B. Petersen, "A Novel Approach to Prediction of the 3-Dimensional Structures of Protein Backbones by Neural Networks", *FEBS Lett.* **261** (1990) 43 – 46.

17-6. H. Andreassen, H. Bohr, J. Bohr, S. Brunak, T. Bugge, R. M. J. Cotterill, C. Jacobsen, P. Kusk and B. Lautrup, "Analysis of Secondary Structure of the Human Immunodeficiency Virus (HIV) Proteins p17, gp120, and gp41 by Computer Modeling Based on Neural Network Methods", *J. Acquired Immune Defic. Syndr.* **3** (1990) 615 – 622.

17-7. S. Brunak, J. Engelbrecht and S. Knudsen, "Neural Network Detects Errors in the Assignment of MRNA Splice Sites", *Nucleic Acid Res.* **18** (1990) 4797 – 4801.

17-8. P. Y. Chou and G. D. Fasman, "Conformational Parameters for Amino Acids in Helical, β-Sheet, and Random Coil Regions Calculated from Proteins", *Biochemistry* **13** (1974) 211 – 222.

17-9. P. Y. Chou and G. D. Fasman, "Prediction of Protein Conformation", *Biochemistry* **13** (1974) 222 – 241.

17-10. B. Rost, "PHD: Predicting one-dimensional protein structure by profile based neural networks", *Meth. Enzymol.* **266** (1996) 525 – 539.

17-11. B. Rost, "Protein Structure: Prediction in 1D, 2D, and 3D", in *Encyclopedia of Computational Chemistry*, Eds.: P. v. R.

Schleyer, N. L. Allinger, T. Clark, J. Gasteiger, P. A. Kollman, H. F. Schaefer III and P. R. Schreiner, Wiley, Chichester, UK, 1998, pp. 2242 – 2255.

18 Infrared Spectrum-Structure Correlation

---

learning objectives:

- the classification of objects simultaneously into several classes, or hierarchy of classes

- different means of spectra representation by reduced sets of intensities, or by reduced sets of Fourier and Hadamard coefficients,

- the possibility of using different spectrum representations for different spectral regions (i.e., different functional groups)

- use of statistical methods to assist in the interpretation of Kohonen maps

- expanding a study from a Kohonen network to a counter-propagation network

- a mathematical transformation of the 3D structure of a molecule into a fixed length representation

- different ways of selecting a training set

- obtaining a 3D structure from an infrared spectrum

18.1 The Problem

Previously, we have seen classification problems where an object has to be assigned to **one** of several categories. Now, we will look at an example where the object has to be assigned **simultaneously to several** classes out of many possible ones.

After assigning compounds to various structure classes on the basis of their infrared spectra, a modeling of the relationships between structure and infrared spectra is presented that leads to the simulation of high quality infrared spectra. By carrying this work further it will be demonstrated how the 3D structure of a molecule can be derived from its infrared spectrum.

The objects of our present example are the infrared spectra of various compounds. The output of the neural network should be a series of substructures that are contained in the compound whose infrared spectrum is being investigated. Except for the immediately preceding example (Chapter 17 – the secondary structure of proteins), our previous examples have had rather simple neural networks with small numbers of weights; in this application, we will meet much larger multilayer neural networks containing 10,000 or more weights.

The elucidation of the structure of organic compounds relies heavily on spectroscopic methods. However, the relationships between structure and spectral data are usually too complex to be expressed as explicit equations. As in many complex associational problems (medical diagnosis, for example), a series of empirical rules has been developed. The search for structure/spectra methods can, fortunately, build on a host of experimental data, much of which is now available in computerized databases.

Today, along with high-resolution full-curve spectrum, the databases also contain chemical structure coded as a connection table. The clear (though complicated!) relationships between structure and spectra, and the availability of large computerized datasets (50,000 spectra now, and more every year) make this field an ideal – and important – area of application for neural networks.

The work presented in this chapter stresses the importance of structure representation. Both, the investigations reported in Sections 18.2 - 18.3 and those in Sections 18.4 - 18.6 represent structures by a set of functional groups. In these studies, the objective is to predict the presence or absence of functional groups from the information contained in an infrared spectrum. The two different studies allow a comparison of what can be achieved with a back-propagation network against results from a Kohonen network. On the other hand, a representation of structures by functional groups is hopelessly inadequate for the reverse problem, for the simulation of an infrared spectrum over the entire frequency domain. A break-through in this area could only be achieved by the use of a novel molecular transform of the 3D structure. This structure representation then allowed even the derivation of the 3D structure from the information present in an infrared spectrum (Figure 18-1).

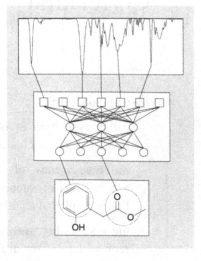

Figure 18-1: The problem: derivation of substructural features from the infrared spectrum.

This chapter only deals with the application of neural networks to structure-infrared spectrum correlations. However, work on finding correlations between structure and data from other spectroscopic methods like **mass spectra** or **^{13}C NMR spectra** with neural networks has already been done. Some of these investigations are mentioned in Section 18.7 (References).

18.2 The Representation of Infrared Spectra as Intensities

Munk, Madison, and Robb, (Reference 18-1) represented infrared spectra in the following way. As a first approach, the range of a spectrum from $4000 - 400 \ cm^{-1}$ was divided into 640 intervals of width $5.6 \ cm^{-1}$. The transmission intensity value of one interval was then scaled according to the equation:

$$x_i = 1.00 - (\%t) / 100.0 \qquad (18.1)$$

where $\%t = \%$ transmission

Thus the neural network would need 640 input units; but this large number of input units caused some spurious results, and it was reduced to 256. At the same time, they adjusted the widths of the intervals to be narrowest at low frequencies and broadest at the high frequency end of the spectrum (to take into account the varying discrimination from one end of the spectrum to the other). The formula that makes the **length** of the interval i dependent on the frequency is:

$$i = 6.0 \, (frequency)^{0.5} - 120.0 \qquad (18.2)$$

rounded to the nearest integer. This equation assigns a frequency interval of $10 \ cm^{-1}$ (from $400 - 410 \ cm^{-1}$) to input unit 1; on the other end of the spectrum, it assigns a frequency interval of $20 \ cm^{-1}$ (from $3928 - 3948 \ cm^{-1}$) to input unit 256. The assigned frequency interval for each unit is then scanned for peaks; if a peak is found, its intensity (scaled to lie between 0.000 and 1.000) is the input to this unit, otherwise the input to the unit is zero.

The structure of the compound is described in terms of 36 functional groups (primary alcohol, phenol, tertiary amine, ester, etc.), each represented by one output unit. Hence, the target vector is a 36-

variate binary vector in which each 1 indicates the presence of the associated functional group, and zero indicates its absence.

In general, a structure can have several such functional groups, and thus several output units might be simultaneously active. After trying 14 different networks varying from fewer than ten to more than 60 neurons in the hidden layer, 34 were found to be appropriate. Thus, they used a neural network having just under 10,000 $((256 + 1) \times 34 + (34 + 1) \times 36 = 9{,}998)$ weights. Figure 18-2 shows the network used in this example.

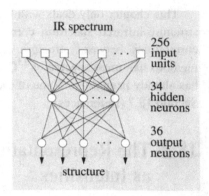

Figure 18-2: The network for the infrared spectra-structure correlation problem.

18.3 The Dataset, and Learning by Back-Propagation

If the back-propagation algorithm is used to train a large neural network such as the one described in the previous section, we must have a large training set. A good rule of thumb is that the number of data values taken for training should be equal to or greater than the number of weights to be determined in the network. Here, the data matrix contains about 640,000 values (2,499 spectra times 256 intensities), which is about 60 times larger than the number of weights. An additional 416 spectra were set aside to test the prediction ability of the trained neural network.

A relatively small learning rate η of 0.083 was used. One epoch of training using the entire dataset of 2500 spectra required about 10 min of CPU time on a VAX 3500; typically 100 epochs were needed to stabilize the network.

The actual outputs of the network are seldom exactly zero or one. Therefore, predicting the presence or absence of a functional group strongly depends on where the threshold is set. For example, in this network only 30% of the 265 primary alcohols contained in the data set produce an output equal to 1.0. However, if the threshold is lowered to 0.86, 50% (132) of the primary alcohols are classified correctly. But, unfortunately, 34 compounds not having this group produce a value higher than 0.86 **on this output unit** (false positives).

Any given functional group represents **only a small fraction** of the entire training set; lowering the threshold value would simply generate an excessive number of false positives.

Quite often, such false positives are not considered in the figures of merit calculated for multicategory classifications. Therefore, they have to find a balance between a small percentage of reliable correct predictions and a higher percentage of slightly less reliable ones.

To account for both correct and incorrect assignments, a reliability index called the *A50* value is taken as a measure of the reliability of the prediction for a given functional group *j*:

$$A50 = \frac{0.5\,n_j}{0.5\,n_j + n_{wrong}} \qquad (18.3)$$

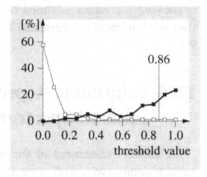

Figure 18-3: Percentage of correct classifications of primary alcohols depending on the threshold value (vertical line); the thin curve shows the percentage of false positives.

where n_j is the number of compounds having functional group *j* in the training set, and n_{wrong} is the number of false positives involving group *j*.

NOTE: the *A50* value defines the reliability of the predictions and not the prediction ability. An *A50* value of 100% would mean that at the current threshold level for prediction, only **half** the objects from class *j* are correctly classified, with **no** false positives. This means that if the prediction is positive it is extremely reliable (Figure 18-3), but half of the compounds having this functional group are not identified at all (false negatives).

At a threshold value of 0.86, 132 of the 265 primary alcohols in the training set are correctly identified, and there are 34 false positives. This gives an *A50* value of 132/(132 + 34) = 79.5%. This is considered a good reliability for predictions. Thirty out of the 36 functional groups are determined with an equally good or better reliability.

These results, obtained with a two-layer neural network (with one hidden layer of 34 neurons), were also compared to an earlier investigation of the same group of authors (Reference 18-2), in which they used only one active layer of neurons, and no hidden layer. It turned out that the hidden layer of neurons leads to remarkable improvements in the reliability of the predictions.

A combination of methods is often more powerful than either one taken separately. Thus, a neural network can be incorporated into an expert system for structure elucidation, where it could help to identify those structures for which functional group identification can be made with a rather high reliability. The search space for the expert system could be narrowed to include only those compounds for which no reliable prediction can be made by the neural network.

A further merit of this work is that even such simple representations of the spectrum and the molecular structure lead to practical results. This opens the door to further investigations aimed at improving the representations of spectral and structural information (See the comment by Bernhard Widrow quoted at the End of Chapter 9.).

18.4 Adjustable Representation of an Infrared Spectrum

A further refinement of the investigation discussed above is to build a number of neural network decision modules and arrange them in a hierarchical manner. In this way, a wide variety of decision possibilities can be achieved, while maintaining a short decision path. Such a work has already been initiated by Kateman and coworkers of the Analytical Department of the Nijmegen Catholic University (see Reference 18-6).

The problem of priorities associated with setting up a hierarchy of decisions was addressed in Section 9.2: how to determine which decision or decisions to put onto the top level of the decision hierarchy, which in the next one, etc. But another, even more important problem is the choice of representation for each of these decision modules.

Since the first attempts to build automated interpretation systems for infrared spectra on the basis of the **full spectral curve**, all authors have stressed the fact that different spectral regions are actually used for each decision, so that ideally **a different spectral representation** should be used for each decision. The closest approach to this is offered by a rule-based expert system, which requires a set of rules for each structural feature in the form "**if**-spectral-feature **then** functional-group". Unfortunately, not all the rules necessary to interpret the infrared spectrum have been worked out.

Although most workers are aware of the need for different representations and have stressed it many times, a system that would actually **adapt** spectral input to **each** decision separately has not yet been developed.

In the following two Sections we would like to show a way that possibly enables us to handle both problems: setting the priorities of functional groups (i.e., their positions in the decision hierarchy), and how to select the most suitable spectral representation for a particular decision.

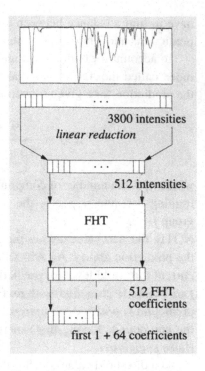

3800 intensities
linear reduction

512 intensities

FHT

512 FHT coefficients

first 1 + 64 coefficients

Figure 18-4: The way from the full spectral curve to a reduced set of Hadamard coefficients.

18.5 Representing Spectra using Truncated Sets of Fourier or Hadamard Coefficients

We will discuss how to map spectra from a 512-dimensional spectral space into a two-dimensional structure feature space. We will follow the example of Novic and Zupan of the National Institute of Chemistry in Ljubljana (see Reference 18-3). Their initial representation of an infrared spectrum does not follow the reduction of Equation (18.2); rather, it tries to capture the entire shape of the spectral curve by first making a Hadamard (or Fourier) transformation and then using as input only a **truncated** set (e.g., the first half or quarter) of the coefficients (Figure 18-4).

The Fast Fourier Transformation (FFT) uses as a basis set a set of sines and cosines of different frequencies, while the Fast Hadamard Transformation (FHT) uses box (square wave) functions with different frequencies (Figure 18-5). For the reduction of a measurement space and for recovery of the original information, both transformations have about the same merits and deficiencies (see Reference 18-17).

Novic and Zupan preferred the Fast Hadamard Transformation because it is 4 to 8 times faster (depending on the hardware) and because it does not use complex coefficients (for a comparison of the algorithms for these two fast transformations see Reference 18-4).

First, the infrared spectrum is divided into 512 intervals, in each of which the corresponding intensity is taken. The intervals are of two different lengths: larger (20 cm^{-1}) at the higher wave numbers (4000 to 2000 cm^{-1}) and narrower (4 cm^{-1}) in the remaining part of the spectrum (2000 to 352 cm^{-1}). Applying the fast Hadamard transformation produces 512 Hadamard coefficients; the first 64 of these are taken as a representation of the spectrum.

This gives us: first, a considerably shorter representation (64 variables compared to 512, a factor of 8), and second, a reasonably good reproduction of the original spectrum. Figure 18-6 shows the same infrared spectrum after its 512 intensities are transformed with the Fast Hadamard Transformation, then reduced, and finally transformed back to the "original".

The **first** of the two goals in this example is to find out which functional groups are so characteristic in the infrared spectra that they

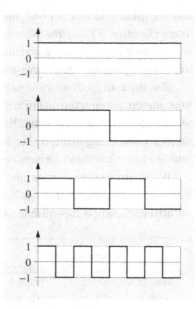

Figure 18-5: A set of square wave (Walsh) functions used in the Hadamard transformation.

can be placed at the top of the decision hierarchy. First, Kohonen maps (Section 7.6) of the 64-variate infrared spectra are made. Then, an attempt is made to associate clusters (of spectra) with common functional groups in the corresponding molecules.

The **second** goal is to obtain guidelines for the **selection of the best spectral representation** for different structures. This task aims to obtain rules (or at least some hints and suggestions) for making new representations appropriate for infrared spectra of compounds having one or more functional groups in common.

As a preliminary example, a modestly large (11 x 11 x 64) Kohonen network (Figure 18-7) was trained with 150 infrared spectra of different compounds (Table 18-1).

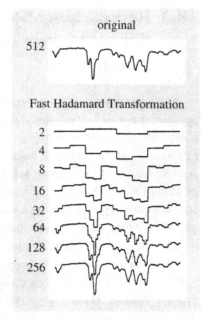

Figure 18-6: Reproduction of the same infrared spectrum from different forms of compression. First the 512 intensities are transformed with the Fast Hadamard Transformation, then reduced, and finally transformed back to the "original".

functional group		no. of compounds	label
–O–CO–	*(ester)*	25	E
–CO–	*(ketone)*	21	K
–C–O–C–	*(ether)*	17	O
–COOH	*(acid)*	15	A
S (thiophene)	*(thiophene)*	9	T
ketone + ether		4	k = K + O
ketone + thiophene		3	v = K + T
ketone + acid		2	b = K + A
ketone + ester		2	e = K + E
ether + ester		2	c = O + E
acid + ether		1	a = A + O

Table 18-1:   Most common functional groups in the training set, and some of the compounds having two functional groups.

The training set contained only compounds having one or two functional groups known to have clearly distinguishable spectral features. Table 18-1 lists some of the functional groups in the compounds whose spectra are input to the Kohonen network.

NOTE: in this application Kohonen maps were obtained **without** toroidal boundary conditions; see Section 6.2 for more details.

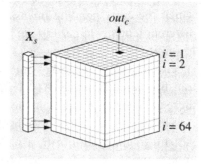

Figure 18-7: Kohonen network for grouping infrared spectra.

18.6 Results of Kohonen Learning

One epoch represents input of all 150 sets of 64 Hadamard coefficients into the (11 x 11 x 64) network. Different numbers of epochs, from 20 to 140, were tried; as a measure of how well the network had adapted, the error produced by all 150 spectra was calculated from all weights of the excited neuron:

$$\varepsilon^{tot} = \frac{\sqrt{\sum_{s=1}^{150} \sum_{i=1}^{64} \left(x_{si} - w_i^{excited} \right)^2}}{\sqrt{150}} \qquad (18.4)$$

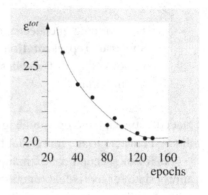

Figure 18-8: Total error ε^{tot} as a function of epochs used in training.

About 100 epochs are needed before the network stabilizes. Figure 18-8 shows a plot of the error at the end of training vs. the number of epochs used. So, no essential improvement can be obtained by going beyond 100 epochs.

In training, the learning rate constant η (Equation (6.6)) was changed linearly, from 0.5 at the beginning to 0.1 at the end, regardless of how many epochs were used.

After the end of training each neuron in the (11 x 11) map was labeled with the letters for functional groups in the compound corresponding to the spectrum that excited it. If **several** spectra excite the same neuron, **all** their labels are attached to its location on the map. The labeled map is shown in Figure 18-9.

Kohonen learning did produce a map with distinct clusters of some of the labels, which suggests that we can decide how to place the functional groups in a decision hierarchy (of networks) by inspecting these clusters.

The most compact clusters are esters (E) and acids (A); the other two largest sets of compounds, ketones (K) and ethers (O), also form clusters, but have a few outliers; that is, a few spectra labeled "K" or "O" excite neurons outside (but near to) the respective clusters.

Addressing our first goal, the selection of functional groups for the top of a decision hierarchy, E (esters) and A (acids) would seem to be natural choices.

It is interesting to observe the locations of spectra from compounds having **two** functional groups. For example, the two compounds labeled "k" contain both a ketone and an ether, and those labeled "b" contain both the ketone and an acid. These compounds are

Figure 18-9: The (11 x 11) Kohonen map with some of the 150 labels (each label marks one compound) whose spectrum (via the functional group identified by the label) excited the neuron at that position.

> For building the decision hierarchy with **adjusted spectral representations** both types of spectral regions, those of smallest and of largest similarity, are important.

placed between the corresponding clusters, right on the borders where the two clusters of the individual functional groups meet.

As for our second goal, finding guidelines for representing spectra appropriate for individual functional groups, this information is stored in the weights of the 121 neurons.

The 64 weights of each neuron were obtained during training from the truncated set of 64 Hadamard coefficients used to represent the infrared spectrum. Therefore, the "adjusted" weights can be regarded as *"adjusted" Hadamard coefficients* and can consequently be transformed back with the inverse Hadamard transformation to *"adjusted" infrared spectra*.

(The inverse Hadamard transformation of a **truncated** set of 64 coefficients yields 512 intensities; however, there are only 64 different ones: each intensity is repeated eight times in a row (64 x 8 = 512). Hence, the "adjusted" infrared spectrum has only 64 different intensities over the entire region.)

Note that these 121 adjusted spectra have now replaced the weights of the neurons, i.e., they can be stored in hypercolumns (as neurons) in the Kohonen network. Because the learning procedure adjusts **all weights in all neurons**, the adjusted spectrum can be obtained even **at positions where the neuron was not excited by any of the training spectra**, i.e., from "empty spaces" (see Section 10.4). One such position is marked by a small square in Figure 18-10.

Consider the group of acids (A) in Figure 18-9. The question now is how to obtain the **spectral** features that contribute most to the identification of the **structural** cluster marked with A's.

The spectra of compounds having a specific functional group in common have similar spectral features in certain regions. The hypothesis is that the adjusted spectra forming the region labeled "A" are more similar to each other than they are to the rest of the adjusted spectra in the network, and our goal is to find the regions where the similarity is smallest, and those where it is largest.

The spectral region with a high degree of similarity should be responsible for deciding whether a particular functional group is

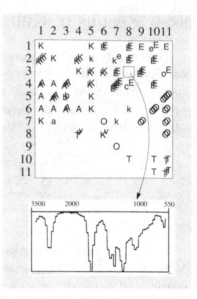

Figure 18-10: Top: Empty space (neuron) which was not excited by any of the training spectra in the "ester" region. Bottom: the corresponding adjusted spectrum stored as a weight vector in this neuron.

present, while the parts of the spectrum with lower similarities are responsible for other decisions **afterwards**.

Figure 18-11 shows a comparison between one adjusted spectrum from the left-hand side of the map (acid region) and one from the right-hand side (ether region).

Recall that all 121 adjusted spectra are stored as weights in 64 levels (as hypercolumns) of the (11 x 11) Kohonen network; therefore, it is possible to inspect the **maps of all weight levels**.

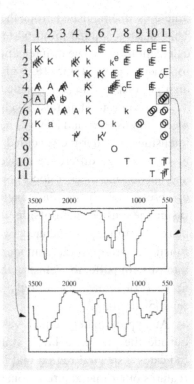

> Remember that a given level corresponds to a certain wavelength region of the spectrum.

This is done by cutting through the Kohonen network at a certain level (of weights), and plotting a contour map of the intensities at this wavelength region (Figure 18-12 (a)). Figure 18-12)b) shows the map obtained when a cut is made across the network at weight levels 6 and 48.

Taking a closer look at all 64 contour maps, we find that some of the contour lines in certain maps almost exactly coincide with the borders of labeled clusters. Figure 18-13 (b) shows that on level 6 the iso-intensity contours coincide with the cluster labeled A.

Recall that the levels correspond to different spectral regions of "adjusted" spectra. Hence, by finding the levels having intensity contours most similarly distributed to the cluster of a functional group, we will have found the most important spectral regions of this functional group.

If a certain spectral region is highly selective for one particular functional group then, consequently, it is less important to other functional groups. In other words, the **remaining** parts of the spectrum should play a more important role in the next steps, where decisions about **other** functional groups are made.

However, visual comparison of intensity contours using the functional group map is only a preliminary way to find the most relevant spectral regions for a particular decision. A more unbiased way to determine the relevant levels is to compare statistical quantities such as means and standard deviations of weight values (intensities) between the regions inside and outside the cluster.

First, we should determine the area on the map for which a statistical comparison with the outside area has to be made. Figure 18-13 (c) shows the (11 x 11) mask that identifies the "acids" region

Figure 18-11: Two adjusted spectra from opposite sides of the (11 x 11) Kohonen network.

(ones) as distinguished from the region "outside" the acids (zeros). This mask is applied in all 64 levels of weights, to select spectra for calculation of the mean intensities and their standard deviations, one for each region. Table 18-2 gives the statistical data for intensity distributions within and outside the "acids" region for 18 levels.

According to statistics, levels with large differences between the mean intensities and levels having large standard deviations of intensities outsider the specified cluster are the most significant. That is, first, a large difference between the mean intensity values inside and outside indicates a good possibility for determining the presence or absence of a particular functional group; a larger mean intensity **inside** the cluster than **outside** means that the presence (absence) of a peak is strongly correlated with the presence (absence) of the functional group in the compound. On the other hand, a larger mean intensity **outside** the cluster than **inside** it would mean that the absence of a peak is strongly correlated with the presence of the functional group.

At approximately equal values of mean intensities within and outside the cluster, a **large standard deviation outside** the cluster indicates a spectral region that is rich in information about structural features other than the functional group specified by the cluster.

From Table 18-2 it can be seen that levels 3 to 7 ($3310 - 2990$ cm^1) can be used for recognizing the acids because the mean intensity inside the A region is large compared to the mean intensity outside. On the other hand, the spectral regions represented by levels 26 to 30 (1788 to 1660 cm^{-1}), 36 to 39 (1468 to 1372 cm^{-1}) which have means of intensities and standard deviations of comparable sizes, not too much different to each other, are quite irrelevant for the decision about the acid functional group.

Additional statistical calculations and visual inspection can yield a great deal of information about the correlation between functional groups (represented by clusters in the Kohonen map) and spectral regions (represented by the levels across which the Kohonen network was cut).

With a larger Kohonen map and thousands of spectra of compounds having a larger assortment of structures, such a study can extend the decision sequence to much deeper hierarchical levels.

In this short example, we have seen yet another way to extract information from a trained neural network. The trouble with textbook examples, of course, is that they make it all seem so easy (!).

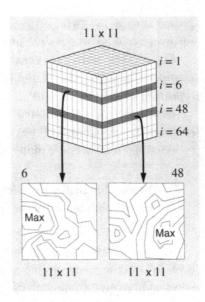

Figure 18-12: Above: if the Kohonen network is cut across different levels, contour plots of intensities can be obtained. Below: these contours are obtained when inspecting levels 6 (b) and 48 (c) corresponding to wave number regions $3190 - 3110$ and $1100 - 1068$ cm^{-1}.

level	wave no.	w_{av}^{in}	σ_{av}^{in}	w_{av}^{out}	σ_{av}^{out}	Δw_{av}^{in-out}
1	3470	3.22	1.28	1.10	1.19	2.11
2	3390	4.99	1.63	1.35	1.85	3.64
3	3310	6.46	1.58	1.28	1.91	**5.18**
4	3230	7.32	1.22	1.31	1.72	**6.01**
5	3150	7.67	0.93	1.90	1.64	**5.77**
6	3070	7.87	0.74	2.44	1.42	**5.43**
7	2990	8.89	0.57	2.45	1.96	**6.44**
8	2910	7.95	0.67	5.39	2.40	2.56
26	1788	3.73	1.16	1.78	1.52	2.21
27	1756	7.23	0.86	3.86	3.15	3.37
28	1724	9.84	0.51	5.44	3.69	4.40
29	1692	8.13	1.44	4.22	2.70	3.91
30	1660	5.10	1.49	3.37	2.24	1.73
36	1468	4.77	1.34	5.79	1.64	−1.02
37	1276	6.53	0.87	5.97	1.57	0.56
38	1244	6.51	1.13	5.53	1.56	0.98
39	1372	5.22	1.66	5.98	1.75	−0.76
48	1084	2.78	1.27	4.86	2.07	0.71
49	1052	2.15	1.60	4.76	2.10	−2.61
50	1020	1.61	1.23	4.92	2.27	−3.31
51	988	1.79	0.55	3.26	1.22	−1.47
61	668	2.70	0.93	1.68	1.29	1.02
62	636	2.37	0.69	1.50	1.08	0.87
63	604	1.75	0.80	1.85	1.06	−0.10
64	572	0.72	0.85	1.24	0.79	−0.52

Table 18-2: Means of weights, w_{av}, and standard deviations, σ, of intensities inside and outside the masked area (see Figure 18-13 (c)) for some of the levels in the (11 x 11 x 64) Kohonen network of infrared spectra.

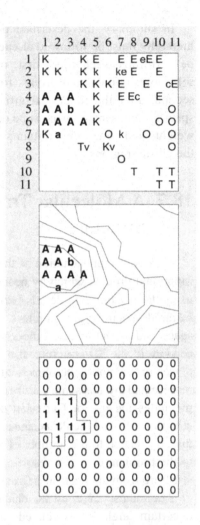

Figure 18-13: The resulting Kohonen map of functional group labels (a), the map of weights on level six (b) and the mask (c) for sampling the intensities from inside (ones) and outside (zeros) the acids region A (see Figure 18-9).

However, be warned that exquisite care must be taken when applying this procedure for finding distinguishable functional groups and the corresponding spectral regions relevant to adjusting the representation. For each node in the decision hierarchy, a new Kohonen network, new (carefully selected) data, and a new learning procedure are required.

In summary, the described procedure for setting up a decision hierarchy must be followed at **each** decision node. Two steps have to be performed each time: first, on the basis of a Kohonen map, a selection of functional groups for the next decision must be made; second, after the functional groups are selected, the corresponding spectral representations must be determined by inspecting the maps of weights and comparing certain statistical features within and outside the cluster regions.

18.7 A Molecular Transform of the 3D Structure

Clearly, a representation of the structure of a molecule by a set of functional groups, which by necessity is limited, must be insufficient to model the intricate details between structure and infrared spectra. Infrared spectroscopy monitors the vibrations of a molecule in 3D space and a sophisticated modeling of these relationships must take account of the 3D structure of a molecule. As emphasized at several places in this book, the objects of a neural networks study have to be represented by the same number of descriptors (Figure 18-14). This precludes the use of Cartesian coordinates for representing the 3D structure of a molecule because, then, the number of descriptors is directly related to the number of atoms in this molecule. The solution to the problem came by resorting to equations used to derive the 3D structure from an electron diffraction experiment (Figure 18-15).

The intensity, $I(s)$, of the electron beam scattered by a molecule at a certain angle, s, is related to the 3D structure of a molecule, represented by the vectors, r_{ij}, of the interatomic distances between atoms i and j, as given by Equation (18.5):

$$I(s) = \sum_{j=i+1}^{n} \sum_{i=1}^{n-1} a_i a_j \frac{\sin(sr_{ij})}{sr_{ij}} \qquad (18.5)$$

In this equation, s measures the scattering angle given by

$$s = 4\pi \sin(\theta/2)/\lambda \qquad (18.6)$$

with θ being the scattering angle and λ the wave length. a_i and a_j are the atomic numbers of atoms in the molecule. Equation (18.5) is the simplified form of the original equation obtained by assuming the atoms to be point scatterers and the molecule to be rigid. All

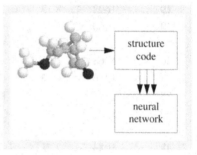

Figure 18-14: Transforming the 3D structure of a molecule into a fixed-length representation.

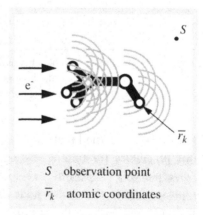

S   observation point

$\bar{r}_k$ atomic coordinates

Figure 18-15: An electron diffraction experiment.

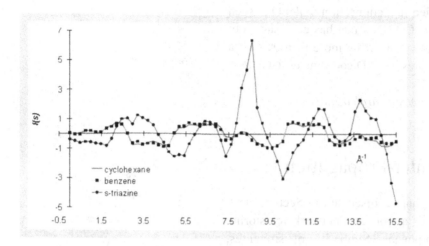

instrument variables were collected into a single constant which was set to 1.

In electron diffraction, the scattered intensity, $I(s)$, is measured, and the 3D structure of a molecule, given by r_{ij}, is derived therefrom. We, however, turned Equation (18.5) around by inputting the 3D structure of a molecule and calculating $I(s)$. In addition, $I(s)$ was calculated only for a fixed number (e.g. 32 or 64) of discrete, equidistant values of s.

Thus, the 3D structure was transformed into a fixed number of descriptors. Furthermore, it is possible to use atomic properties, a_i other than atomic numbers in Equation (18.5). In most applications for the simulation of infrared spectra we use partial atomic charges for a_i as calculated by the PEOE method. Figure 18-16 shows such a representation of a molecule. As this molecular representation was derived from an analysis of electron diffraction experiments, we named it 3D-MoRSE code (3D-Molecule Representation of Structure derived from Electron diffraction).

Clearly, such an approach requires access to the 3D coordinates of a molecule. Although X-ray structures have been determined for about 140,000 organic and organometallic compounds, this number is still very small in comparison to the about 14 million known organic compounds. In order to allow the study of the relationships between infrared spectra and the 3D structure of molecules on a broad range, a more universal access to the 3D structure of a molecule is required.

Fortunately, in recent years, automatic 3D structure generators have been developed that can build a 3D model of a molecule from constitutional information embodied in a connection table. One such automatic 3D structure generator is CORINA that has been shown to have a broad scope by automatically converting more than 99% of a database of over 6.5 million structures into 3D coordinates. CORINA can be accessed on the internet

(*http://www2.ccc.uni-erlangen.de/software/corina/*);
see the Appendix for further details.

18.8 Learning by Counter-Propagation

The previous example in this chapter, discussed in Sections 18.5 and 18.6, has shown the merits of a Kohonen network for storing infrared spectra of similar compounds. Similarity of structures, however, was measured rather rudimentary, by a small number of functional groups, only. With the structure coding developed in section 18.7 we have a much more sophisticated structure representation and are now in a position to model the relationships between infrared spectra and structure. We must therefore use a supervised learning technique. In addition, we wanted to retain the advantages of the learning technique embodied in a Kohonen network. The answer is therefore: use a counter-propagation neural network, as this is a supervised learning method using the competitive learning technique also contained in a Kohonen network.

Figure 18-17 shows the architecture of the counter-propagation network used in this study. The upper block of the network contains the weights that are adjusted based on the intensity descriptors, $I(s)$, derived from the 3D structure of the molecules according to Equation (18.5). The lower block contains the weights that are adjusted from the infrared spectra represented as detailed in Section 18.5.

In a typical example, all mono-, di-, and tri-substituted benzene derivatives carrying substituents with no more than eight consecutive bonds and consisting of the atoms C, H, N, O, F, Cl, and Br were retrieved from the SpecInfo$^©$ database. This provided a data set of 871 benzene derivatives and their infrared spectra. In order to split this data set into a training and a test set, a planar counter-propagation network of size 30 x 30 was trained with the entire data set. From each of the occupied neurons, that molecule was selected for the training set that had a structure code most similar to the weights of this neuron. This provided a training set of 487 molecules. All other molecules (384) were transferred into the test set. The 3D structure of all

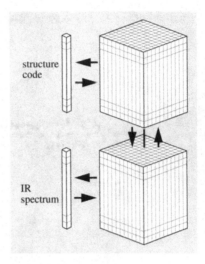

Figure 18-17: Architecture of the counter-propagation network for learning the relationships between 3D structure and infrared spectra.

molecules was generated by CORINA and converted into 32 intensity coefficients according to Equation (18.5) using partial atomic charges as calculated by the PETRA package as atom descriptors, a_i. The infrared spectrum was reported by 128 intensity values after Hadamard transformation as described in Section 18.5.

A counter-propagation network with toroidal topology consisting of 25 x 25 neurons was trained with these 487 molecules and their infrared spectra.

The similarity between the experimental and the simulated infrared spectra was measured by the correlation coefficient. Figure 18-18 gives the distribution of the correlation coefficients for the entire training set of 487 molecules.

Clearly, the performance of this approach has to be measured by the test set. Figure 18-19 shows the distribution of the correlation coefficient for the entire test set of 384 compounds.

Figure 18-20 compares one of the higher quality simulated infrared spectrum with the experimental one from the test set. The most important result is that good correspondence can be obtained over the entire frequency range, not only in the region of valence bond vibrations but also in the fingerprint region. This attests to the potential of the 3D-MoRSE code to represent the entire structure of a molecule in its coefficients, not only parts of the structure such as functional groups.

As with any learning procedure, the quality of prediction is dependent on the availability of information. We have found that the cases of predictions of poorer quality can, by and large, be attributed to a lack of data for these types of compounds. However, all cases with correlation coefficients of 0.7 and higher can be considered for many applications as satisfactory.

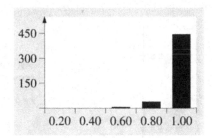

Figure 18-18: Distribution of correlation coefficients of the experimental with the simulated infrared spectra of the 487 molecules of the training set of benzene derivatives.

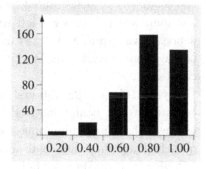

Figure 18-19: Distribution of the correlation coefficients of the experimental with the simulated infrared spectra of the 384 molecules of the test set of benzene derivatives.

18.9 Different Strategies for the Selection of a Training Set

In the example given in Section 18.8, a certain group of compounds, in this case, mono-, di-, and tri-substituted benzene derivatives, was selected as data set to train a counter-propagation network for the simulation of infrared spectra.

It is also possible to train a single counter-propagation network for the simulation of infrared spectra over the entire range of organic

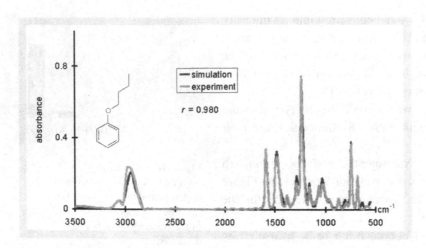

Figure 18-20: Comparison of a simulated with an experimental infrared spectrum for a molecule of the test set of benzene derivatives.

chemistry. The SpecInfo database Version 2.0 contained approximately 15,000 infrared spectra. After elimination of duplicates and ionic species we were left with 9,850 different structures. This data set was split into 3,244 compounds for training a counter-propagation network with 70 x 70 neurons, and two test sets of about equal size.

By and large the infrared spectra simulated with this large comprehensive neural network were quite acceptable. Clearly, however, the computation times for training such a huge network were quite high and the results with smaller, dedicated networks were often of higher quality. Thus, we rather prefer smaller networks.

In both approaches, working with a large network encompassing the entire range of organic chemistry or when training a network for a certain class of compounds, the training of the network can be done once and for all; predictions are then rather rapid, indeed.

However, it must be realized that the selection of a compound into a class of compounds might be quite arbitrary. Is the compound shown in Figure 18-21 a quinoline derivative or a substituted furane? (Clearly, it is both!) For such cases, we have developed an alternative for the selection of a data set for training a network: The structure for which an infrared spectrum should be simulated, the query structure, determines its own training set: From the database those 50 molecules – and their associated infrared spectra – are selected for the training of a counter-propagation network, that have a structure code that is most similar to the code of the query structure. We have found this approach to give the most satisfactory results although in many cases,

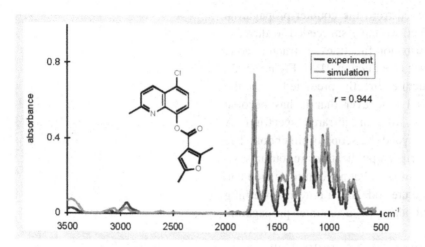

Figure 18-21: Comparison of an experimental infrared spectrum with the simulated spectrum obtained by the query-directed approach from a counter-propagation network trained with the 50 most similar structures.

where the definition of a class of compounds is quite clear, the improvements are only minor.

Figure 18-21 gives the results of such a query-directed simulation of an infrared spectrum. The disadvantage of this approach is that each new query structure requires the training of a specific network, one cannot work with pre-trained networks. However, with training sets of 50 compounds the training times are quite acceptable with about a minute on a PC.

18.10 From the Infrared Spectrum to the 3D Structure

The architecture of Figure 18-17 shows that there is a direct relationship between the block of weights obtained from the structure code and the block of weights from the infrared spectra. As of now, we have used such a counter-propagation network in a single direction, inputting a structure code and outputting an infrared spectrum.

However, it should be possible to also operate such a counter-propagation network in reverse mode, inputting an infrared spectrum and outputting a structure code (Figure 18-22). Now, remember, that the structure code as calculated by Equation (18.5) is nothing else than a discrete form of the electron diffraction pattern, the very information that is used – in its full form – to derive a 3D structure from an electron diffraction experiment. Thus, it should be possible to

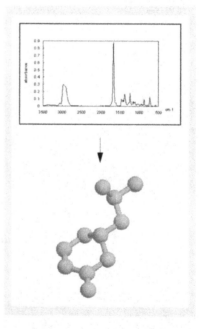

Figure 18-22: 3D structure derived from an infrared spectrum.

transform the structure code obtained from the counter-propagation network into a 3D structure. In fact, we have succeeded to develop such a method based on radial distribution functions, a structure code quite similar to the 3D-MoRSE code (see Section 21.3). Figure 18-22 shows an example of a 3D structure directly predicted from the infrared spectrum. This is the first time ever that is has become possible to derive a 3D structure from an infrared spectrum. A discussion of this procedure goes beyond the scope of this book. The interested reader is referred to the original publications in journals.

The important message to carry away is that novel information can be gained with a sophisticated structure code and a powerful learning algorithm such as the one embodied in a counter-propagation network.

The methods explained in Sections 18.7 – 18.10, both for the simulation of infrared spectra and for the derivation of the 3D structure from an infrared spectrum, are made available for the general public to use on the internet through the project TeleSpec at

http://www2.ccc.uni-erlangen.de/research/ir/.

Furthermore, the scientific community is invited to use and cooperate in building a freely accessible database of infrared spectra. See the Appendix for further details.

18.11 References and Suggested Readings

18-1. M. E. Munk, M. S. Madison and E. W. Robb, "Neural Network Models for Infrared Spectrum Interpretation", *Mikrochim. Acta* [Wien] 1991 II, 505 – 514.

18-2. E. W. Robb and M. E. Munk, "A Neural Network Approach to Infrared Spectrum Interpretation", *Mikrochim. Acta*, [Wien], 1990 I, 131 – 155.

18-3. M. Novic and J. Zupan, "2-D Mapping of Infrared Spectra Using Kohonen Neural Network", *Vestn. Slov. Kem. Drus.* **39** (1992) 195 – 212.

18-4. M. Razinger and M. Novic, "Reduction of the Information Space for Data Collection", in *PC's for Chemists*, Ed.: J. Zupan, Elsevier, Amsterdam, NL, 1990, pp. 89 – 103.

18-5. J. R. M. Smiths, P. Schoenmakers, A. Stehmans, F. Sijstermans and G. Kateman, "Interpretation of Infrared Spectra with Modular Neural Network Systems", *Chemom. Intell. Lab. Syst.* **18** (1993) 27 – 39.

18-6. W. J. Melssen, J. R. M. Smits, G. H. Rolf and G. Kateman, "Two-dimensional Mapping of Infrared Spectra Using Parallel Implemented Self-organising Feature Map", *Chemom. Intell. Lab. Syst.* **18** (1993) 195 – 204.

18-7. J. Zupan and M. Novic, "Hierarchical Ordering of Spectral Data", in *Computer Supported Spectroscopic Data Bases*, Ed.: J. Zupan, Ellis Horwood, Chichester, UK, 1986, pp. 42 – 63.

18-8. M. Tusar and J. Zupan, "Neural Networks", in *Software Development in Chemistry 4*, Ed.: J. Gasteiger, Springer Verlag, Berlin, FRG, 1990, pp. 363 – 376.

18-9. M. Otto and U. Hörchner, "Application of Fuzzy Neural Networks to Spectrum Identification", in *Software Development in Chemistry 4*, Ed.: J. Gasteiger, Springer Verlag, Berlin, FRG, 1990, pp. 377 – 384.

18-10. J. R. Long, V. G. Gregoriou and P. J. Gemperline, "Spectroscopic Calibration and Quantization Using Artificial Neural Networks", *Anal. Chem.* **62** (1990) 1791 – 1797.

18-11. B. Curry and D. E. Rumelhart, "MSnet: A Neural Network That Classifies Mass Spectra", *Tetrahedron Comput. Methodol.* **3** (1990) 213 – 237.

18-12. J. U. Thomsen and B. Mayer, "Pattern Recognition of the ^{1}H NMR Spectra of Sugar Alditols Using a Neural Network", *J. Magn. Res.* **84** (1989) 212 – 217.

18-13. H. Lohninger, "Classification of Mass Spectral Data Using Neural Networks", in *Software Development in Chemistry* **5**, Ed.: J. Gmehling, Springer Verlag, Berlin, FRG, 1991, pp. 159 – 168.

18-14. B. J. Withoff, S. P. Levine and S. A. Sterling, "Spectral Peak Verification and Recognition Using a Multilayered Neural Network", *Anal. Chem.* **62** (1990) 2709 – 2719.

18-15. P. J. Gemperline, J. R. Long and V. G. Gregoriou, "Nonlinear Multivariate Calibration Using Principal Component Regression and Artificial Neural Networks", *Anal. Chem.* **63** (1991) 2314 – 2323.

18-16. V. Kvasnicka, "An Application of Neural Networks in Chemistry. Prediction of ^{13}C NMR Chemical Shifts", *J. Math. Chem.* **6** (1991) 63 – 76.

18-17. J. Zupan, *Algorithms for Chemists*, Wiley, Chichester, UK, 1989, Chapter 5.

18-18. J. Gasteiger, X. Li, V. Simon, M. Novic and J. Zupan, "Neural Nets for Mass and Vibrational Spectra", *J. Mol. Struct.* **292** (1993) 141 – 160.

18-19. J. Sadowski and J. Gasteiger, "From Atoms and Bonds to Three-Dimensional Atomic Coordinates: Automatic Model Builders", *Chem. Reviews* **93** (1993) 2567 – 2581.

18-20. J. Sadowski, J. Gasteiger and G. Klebe, "Model Builders Using 639 X-Ray Structures", *J. Chem. Inf. Comput. Sci.* **34** (1994) 1000 – 1008.

18-21. M. Novic and J. Zupan, "Investigations of Infrared-Spectra-Structure Correlation Using a Kohonen and Counterpropagation Neural Networks", *J. Chem. Inf. Comput. Sci.* **35** (1995) 454 – 466.

18-22. J. H. Schuur, P. Selzer and J. Gasteiger, "The Coding of the Three-dimensional Structure of Molecules by Molecular Transforms and Its Application to Structure - Spectra Correlations and Studies of Biological Activity", *J. Chem. Inf. Comput. Sci.* **36** (1996) 334 – 344.

18-23. J. Schuur and J. Gasteiger, "Infrared Spectra Simulation of Substituted Benzene Derivatives on the Basis of a Novel 3D Structure Representation", *Anal. Chem.* **69** (1997) 2398 – 2405.

18-24. The SpecInfo© database was provided by Chemical Concepts GmbH, Weinheim, Germany.

18-25. M. C. Hemmer, V. Steinhauer and J. Gasteiger , "The Prediction of the 3D Structure of Organic Molecules from Their Infrared Spectra", *Vibrat. Spectroscopy* **19** (1999) 151 – 164.

19 Properties of Molecular Surfaces

learning objectives:

- color-coding of electrostatic potentials on the van der Waals surface of a molecule

- electrostatic potentials, like any other molecular surface property, can be mapped onto a 2-D plane by a Kohonen neural network

- different conformations will show different Kohonen electrostatic potential maps

- the Kohonen maps of the electrostatic potentials can be used for finding molecular similarities, e.g., in substrates binding to the same receptor

- **artificial** neural networks can be used to eludicate chemical processes in **biological** neural networks

- Kohonen networks can be used to study the shape of molecules

- Kohonen networks can help in the search and optimization of pharmaceutical lead structures

19.1 The Problems

The shape of a molecule, and properties on molecular surfaces such as hydrophobicity, hydrogen bonding potential, and the electrostatic potential are of profound influence on many physical, chemical, or biological properties. A ligand must have a certain shape in order to fit into a receptor, and the properties on the surface of a ligand must correspond to those in the pocket of a receptor.

The study of the geometry of molecular surfaces and of the distribution of electronic and other properties on molecular surfaces may, therefore, give important insights into mechanisms of interactions of molecules and their influence on the properties of compounds.

We will largely concentrate our discussion on the molecular electrostatic potential (MEP) although much of what is being said is also applicable to other molecular surface properties. After the analysis of the molecular electrostatic potential we will turn our attention to the geometric shape of molecular surfaces.

Molecular electrostatic potentials give detailed information for studies on chemical reactivity or pharmacological activity of a compound. The spatial distribution and the values of the electrostatic potential determine the attack of an electrophilic or nucleophilic agent as the primary event of a chemical reaction. By the same token, the three-dimensional distribution of the electrostatic potential is largely responsible for the binding of a substrate molecule at the active site of a biologically active receptor.

The *electrostatic potential* at a certain point near a molecule is the work involved when a unit positive charge is brought from infinity to this point. It can be calculated by quantum mechanical procedures of various degrees of sophistication, or by a simple point-charge model if the atoms of a molecule have been assigned partial charge values. Negative values of the electrostatic potential indicate attraction for a positive charge, positive values stand for repulsion.

The three-dimensional nature of the electrostatic potential makes it difficult to simultaneously visualize its **spatial distribution** and its **magnitude**.

To simplify matters, in the rest of the study we will deal with only the electrostatic potential on the van der Waals surface of a molecule. This part of the total electrostatic potential is the most important part as molecules come into contact with other molecules at the van der Waals surface. It is calculated by a point-charge model at points evenly distributed over this surface. The partial charges on the atoms are calculated by iterative partial equalization of orbital electronegativity (PEOE), a well-established empirical method for the rapid calculation of charge distributions (see References 19-1 and 19-2).

It has become customary to color-code the **magnitude** of the electrostatic potential, showing values from negative to positive as colors from red through yellow and green to blue.

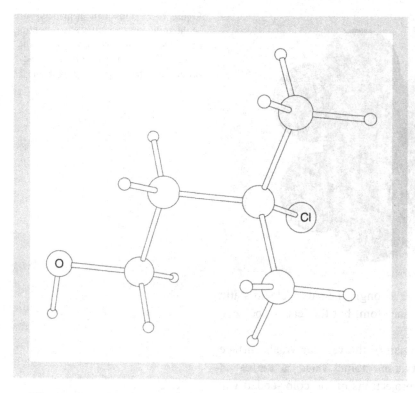

Figure 19-1: Ball and stick model of
3-chloro-3-methylbutan-1-ol.

The next problem is how to represent the **spatial distribution** of the electrostatic potential. This is usually done by choosing an observation point away from the molecule and showing, on a graphics screen, whatever part of the van der Waals surface can be seen from this point. Basically, this means that a parallel **linear projection** of the molecular electrostatic potential is performed, indicating only that side of the molecule visible from the observation point.

Figure 19-1 shows a three-dimensional model of *3-chloro-3-methylbutan-1-ol* viewed from an observation point from which the chlorine atom is hardly visible.

The electrostatic potential for this molecule is then calculated by moving a unit positive charge across the van der Waals surface and calculating at evenly distributed points the electrostatic interactions with all the atoms that bear partial charges.

The electrostatic potential on the van der Waals surface is shown in Figure 19-2 from the same observation point as the one chosen (Figure 19-1) for looking at the ball and stick model. The positive charge-attracting (nucleophilic) site corresponding to the oxygen atom can clearly be seen by the red color (negative electrostatic potential).

Figure 19-2: View (linear projection) of the molecular electrostatic potential of *3-chloro-3-methylbutan-1-ol* on the van der Waals surface from the same observation point as in Figure 19-1.

On the other hand, there is also a strongly negative electrostatic potential at the surface of the chlorine atom, but this cannot be seen from this viewpoint.

With a linear projection a large part of the van der Waals surface **cannot** be seen from the observation point. Thus, a **series** of observation points and associated projections of the color-coded van der Waals surface have to be chosen to get the overall picture of the electrostatic potential.

It would be quite helpful if the entire electrostatic potential of the molecule on the van der Waals surface could be shown in one picture. Clearly, this requires a **nonlinear** projection method; a Kohonen network can do this for us.

19.2 The Network Architecture and Training

In Section 6.4, we saw how a two-dimensional map of the surface of a sphere can be obtained by training a Kohonen network with the three Cartesian coordinates of a series of points taken from that surface.

Basically, the same procedure is followed in this example to generate a map of the distribution of the electrostatic potential; 20,000 points are randomly selected from the van der Waals surface and a (60 x 60) Kohonen network is trained with the three Cartesian coordinates of each of these points. Thus, each neuron in the network has three weights.

Recall that a Kohonen network undergoes unsupervised training. Hence, the value of the electrostatic potential is not used in the training phase.

The plane of projection is the surface of a torus. For any point of the van der Waals surface that enters the learning phase, the neuron with the three weights **most similar** to the input coordinates of this particular point is selected as the winning neuron, and its weights and those of neurons in an appropriate neighborhood are corrected according to Equations (6.6).

When the Kohonen network stabilizes, all neurons are investigated to see which points from the van der Waals surface they contain.

It is found that points which are very close together on the van der Waals surface (and therefore have nearly the same potential) map to the same neuron, and adjacent neurons are excited by points with **similar** electrostatic potentials. Thus, the magnitude of the electrostatic potential of the points that map to a neuron can be used to decide its color. This leads to the color-coded map of the electrostatic potential shown in Figure 19-3.

In this map of the electrostatic potential, the effect of the oxygen and of the chlorine atom can both be seen simultaneously. The oxygen atom corresponds to the yellow and orange spot whereas the chlorine atom is indicated by the large blue-green area. Thus, the Kohonen network has achieved a nonlinear projection of the entire molecular surface and can, therefore, indicate all features of a property, such as the electrostatic potential, on a molecular surface.

As discussed in Section 6.2, the map is projected onto the surface of a torus. In fact, the learning process adjusts the form of the torus as well as possible to the three-dimensional structure of the molecule. This is shown in Figure 19-4.

The torus can be cut open and flattened into a square arrangement of neurons at any two perpendicular lines on the torus. Thus, the map can be shifted right, left, up or down. The two maps shown in Figure 19-5 are just as valid representations of the electrostatic potential as the map of Figure 19-3.

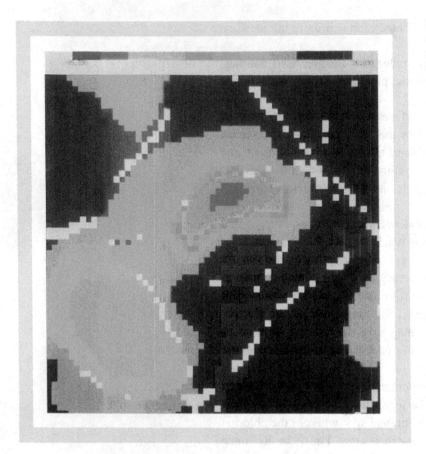

Figure 19-3: Color-coded Kohonen map of the electrostatic potential of *3-chloro-3-methylbutan-1-ol* in the conformation shown in Figure 19-1.

Figure 19-4: Adjustment of the torus used for projection of the surface of the molecule.

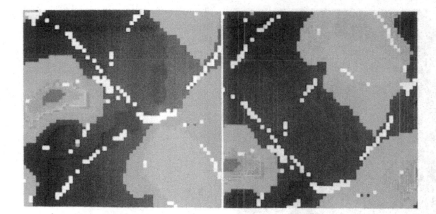

Figure 19-5: Shifted maps of the electrostatic potential of *3-chloro-3-methylbutan-1-ol.*

19.3 Tiling with Kohonen Maps; Conformational Effects

In order to show all of the features of Kohonen maps of the electrostatic potential, it is helpful to put several identical maps together like tiles (see Section 6.4). Figure 19-6 shows the composite of six identical Kohonen maps of the electrostatic potential of *3-chloro-3-methylbutan-1-ol* as shown in Figure 19-3.

The electrostatic potential of a molecule depends on the relative arrangement of the individual atoms. If a molecule has conformational flexibility, the different conformations will result in different distributions of the electrostatic potential on the van der Waals surface; consequently, different conformations give different Kohonen maps of the electrostatic potential.

The Kohonen map of Figure 19-3 was obtained from a conformation of *3-chloro-3-methylbutan-1-ol* having a torsional angle for the sequence C1–C2–C3–Cl of 90° (see Figure 19-1). Figure 19-7 shows a conformation of *3-chloro-3-methylbutan-1-ol* that has a torsional angle of 0° for the sequence C1–C2–C3–Cl and furthermore a hydrogen bridge between the OH group and the chlorine atom.

The molecular electrostatic potential is influenced by the three-dimensional arrangement of atoms and thus also by the conformation of a molecule. This should also be reflected in the Kohonen maps of the molecular electrostatic potential. Figure 19-8 shows the Kohonen map of the electrostatic potential of *3-chloro-3-methylbutan-1-ol* calculated for the conformation shown in Figure 19-7.

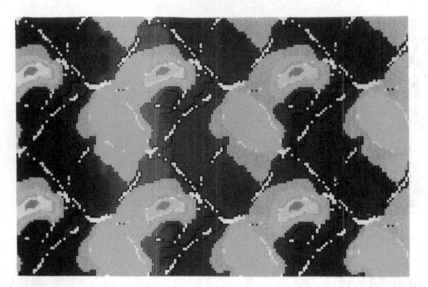

Figure 19-6: Tiling six identical Kohonen maps equivalent to Figure 19-3.

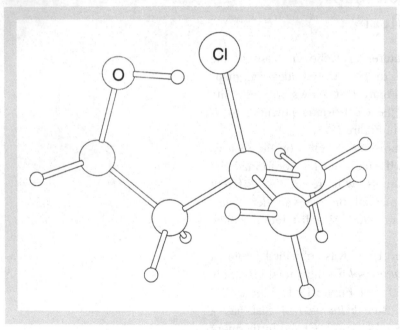

Figure 19-7: Ball and stick model of *3-chloro-3-methylbutan-1-ol* in a conformation different from the one in Figure 19-1.

It can be clearly seen that the maps shown in Figure 19-3 and 19-8 are quite different. In particular, the areas of negative electrostatic potentials resulting from the oxygen and the chlorine atom have now merged due to the close proximity of these two electronegative atoms. Thus, indeed, Kohonen maps of molecular electrostatic potentials give information on the influence of conformation on this property.

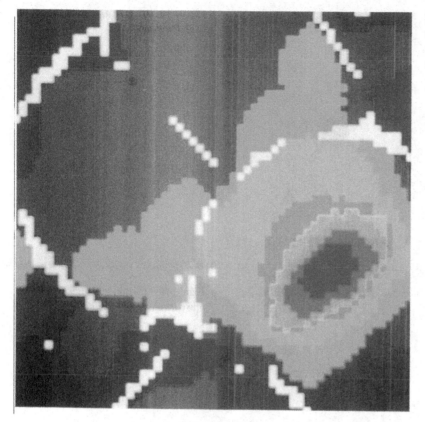

Figure 19-8: Kohonen map of the electrostatic potential of *3-chloro-3-methylbutan-1-ol* in the conformation shown in Figure 19-7.

19.4 Investigation of Receptors of Biological Neural Networks

In the following Section we will show how these Kohonen maps can be used to draw chemical inferences. Since the electrostatic potential of a biologically active molecule plays a major role in substrate binding, Kohonen maps of electrostatic potentials can be valuable for finding similarities in structures that bind to the same receptor. In fact, we will now use an **artificial** neural network (Kohonen network) to shed some light onto chemical compounds involved in **biological** neural networks.

Signals are transmitted **within** a biological neuron electrically, but **between** two different neurons – across the synaptic gap **chemically**. A compound called a *neurotransmitter* is released at the end of an axon, crosses the synaptic gap and initiates another electric signal in the dendrite of a second neuron (Figure 19-9).

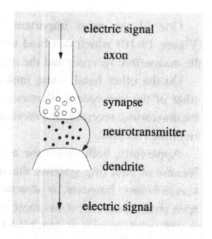

Figure 19-9: The synapse between two neurons.

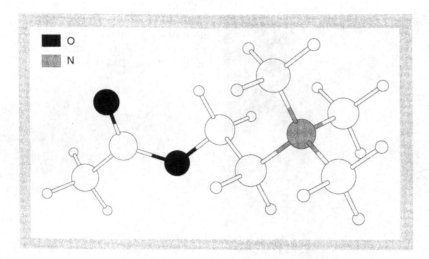

Figure 19-10: Ball and stick model of *acetylcholine*.

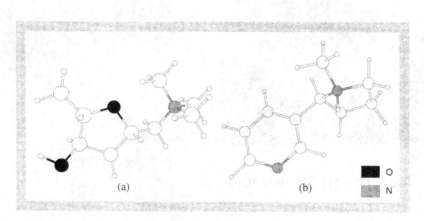

Figure 19-11: Ball and stick model of *muscarine* (a) and of *nicotine* (b).

One of the most important neurotransmitters is *acetylcholine* (Figure 19-10) which can bind to at least to two different receptors, the *muscarinic receptor* and the *nicotinic receptor*.

On the other hand, some molecules can only bind to one or the other of the two receptors: *muscarine* (Figure 19-11a) binds only to the muscarinic receptor, and *nicotine* (Figure 19-11b) binds only to the nicotinic receptor.

Apparently, both *muscarine* and *nicotine* are structurally so rigid, because of their ring systems, that they fit only in one of the receptors. *Acetylcholine*, however, is structurally more flexible; because of its open chain character, it has more conformational degrees of freedom: in one conformation it can bind to the muscarinic receptor, and in another one, to the nicotinic receptor.

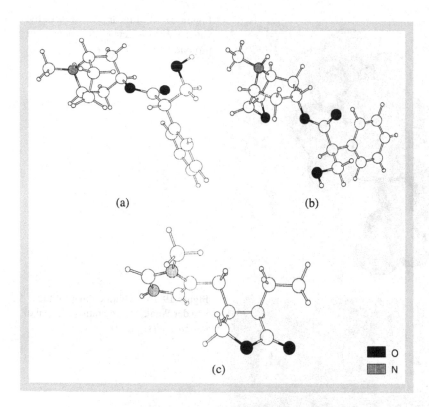

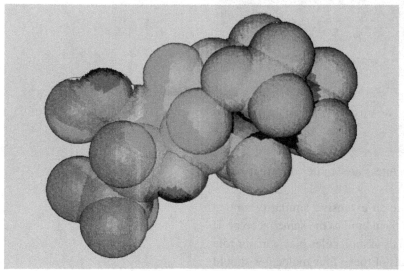

Figure 19-12: Molecules that bind to the muscarinic receptor: *atropine* (a), *scopolamine* (b), and *pilocarpine* (c).

Figure 19-13: Linear projection of the van der Waals electrostatic potential of *muscarine*.

Figure 19-12 shows three more molecules that bind to the muscarinic receptor: *atropine*, *scopolamine*, and *pilocarpine*. Clearly, there is a close structural relationship between *atropine* and

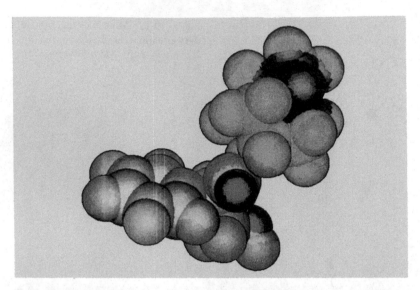

Figure 19-14: Linear projection of the van der Waals electrostatic potential of protonated *scopolamine*.

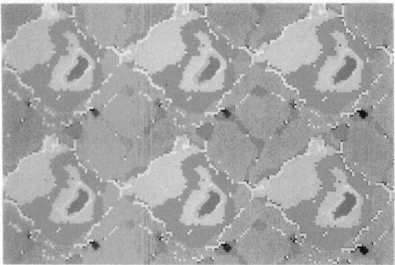

Figure 19-15: Kohonen map of the van der Waals electrostatic potential of *muscarine* (six fold).

scopolamine, but these two structures are quite different from *muscarine* and from *pilocarpine*.

Nevertheless, there must be quite an extensive similarity among these molecules, because all four of them bind to the same receptor. If it is true that the electrostatic potentials of molecules play a major role in binding, the electrostatic potentials of these four molecules should be quite similar. *Acetylcholine* and *muscarine* both have a quaternary nitrogen atom, i.e., a nitrogen atom bearing a positive charge. To enable a correct comparison, the other uncharged molecules have been protonated at the most basic nitrogen before calculating the

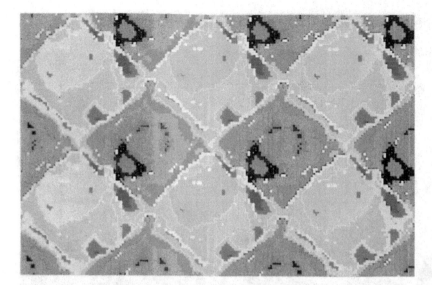

Figure 19-16: Kohonen map of the van der Waals electrostatic potential of *scopolamine* (six fold).

electrostatic potentials. This also etter reflects the situation at the receptor where the molecule surely is protonated. Figure 19-13 and Figure 19-14 show the electrostatic potentials of *muscarine* and of *scopolamine* (in its protonated form).

Clearly, a comparison of the electrostatic potentials of the two molecules suffers from the deficiencies of a linear projection, which shows only that part of the electrostatic potential that can be seen from some chosen viewpoint. In order to compare the entire electrostatic potentials of the two molecules, we have generated the Kohonen maps of both. These are shown in Figures 19-15 and 19-16.

A comparison of Figures 19-15 and 19-16 shows certain similarities in the gross features of the two Kohonen maps. These similarities will become even clearer when the Kohonen maps of the four muscarinic compounds (*muscarine*, *atropine*, *scopolamine*, and *pilocarpine*) are compared with those of the four compounds that bind at the nicotinic receptor (*nicotine*, *anatoxine*, *mecamylamine*, and *pempidine*). This is done in Figure 19-17.

A detailed discussion of the features in these eight Kohonen maps is beyond the scope of this book (for further details see References 19-5 to 19-7). However, it is hoped that the reader realizes and appreciates the similarities in the four Kohonen maps in the top and bottom row, respectively, and perceives the differences between the two groups of maps.

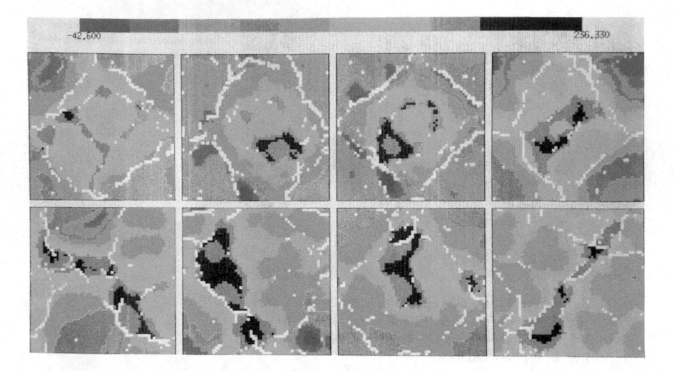

The mapping of molecular electrostatic potentials by Kohonen networks is apparently able to extract essential features responsible for the binding of biologically active compounds to their receptors..

We hope that this application of an **artificial neural network** to a study from the field of **biological neural networks** underscores the broad applicability of neural networks. Again, applications have their limits only in the imagination of the potential user.

Figure 19-17: Kohonen maps of the electrostatic potentials of four muscarinic compounds (*muscarine*, and protonated *atropine*, *scopolamine* and *pilocarpine*) (top) and four protonated nicotinic compounds (*nicotine*, *anatoxine*, *mecamylamine*, and *pempidine*) (bottom).

19.5 Comparison of Kohonen Maps

The example in the previous section has shown the benefits in comparing Kohonen maps of the molecular electrostatic potential. Particularly, the eight maps assembled in Figure 19-17 emphasize how the maps of the molecular electrostatic potential can show similarities of ligands binding to the same receptor. In order to support this comparison, the maps had to be aligned, i.e., the location of the cuts in the torus (cf. Figure 6-7) were chosen such that maximum similarity in the maps could be discerned.

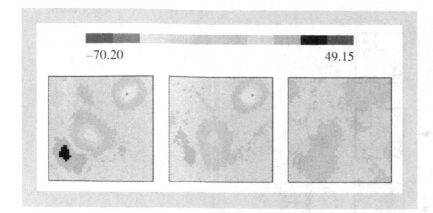

−70.20   49.15

Figure 19-18: Averaged Kohonen maps of the molecular electrostatic potential of steroids binding to the CBG receptor, having (a) high, (b) intermediate, and (c) low binding affinity.

Similarity was assigned, in this case, by human inspection building on the powerful pattern recognition capabilities of the human mind. Clearly, it would be quite desirable to have an automatic procedure for defining or even quantifying similarities in such maps.

We have not yet found a universally convincing method for the calculation of similarity in these maps, as both global and local features of the maps have to be taken into account. Furthermore, the maps may have to be rotated and reflected because the orientation of the features in a map can change with different random initialization of the weights of the network. In one approach, we have analyzed the maps of 31 steroids binding to the *corticosteroid binding globulin* (CBG) receptor (cf. Sections 13.5 – 13.8 and, particularly, Table 13-2). Averaged maps were calculated for the compounds having high, intermediate, or low binding affinities; they are shown in Figure 19-18.

The pattern of the molecular electrostatic potential of the most polar area in the averaged map of the highly active compounds is the most pronounced feature (Figure 19-18a). In the three averaged maps, the distinction of the polar spaces decreases according to decreasing activity of the compounds.

The averaged map of the highly active compounds can be used to build a pharmacophore model. Thus, a comparison of the map of a steroid with the averaged maps allows one to establish whether a molecule belongs to the active or inactive CBG compounds.

The pattern of the molecular electrostatic potential in the Kohonen maps can also be converted into quantitative descriptors. Such descriptors have successfully been used in quantitative structure-activity studies.

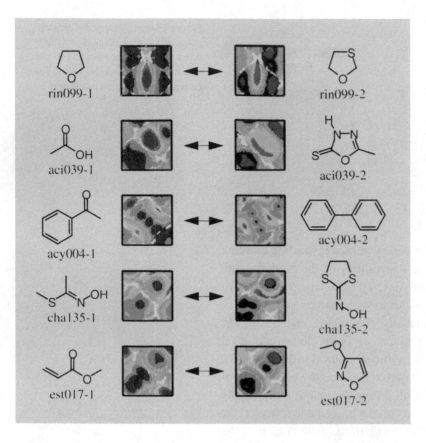

Figure 19-19: Pairs of bioisosteric groups that have quite similar Kohonen maps of their molecular electrostatic potential.

19.6 Bioisosteric Design

In drug design, a large amount of efforts has to be devoted to arrive from an initial lead structure at the final, highly active drug. Many structural variations have to be explored to increase biological activity. In this process, the concept of bioisosteric groups, of structure fragments having similar influence on biological activity has been found quite helpful.

Kohonen maps of the molecular electrostatic potential of several hundred pairs of bioisosteric groups have shown clear similarities in the maps of groups considered to be bioisosteric. Some pairs of bioisosteric groups are shown in Figure 19-19. In fact, this study allowed a more stringent definition of bioisosterism and led to the development of a new, highly active drug.

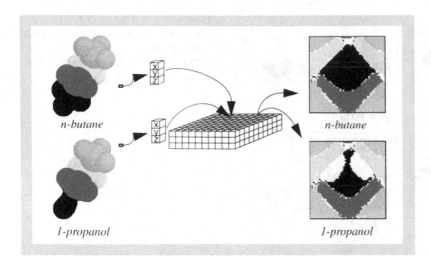

19.7 Molecular Shape Analysis

The projection of molecular surface properties by a Kohonen network as outlined in Section 19.2 involves the training of the Kohonen network by the Cartesian coordinates of points from the molecular surface. Thus, the Kohonen network, in effect, stores the geometry of a molecular surface in its weights (there are three weights for each neuron!). If this is true, such a Kohonen network trained with the coordinates of a molecular surface can be used as a template for the comparison of other molecular surfaces. If the coordinates of a point from the surface of a second molecule (the compared molecule) correspond to the coordinates of a point of the first molecule, the neuron that stores the point from the first molecule (the template) will indicate a hit. Those parts of the surface of the template molecule that have no correspondence in the surface of the compared molecule will provide empty neurons.

Figure 19-20 indicates this process with the surface of *n-butane* as a template and the surface of *1-propanol* as the surface under study (the second molecule). The two areas of empty neurons in the compared Kohonen map correspond to the surface of the two hydrogen atoms that the methyl group of *n-butane* has on top of the atoms of an OH-group propanol. Note, that a proper comparison requires an alignment of the two molecules; in our case, three of the carbon atoms of *n-butane* were superimposed onto the three carbon atoms of *1-propanol*.

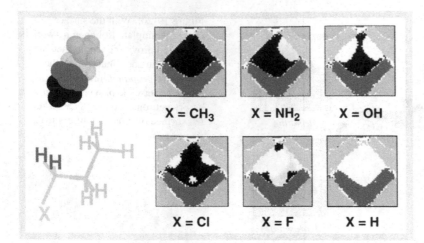

Figure 19-21: A series of 1-substituted *propane* derivatives and the corresponding compared maps obtained from the Kohonen network of *n-butane* as template.

Figure 19-21 shows the results of a more extensive comparison of the surfaces of 1-substituted propane derivatives sent through the template network of *n-butane*. The location of the empty neurons shows where the surface of the reference compound differs from that of the compared compound. An even clearer picture of the location of the differences in the shapes of the two molecular surfaces can be obtained by projecting the Kohonen map of the compared molecule back onto the 3D molecular surface of the template molecule.

Figure 19-21 indicates that the number of empty neurons might be taken as a quantitative measure of the difference in the surfaces of the reference and the compared compound. This has been verified in a number of studies. Thus, Kohonen networks can be used for the comparison of molecular shapes allowing the definition of a quantitative measure of shape difference.

19.8 References and Suggested Readings

19-1. J. Gasteiger and M. Marsili, "Iterative Partial Equalization of Orbital Electronegativity – A Rapid Access to Atomic Charges", *Tetrahedron* **36** (1980) 3219 – 3228.

19-2. J. Gasteiger and H. Saller, "Calculation of the Charge Distribution in Conjugated Systems by a Quantification of the Resonance Concept", *Angew. Chem.* **97** (1985) 699 – 701, *Angew. Chem. Int. Ed. Engl.* **24** (1985) 687 – 689.

19-3. T. Kohonen, "Self-Organized Formation of Topologically Correct Feature Maps", *Biol. Cybern.* **43** (1982) 59 – 69.

19-4. H. Ritter and T. Martinetz, K. Schulten, *Neuronale Netze, Eine Einführung in die Neuroinformatik selbstorganisierender Netzwerke*, Addison-Wesley, Bonn, FRG, 1990.

19-5. J. Gasteiger and X. Li, "Mapping the Electrostatic Potential of Muscarinic and Nicotinic Agonists with Artificial Neural Networks", *Angew. Chem.* **106** (1994) 671 – 674, *Angew. Chem. Int. Ed. Engl.* **33** (1994) 643 – 646.

19-6. X. Li, J. Gasteiger and J. Zupan, "On the Topology in Self-Organizing Feature Maps", *Biol. Cybern.* **70** (1993) 189 – 198.

19-7. J. Gasteiger, X. Li, C. Rudolph, J. Sadowski and J. Zupan, "Representation of Molecular Electrostatic Potentials by Topological Feature Maps", *J. Am. Chem. Soc.* **116** (1994) 4608 – 4620.

19-8. S. Anzali, G. Barnickel, M. Krug, J. Sadowski, M. Wagener, J. Gasteiger and J. Polanski, "The Comparison of Geometric and Electronic Properties of Molecular Surfaces by Neural Networks: Application to the Analysis of Corticosteroid Binding Globulin Activity of Steroids", *J. Comput.-Aided Mol. Design* **10** (1996) 521 – 534.

19-9. J. Polanski, J. Gasteiger, M. Wagener and J. Sadowski, "The Comparison of Molecular Surfaces by Neural Networks and Its Application to Quantitative Structure Activity Studies", *Quant. Struct.-Act. Relat.* **17** (1998) 27 – 36.

19-10. U. Holzgrabe, M. Wagener and J. Gasteiger, "Comparison of Structurally Different Allosteric Modulators of Muscarinic Receptors by Self-organizing Neural Networks", *J. Mol. Graphics* **14** (1996) 185 – 193.

19-11. J. Gasteiger, X. Li and A. Uschold, "The Beauty of Molecular Surfaces as Revealed by Self-organizing Neural Feature Maps", *J. Mol. Graphics* **12** (1994) 90 – 97.

19-12. S. Anzali, G. Barnickel, M. Krug, J. Sadowski, M. Wagener and J. Gasteiger, "Evaluation of Molecular Surface Properties Using a Kohonen Neural Network: Studies on Structure-Activity Relationships", in *Neural Networks in QSAR and Drug Design*, Ed.: J. Devillers, Academic Press, London, UK, 1996, pp. 209 – 222.

20 Libraries of Chemical Compounds

learning objectives:

- separation of structures according to their biological activity

- autocorrelation as a mathematical transformation producing a fixed number of descriptors

- structure coding by autocorrelation considering the constitution of a molecule, different atomic properties, or molecular surface properties

- perception of similarity between chemical structures

- analysis of chemical structure space

- definition of similarity / dissimilarity of chemical compounds

- analysis of diversity and similarity of large chemical libraries

- search for new lead structures

20.1 The Problems

The development of a new drug or agrochemical presently requires, on average, the synthesis of approximately 40,000 new compounds, and this number is still rising. This underscores that the search for a new drug is quite often like the search for the needle in the haystack. On the other hand, biological test systems have become available that allow the testing of many compounds in a short time. The capacity for high-throughput screening (HTS) puts a lot of pressure on rapidly synthesizing many new compounds. These requirements have recently been met by the development of parallel synthesis and combinatorial chemistry that provide large collections, so-called libraries, of chemical compounds.

When synthesizing, or testing, a library of compounds one wants to be sure that the new library is different - dissimilar - to the one previously investigated. Furthermore, in the search for a new lead structure, first, libraries should be synthesized that span the chemical space as broad as possible - are highly diverse - in order to ensure that the kind of compounds that show the desired biological activity are contained in the libraries. In later stages of the search, focussed libraries that center on that part of the chemical space that contains those structures having the desired biological activity should be investigated. Thus, questions of similarity and diversity of chemical structures and libraries become important. Figures 20-1 and 20-2 illustrate the concept of similarity and diversity of chemical libraries.

To answer those question, an appropriate structure coding has to be chosen, a structure coding that is somehow related to the biological activity under investigation. We will first discuss several methods for representing chemical structures and then investigate the capability of one such structure coding method to differentiate between compounds of different biological activity. Furthermore, we will show that this structure representation focusses the structures with the desired biological activity into a restricted part of the chemical space. We will then address questions of similarity and diversity of large chemical libraries as met in combinatorial chemistry.

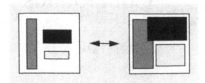

Figure 20-1: Visualization of the concept of diversity: The data sets on the left-hand side are not diverse enough to fill the entire chemical space.

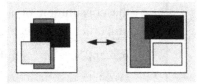

Figure 20-2: Visualization of the concept of similarity: The data sets on the left-hand side are too similar to fill the entire chemical space.

20.2 Structure Coding

The question of similarity can only be defined for a specific purpose, in our case, biological activity: structures are considered similar if they carry similar biological activity. Thus, the structure representation has to consider those properties of a chemical structure that are deemed to be responsible for the biological activity under investigation. Furthermore, the structure coding scheme must produce the same number of descriptors, irrespective of the size of a molecule, the number of atoms in a molecule. For, any automatic learning method such as a neural network has to have a fixed number of input units and, therefore, requires the objects under investigation to be represented by a predetermined, constant number of variables.

Thus, the chemical structure has somehow to be transformed to produce a fixed number of descriptors. In this chapter we will present one such mathematical transformation, autocorrelation, and show how it can consider structure information of various degrees of sophistication, either the constitution only, or molecular surfaces. Furthermore, we will show how various physicochemical properties of atoms, or molecular surfaces can be introduced into the structure coding method. Structure coding by autocorrelation has already been used in Chapter 13. A more extensive discussion of various methods for structure representation is contained in Chapter 21.

The idea of using autocorrelation for the transformation of the constitution of a molecule into a fixed length representation was introduced by Moreau and Broto. The property, p, of an atom, i, is correlated with the same property on atom j and these products are summed over all atom pairs having a certain number of intervening bonds, a certain topological distance, d. An example for the definition of a topological distance is given in Figure 20-3. This gives one element of a topological autocorrelation function $A(d)$:

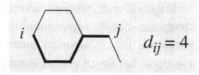

Figure 20-3: Definition of topological distance, d_{ij}, as the number of bonds between two atoms i and j.

$$A(d) = \sum_{j=i+1}^{n} \sum_{i=1}^{n-1} \delta_{ij}\, p(i)\, p(j) \qquad (20.1)$$

with $\delta_{ij} = 1$ if $d_{ij} = d$, otherwise $\delta_{ij} = 0$

The following properties were calculated by previously published empirical methods contained in the program package PETRA for all atoms of a molecule: sigma charge, q_σ, total charge, q_{tot}, sigma-electronegativity, χ_σ, pi-electronegativity, χ_π, lone pair-electro-negativity, χ_{LP} and atom polarizability, α.

In addition to these six electronic variables, the identity function, i.e., each atom being represented by the number 1, was used in Equation (20.1) to only account for the connectivity of the atoms.

The autocorrelation of these variables was calculated for seven topological distances (number of intervening bonds) from two to eight. The basic assumption thus was that the interaction of atoms beyond eight bonds can be neglected. Thus, the descriptor for representing molecular structures is given by Equation (20.2)

$$A(p_k, n) = \sum_{i,j \in M(n)} p_k(i)\, p_k(j) \qquad (20.2)$$

with $M(n) = \{(i,j) \mid \#bond(i,j) = n\}$

$p_k(i)$ is the k-th property on atom i, and #bonds (i,j) is the minimum number of bonds between atoms i and j.

With seven variables and seven distances an autocorrelation vector of dimension 49 was obtained for each molecule, irrespective of its size or number of atoms.

Ligands and proteins interact through molecular surfaces and therefore, clearly, representations of molecular surfaces have to be sought in the endeavor to understand biological activity. Again, we are under the restriction of having to represent molecular surfaces of different size, and, again, autocorrelation was employed to achieve this goal.

First, a set of randomly distributed points on the molecular surface has to be generated. Then, all distances between the surface points are calculated and sorted into preset intervals according to Equation (20.3).

$$A(d) = \frac{1}{m}\sum_{i,j} p(i)\, p(j) \qquad (20.3)$$

with $d_l < d_{ij} < d_u$

where $p(i)$ and $p(j)$ are property values at points i and j, respectively, d_{ij} is the distance between the points i and j, and m is the total number of distances in the interval $[d_l, d_u[$ represented by d. For a series of distance intervals with different lower and upper bounds, d_l and d_u, a vector of autocorrelation coefficients is obtained. It is a condensed representation of the distribution of the property on the molecular surface. This coding was also used in the example contained in Sections 13.6 – 13.8.

20.3 Separation of Benzodiazepine and Dopamine Agonists

In order to investigate the potential of topological autocorrelation functions for the distinction of biological activity, a data set of 112 *dopamine agonists* (DPA) and 60 *benzodiazepine agonists* (BDA) was studied. A Kohonen network of size 10 x 7 was used to project these 172 compounds from the 49-dimensional space spanned by these autocorrelation vectors into two dimensions. The results are shown in Figure 20-4.

It can be seen that the two types of compounds, DPA and BDA, are nearly completely separated in the Kohonen map, underscoring the potential of this molecular representation to model biological activity.

Figure 20-4: Kohonen map of size 10 x 7 neurons obtained for the data set of 112 *dopamine* (black) and 60 *benzodiazepine agonists* (light gray). The separation of the two types of biologically active molecules is nearly complete.

20.4 Finding Active Compounds in a Large Set of Inactive Compounds

To put this potential for comparing data sets of compounds and clustering of compounds with a desired biological activity to a more stringent test, this data set of 112 DPA and 60 BDA compounds was mixed with the entire catalog of a chemical supplier consisting of 8,323 commercially available compounds comprising a wide range of structures from alkanes to triphenylmethane dyestuffs.

The map of Figure 20-5 shows that both sets of compounds, DPA and BDA, occupy only limited areas in the overall map. Furthermore, the areas of DPA and BDA are quite well separated from each other, only one neuron with BDA, intrudes into the domain of DPA and only two neurons have conflicts, obtaining both DPA and BDA. Clearly, the areas of neurons with DPA and BDA are larger than one probably has hoped them to be. For, with the results obtained here the search for new active compounds or new lead structures in a data set of compounds of unknown activity will have to scan a fairly large area and, correspondingly, quite a few compounds. However, compared to the overall size of the network, the areas where DPA and BDA are to be found are distinctly smaller and quite concentrated. Closer analysis of the mapping shows interesting insights that are further discussed in the original publication.

The six electronic factors and the connectivity of a molecule, their encoding into topological autocorrelation vectors and their projection into a two-dimensional map by the self-organizing capability of a Kohonen network provide a powerful means for the detection of similarity in the structure of organic molecules. *Dopamine agonists* can be separated from *benzodiazepine receptor agonists* and this

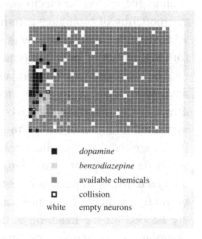

Figure 20-5: Kohonen map of 40 x 30 neurons obtained by training with 112 *dopamine* (DPA), 60 *benzodiazepine agonists* (BDA), and 8,323 commercially available compounds. Only the type of compounds mapped into the individual neurons is indicated. Black identifies DPA, light gray BDA, and dark gray the compounds of unknown activity. Empty neurons are shown in white; the two neurons marked by a black frame indicate conflicts where both DPA and BDA are mapped into the same neuron.

separation is maintained when these two types of compounds are embedded in a larger set of structures.

This opens the way for searching for compounds with a desired biological activity and for discovering new lead structures in large databases of compounds.

Furthermore, this approach can be used for the comparison of libraries of compounds in order to decide whether a commercially offered compound library is distinctly different from the inhouse compound collection.

20.5 Diversity and Similarity of Combinatorial Libraries

The merit of the autocorrelation of molecular surface properties such as the molecular electrostatic potential for the classification and the modeling of biological activity has already been shown in Section 13.6. Here, we will show how this structure representation can be used for the analysis of large combinatorial libraries.

The methods introduced in the previous sections have the advantage that they allow for a rapid visualization of high-dimensional descriptor spaces. The importance of this feature has increased with the advent of the large compound collections that can be generated by combinatorial chemistry and related techniques: small data sets comprising tens or hundreds of compounds can be analyzed using almost any method without reaching the limits of currently available computer hardware. Special techniques, however, are needed for the handling of data sets of hundreds of thousands of compounds. To demonstrate the merits of Kohonen networks and spatial autocorrelation descriptors in handling large data sets, we analyzed three combinatorial libraries that together comprise more than 87,000 compounds.

Rebek et al. published the synthesis of two combinatorial libraries of semi-rigid compounds that were prepared by condensing a rigid central molecule functionalized by four *acid chloride* groups with a set of 19 different *L-amino acids*. This process is summarized in Figure 20-6. In addition to the two published libraries we included a third, hypothetical library with *adamantane* as central molecule into our study.

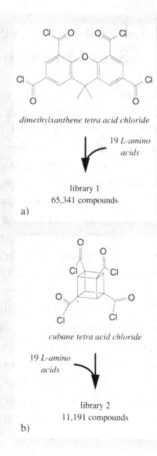

Figure 20-6: Preparation of the *xanthene* and the *cubane* libraries by reaction of four *acid chloride* substituents on these skeletons with four (identical or different) *amino acids* from a set of 19 amino acids. The more symmetric *cubane* skeleton gives less compounds.

A Kohonen network with 50 x 50 neurons was trained with the combined descriptors of the *xanthene* and the *cubane* libraries, each molecule represented by 12 autocorrelation values calculated from the electrostatic potential on the molecular surface by equation 20.3. The resulting map is shown in Figure 20-7. The neurons are colored according to the most frequent central molecule that is mapped into them. All 2,500 neurons of the map are occupied. The compounds of the cubane library form a cluster in the center of the map that is separated from the compounds of the *xanthene* library. The neural network can clearly separate the two libraries quite well - they both cover different parts of chemical space - only 3 per cent of the neurons obtain both *xanthene* and *cubane* derivatives. Consequently, they are remarkably different and, thus, both worthwhile to be considered in a screening program.

In a second experiment we trained the same network with the combined data set of the three libraries of *xanthene*, *cubane* and *adamantane* compounds. This resulted in the Kohonen map shown in Figure 20-8. Again, a distinct cluster that is clearly separated from the *xanthene* derivatives can be seen in the center of the map. The *cubane* and *adamantane* derivatives, on the other hand, cannot be distinguished by the neural network. They are tightly mixed in the central cluster, even more than can be inferred from Figure 20-8 as 88% of the cubane and adamantane compounds are mapped into common neurons.

The *cubane* and *adamantane* libraries, thus, cover the same part of the chemical space - they are so similar to each other that considering both of them in a screening program is a waste of resources and time. The *xanthene* library is evidently different from the other two libraries so that the *xanthene* and one of the *cubane* or *adamantane* libraries should be used for screening.

20.6 Deconvolution of Xanthene Sublibraries

Rebek et al. used their libraries to screen for novel *trypsin inhibitors*. Only the *xanthene* library showed significant *trypsin* inhibition, so that they concentrated further efforts on this library. In the next round of screening they divided the *xanthene* library into six sublibraries by using subsets of only 15 *amino acids* for the generation of the libraries. These subsets were generated by omitting three *amino acids* in turn from a set of 18 *amino acids* (Figure 20-9).

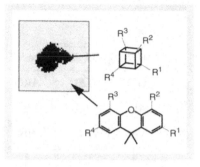

Figure 20-7: Kohonen map obtained from the two libraries of 65,341 *xanthene* derivatives and the 11,191 *cubane* compounds. The area into which the xanthene derivatives are mapped is colored grey whereas the area into which the *cubane* derivatives are mapped is colored black.

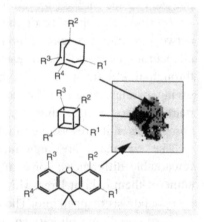

Figure 20-8: Kohonen map obtained from three libraries of 65,341 *xanthene*, 11,191 *cubane* and 11,191 *adamantane* derivatives. The center cluster shows the area into which the *cubane* (colored black) and *adamantane* (colored dark grey) compounds are mapped.

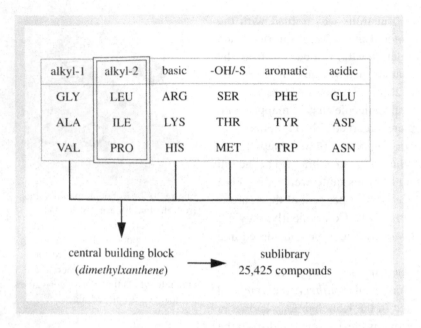

alkyl-1	alkyl-2	basic	-OH/-S	aromatic	acidic
GLY	LEU	ARG	SER	PHE	GLU
ALA	ILE	LYS	THR	TYR	ASP
VAL	PRO	HIS	MET	TRP	ASN

central building block
(*dimethylxanthene*) ➤ sublibrary
25,425 compounds

Figure 20-9: Building six sublibraries by reacting 15 *amino acids* with *dimethylxanthene* having four *acid chloride* group (see Figure 20-6). The three *amino acids* that are omitted in the present case are shown in a double frame. Each one of the six sublibraries contains 25,425 compounds.

This process resulted in six sublibraries each with 25,425 compounds, that were tested for their *trypsin* inhibition.

To study the diversity of the six sublibraries we first trained a network with the complete *xanthene* data set resulting in a map with all neurons occupied. Then, each one of the sublibraries was sent through this template network of the complete library, altogether giving six maps, one each for each sublibrary. In these maps, we then marked, for each sublibrary, only those neurons containing compounds of the respective sublibrary.

The six maps are reproduced in Figure 20-10. They show remarkable differences: some of them are nearly completely filled, some of them exhibit large white areas representing neurons that no compound was mapped into. The larger these white areas are, the less the corresponding sublibrary covers the chemical space of the original *xanthene* library. For example, the omission of the basic or acidic *amino acids* has led to a decreased diversity as shown by the large number of empty neurons. On the other hand, the omission of the larger alkyl *amino acids* or the -OH and -S- substituted *amino acids* from the *xanthene* library does not lead to a remarkable decrease in diversity as there are only small white areas in the corresponding maps.

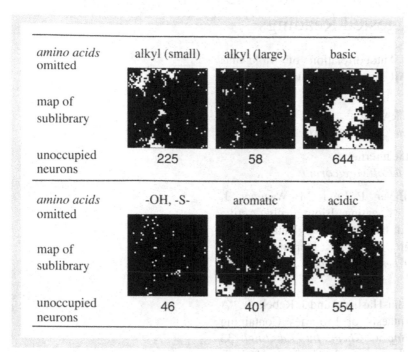

Figure 20-10: Kohonen maps of the six sublibraries built according to Figure 20-9 using the Kohonen network of the complete *xanthene* derivatives library as a template.

On the basis of these Kohonen maps of such sublibraries, strategies for the deconvolution of combinatorial libraries can be developed.

Clearly, these statements about diversity are valid only within the chosen structure representation, i.e., on the basis of the molecular electrostatic potential. However, for compounds containing aminoacids the molecular electrostatic potential is an important factor influencing biological activity.

20.7 References and Suggested Readings

20-1. G. Moreau and P. Broto, "Autocorrelation of Molecular Structures: Application to SAR Studies", *Nouv. J. Chim.* **4** (1980) 757 – 764.

20-2. For further information on PETRA see:
http://www2.ccc.uni-erlangen.de/software/petra/

20-3. CORINA can be accessed on the internet:
http://www2.ccc.uni-erlangen.de/software/corina/

20-4. H. Bauknecht, A. Zell, H. Bayer, P. Levi, M. Wagener, J. Sadowski and J. Gasteiger, "Locating Biologically Active Compounds in Medium-Sized Heterogeneous Data sets by Topological Autocorrelation Vectors: Dopamine and Benzodiazepine Agonists", *J. Chem. Inf. Comput. Sci.* **36** (1996) 1205 – 1213.

20-5. T. Carell, E. A.Wintner, A. Bashir-Hashemi and J. Rebek, Jr, "A Novel Procedure for the Synthesis of Libraries Containing Small Organic Molecules", *Angew. Chem. Int. Ed. Engl.* **33** (1994) 2059-2061; *Angew. Chem.* **106** (1994) 2159-2162; T. Carell, E. A.Wintner, A. Bashir-Hashemi and J. Rebek, Jr, "A Solution-Phase Screening Procedure for the Isolation of Active Compounds from a Library of Molecules" *Angew. Chem. Int. Ed. Engl.* **33** (1994) 2061-2064; *Angew. Chem.* **106** (1994) 2162 – 2165.

20-6. J. Sadowski, M. Wagener and J. Gasteiger, "Assessing Similarity and Diversity of Combinatorial Libraries by Spatial Autocorrelation Functions and Neural Networks", *Angew. Chem. Int. Ed. Engl.* **34** (1995) 2674 – 2677; *Angew. Chem.* **107** (1995) 2892 – 2895.

21 Representation of Chemical Structures

---

learning objectives:

- coding of structure information of various degrees of sophistication

- consideration of physicochemical properties of atoms

- requirements for a good structure representation

- transformation of structure information into fixed-length representations

- molecular transformation of 3D structure information

- encoding of molecular surface properties

- structure encoding methods applicable to large data sets

21.1 The Problem

It has already been mentioned several times in this book that the most important key to success in the application of neural networks lies in the choice of the proper representation of information. In chemical applications, we have often to deal with relationships between chemical structure and physical, chemical, or biological properties. Therefore, the representation of chemical structures is of paramount importance.

The problem of – and solutions to – structure representation has been mentioned in various sections of Chapters 11, 13, 14, and 17–20. This chapter serves to provide an overview and to order the various types of structure representation into a coherent framework.

Chemists have developed a variety of methods for representing and communicating structure information. The most widely used, international language is the structural formula; it is still the method of choice when representing chemical reactions. For a more in-depth analysis, three-dimensional molecular models are built, either by mechanical molecular model kits, or, increasingly by computer modeling. A variety of representations is available, from framework, through ball and stick, to space-filling models. An even more refined analysis of molecules, particularly when studying biological activity, has to consider molecular surfaces, surface properties, and molecular potentials and fields.

All these various representations of chemical structures have to be translated into a form amenable to computer manipulation. A further requirement set by the use of learning methods is that molecules have to be represented by the same number of descriptors, irrespective of their size, of the number of atoms in a molecule. Only then can data sets of different molecules automatically be processed by statistical or pattern recognition methods or by neural networks.

In the following, we will present various techniques for encoding these different forms that the chemists use for structure representation, from the constitution of a molecule, through 3D structures to molecular surfaces. Several of these structure representation have already been introduced in previous chapters. Nevertheless, we mention them here again to collect them in a concise overview. These different encoding methods have been developed for the different requirements made by the intended applications. Furthermore, what kind of coding method will be chosen, will also strongly be dictated by the size of the data sets that have to be studied. Data sets of hundreds of thousands or millions of structures have to rely on rather rapid encoding procedures in order to be handled in a reasonable amount of time.

21.2 Coding the Constitution

The structural formula (Figure 21-1) can be considered as a mathematical graph; graph theory has therefore played a major role in the computer handling of structure information. However, the representation of a molecule as a graph, as a list of atoms and bonds does not fulfill the requirement for a fixed number of descriptors, irrespective of the size of a molecule. In many applications, molecules are represented by lists of fragments or substructures, in the form of bit strings; the presence or absence of a certain functional group is

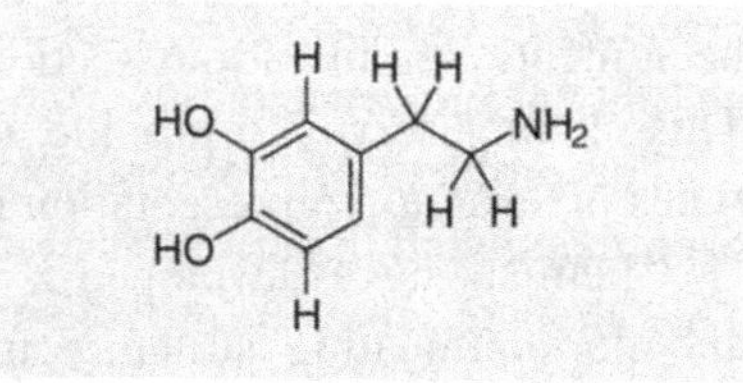

Figure 21-1: The constitution of a molecule.

indicated by a 1 or 0. Thus, there is a clear correspondence between the position of a bit and the substructure present or absent. Such representations of a structural formula are often called 2D descriptors. However, they do not carry any direct 2D information; they are only a reflection of the constitution of a molecule, and therefore should be called topological descriptors, at most.

With a predefined number of substructures one does, indeed, arrive at a structure representation with a fixed length. However, the choice of substructures to be considered will always be arbitrary because, in principle, the number of substructures in organic compounds is unlimited.

In order to be able to consider large sets of substructures an extension of this structure coding by fragments has been introduced, the so-called fingerprints. The occurrence of a large set of fragments is hash-coded into a bit-string representation in order to arrive at a more concise representation. Nevertheless, such a fingerprint representation may consist of a string of 500 – 1,000 bits. In contrast to the former representation no inference on the kind of substructures present can any more be made. Because of the hash-code algorithm no direct relationships between a bit position and a substructure exists any more.

We have sought for methods that allow one to encode various physicochemical properties of the atoms in a molecule, such as partial charges, polarizability, etc. Our approach rests on autocorrelation functions as outlined in Section 13.6 and Chapter 20 and given by Equation (21.1):

$$A(d) \ = \ \sum_{j=i+1}^{n} \sum_{i=1}^{n-1} \delta_{ij} \ p(i) \ p(j) \qquad (21.1)$$

With a topological autocorrelation vector only the constitution of a molecule is considered.

A value for the autocorrelation function A, at a certain topological distance (number of bonds), d, is calculated by summation over all products of a certain property, p, of atoms i and j having the required distance, d.

A range of properties such as partial atomic charges, measures of the inductive effect, resonance, or polarizability effect were calculated by rapid empirical methods contained in the program package PETRA (Parameter Estimation for the Treatment of Reactivity Applications).

Various applications have shown the merit of such a representation. Of particular importance is the possibility of choosing such physicochemical properties of the atoms that are deemed responsible for the effect under investigation. The example given in Section 20.3 – 20.4 is a case in point. The representation of structures by topological autocorrelation of a variety of electronic properties of the atoms in a molecule allows one to distinguish structures having different biological activity, to limit the search space in lead compound search, and to compare different libraries of compounds.

21.3 Coding the 3D Structure I

The study of the relationships between biological activity and the 3D structure (Figure 21-2) of a molecule on a broad scale has been made possible by the advent of universal and efficient automatic 3D generators. Programs are available such as CORINA which can convert large databases such as the Beilstein file with about 7 million structures. With a 3D structure accessible for practically any organic molecule, the problem is then, how to encode the 3D structure under the restriction of having to come up with a fixed number of variables, independent of the number of atoms in a molecule. Clearly, again, autocorrelation of atomic properties as given by Equation (21.1), now inserting genuine spatial distances can be used.

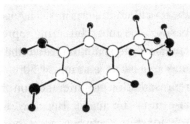

Figure 21-2: The 3D structure of the molecule shown in Figure 21-1.

In fact, useful applications of such a structure representation were made. We were, however, also seeking for a structure representation that offers the possibility of regaining the 3D structure from the molecular code. Building on equations used for obtaining the 3D structure of a molecule from electron diffraction experiments the encoding procedure embodied in Equation (21.2) was developed (cf. Section 18.7).

$$I(s) = \sum_{j=i+1}^{n} \sum_{i=1}^{n-1} a_i a_j \frac{\sin(sr_{ij})}{sr_{ij}} \qquad (21.2)$$

In this equation, $I(s)$ is the intensity of the scattered electron beam at observation angles s, a_i and a_j are atomic properties such as atomic number, or partial charges, and r_{ij} is the distance between the atoms i and j; n is the number of atoms in the molecule.

In electron diffraction, the intensity is measured and the 3D structure as given by all distances r_{ij} is derived from the intensities on the basis of Equation (21.2). In our approach, we have turned the equation around, inputting the 3D structure of a molecule in the form of the distances r_{ij} and calculating $I(s)$. Furthermore, these values of $I(s)$ are calculated only at discrete, equidistant values of s, providing a fixed, predefined number of values of $I(s)$ which are then used as an encoding of the 3D structure of a molecule. This molecular representation was called 3D-MoRSE Code (3D Molecule Representation of Structures based on Electron diffraction).

This 3D-MoRSE code was mainly used for the simulation of infrared spectra. However, it has also be demonstrated that this code shows great promise for correlating structure with biological activity. Dopamine D1 agonists could be separated from dopamine D2 agonists on the basis of the 3D-MoRSE code by a Kohonen network.

In more recent work, a structure code quite similar to the 3D-MoRSE code was developed because it could be more easily transformed back into a 3D structure. This novel encoding scheme is based on atom radial distribution functions and is therefore called RDF code. Equation 21.3 gives the basis of this code:

$$G(r) = \sum_{j=i+1}^{n} \sum_{i=1}^{n-1} a_i a_j e^{-b(r-r_{ij})^2} \qquad (21.3)$$

As in Equation 21.2, a_i and a_j are arbitrary atomic properties, n is the number of atoms in a molecule, and r_{ij} is the distance between the atoms i and j. b is a factor, the so-called temperature factor, that determines the accuracy of position of atoms. r is the running variable and is made discrete, to arrive at a fixed length representation of a molecule by $G(r)$.

Use of the RDF code allows the simulation of infrared spectra of similar quality than when using the 3D-MoRSE code. On top of this, it has become possible to transform this structure representation back into 3D space and, thus, develop a method to derive the 3D structure from an infrared spectrum.

21.4 Coding the 3D Structure II

To repeat once more, a good structure representation should satisfy four conditions that will make it universal:

- uniqueness – one and only one code for each compound and different codes for different structures,

- uniformness – each compound should be represented by the same number and type of variables,

- reversibility – the structure should be retrievable back from the code, and

- translational and rotational invariance – for rotated and/or translated structures the code should remain unchanged.

In the present Section we are discussing a method for representing chemical structures which is uniform, unique and reversible, but lacks the translational and rotational invariance. Although in general, the lack of the origin invariance may be regarded as a drawback it will be shown that exactly this property, namely dependence of the representation on the choice of coordinate system, might in special cases offer some advantages.

The representation of a structure (having n atoms) described by n $[x_j, y_j, z_j]$ triplets does not fulfill the condition of uniformity. The most important feature of the described transformation from a 3N-dimensional representation into a unique m-dimensional *spectrum-like* representation is its reversibility. The new representation is based on the projection of constituent atoms onto an imaginary spherical surface large enough for a molecule under consideration to be accommodated within it. The projection of atoms is not made onto the entire sphere, but onto three perpendicular equatorial trajectories of this sphere. For the representation of n triplets $[x_j, y_j, z_j]$ the coordinates z_j, y_i and x_i, of all atoms are set to zero in sequence, hence, defining three "planar molecules" described by the $3n$ two-plets $[x_j, y_j]$, $[x_j, z_j]$, and $[y_j, z_j]$. The obtained three planar molecules are projected onto circles in the (x,y), (x,z), and (y,z) planes, respectively, forming three equivalent sets of the new representation S.

Each component of the representation S in one of the planes is defined as a cumulative intensity, s_i, at a given point i on the circle at angle φ_j (Figure 21-3) as a sum of n contributions from each atom j in the molecule.

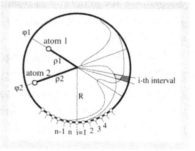

Figure 21-3: Contribution of atoms no. 1 and no. 2 (at positions (ρ_1, φ_1) and (ρ_2, φ_2) to the intensity s_i at the interval (position) i on the circle with radius R, shown as shaded areas of the corresponding Lorentzian bell-shape curves. The circle is divided into m intervals.

$$s_i = \sum_{j=1}^{n} \frac{\rho_j}{(\varphi_j - \varphi_i)^2 + \sigma_j^2} \qquad (21.4)$$

$$\text{for } i = 1...m$$

The contribution under the sum associated with each atom j can be any bell-shaped function. In the case of a *spectrum-like* representation, the Lorentzian shape was chosen for this purpose. For the explanation of the evaluation of the Lorentzian shapes of the projection of atoms onto the circle, the polar coordinates are more plausible than Cartesian ones. In the actual calculations when the atoms are described by two-plets of Cartesian coordinates, Equation (21.4) is used in the rewritten form with Cartesian coordinates (cf. Equations (21.5) and (21.6)).

Each atom j is represented by one Lorentzian curve with the maximum located at angle φ_j. The intensity at the maximum point ($\varphi_i = \varphi_j$) is proportional to the radii-vector ρ_j. The parameter σ_j describes the width of the bell-shaped curve and can bear any information about the nature of the atom (atomic number, van der Waals radius, charge, etc.). If the geometry and the shape of a molecule is to be described only, the parameter σ_j should be set to one for all atoms.

The dimensionality of the new representation is determined by the number of equidistant points on the circles to which the atoms are projected. In principle, the division of the circle does not depend on the number of atoms n in the molecule. However, because the division of the circle determines the resolution, i.e., the quality of the representation and consequently the quality of the inverse process of recovering the structure, this is not entirely true. The larger the division, the more precise is the description of the atoms in the molecule. In general, the division of the circles should be adapted to the number of atoms n in the *largest* molecule of the study. After the division m is chosen, the new representation should be able to map each molecule, regardless of the number of its constituent atoms, into the same *3m*-dimensional space. In many studies where only approximate positions of the substituents with respect to the skeleton are sought m can be as low as 36 or even $m=18$. On the other hand, in cases when small differences in space positions of atoms are important, or for precise recoveries of structures back from the *spectrum-like* representations m can be as large as 720 (division by 0.5^o), thus making the full *spectrum-like* representation 2160 intensities (variables) long.

For actual evaluations of spectral intensities from Cartesian coordinate two-plets, the following transformations are substituted into Equation (21.4)

$$\rho_j = \sqrt{x_j^2 + y_j^2} \quad \text{and} \quad \cos\varphi_j = \frac{x_j}{\sqrt{x_j^2 + y_j^2}} \qquad (21.5)$$

354   *Representation of Chemical Structures*

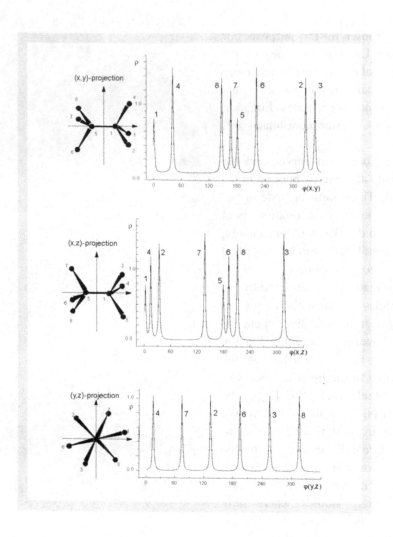

Figure 21-4: Structure of *ethane* in the three perpendicular projections together with the corresponding *spectrum-like* representations in the *(x,y)*, *(x,z)*, and *(y,z)*. The assignments of peaks to the eight *ethane* atoms is shown with figures standing next to the peaks. Atoms at the coordinate origin **do not** produce peaks.

hence:

$$s_i = \sum_{j=1}^{n} \frac{\sqrt{x_j^2 + y_j^2}}{\left[\mathrm{acos}\left(\frac{x_j}{\sqrt{x_j^2 + y_j^2}} \right) - \varphi_i \right]^2 + \sigma_j^2}$$

(21.6)

and equivalent forms for the *(x,z)* and *(y,z)* projections. Figure 21-4 shows how a *spectrum-like* representation of all three projections of the *ethane* molecule is obtained.

The *spectrum-like* representations as described by Equation (21.6) depend on the choice of the coordinate origin. Therefore, such description of molecular structures can be used only for comparative studies on **sets** of structures that can somehow be aligned to each other either by overlapping the skeletons or some other larger structural parts. The liberty to chose the coordinate origin from which the *spectrum-like* representations for an entire set of structures is calculated offers a possibility to search for a position of the origin from which the representation is most useful for a given task, i.e., the position from which the most relevant structural features could be characterized or "seen" best. In the example given in Section 13.10 the optimal coordinate origin for a set of flavonoids was determined (the point –1.7, 3.9, –0.5 in the benzene ring, relative to the atom 2 on the benzopyran ring system) in a preliminary screening by selecting the broadest variance distribution among all representations calculated from each of the tested origins.

21.5 Coding Molecular Surfaces

Molecules interact with each other at molecular surfaces (Figure 21-5). This is particularly true for the interaction of a ligand binding to its receptor. The investigation of molecular surfaces, the coding of surface properties, is therefore of primary importance.

Again, autocorrelation can be used to encode surface properties. Equation (21.1) is now modified such that the properties, p, are sampled on a molecular surface; for the distance parameter, d, all distances within a certain range, e.g., between 3 and 4 Å are collected in one autocorrelation value (cf. Equation 13.3).

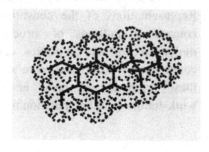

Figure 21-5: The van der Waals surface of the molecule shown in Figure 21-1.

We have shown in Section 13 that autocorrelation of the electrostatic potential on the van der Waals surfaces into 12 descriptors provides excellent descriptors for modeling the affinity of steroids for binding to the corticosteroid binding globulin (CBG) receptor. In a similar manner, autocorrelation of the hydrophobicity potential of a series of 78 polyhalogenated aromatic compounds can quantitatively model the binding to the cytosolic Ah receptor. The same encoding method, autocorrelation of the molecular electrostatic potential (MEP) into 12 descriptors, was used for the definition of diversity and similarity of combinatorial libraries (Section 20).

Furthermore, we have shown in Chapter 19 that a Kohonen network can directly be used to encode a molecular surface and produce maps of surface properties such as the molecular electrostatic potential.

It was shown that such maps of the electrostatic potential on a molecular surface can be used to distinguish between muscarinic and nicotinic agonists. The Kohonen network, in effect, stores the three-dimensional coordinates of points on the molecular surface. Such a network can, therefore, be used as a template for quantifying shape similarities in a series of compounds.

21.6 A Hierarchy of Representations

Approaches to the encoding of molecular structures have been developed that allow the investigation of data sets of diverse molecules by learning methods. These structure representations form a hierarchy of increasing sophistication. At the lowest level, only the constitution of a molecule is taken into account. Next, the 3D structure is considered. For more sophisticated applications, properties of molecular surfaces have to be encoded. The level used will largely be dictated by the size of the data set to be investigated. Representations of the constitution will be applied to data sets comprising millions of structures, whereas representations of molecular surface properties can still be chosen for data sets comprising 100,000 and more structures. Even with large data sets these methods are rapid enough to be performed on small workstations with computation times of a few hours.

21.7 References and Suggested Readings

21-1. G. Moreau and P. Broto, "Autocorrelation of molecular structures: Application to SAR studies", *Nouv. J. Chim.* **4** (1980) 757 – 564.

21-2. J. Gasteiger and M. Marsili, "Iterative Partial Equalization of Orbital Electronegativity - A Rapid Access to Atomic Charges", *Tetrahedron* **36** (1980) 3219 – 3228.

21-3. J. Gasteiger and H. Saller, "Berechnung der Ladungsverteilung in konjugierten Systemen durch eine Quantifizierung des Mesomeriekonzeptes", *Angew. Chem.* **97** (1985) 699 – 701; J. Gasteiger and H. Saller, "Calculation of the Charge Distribution in Conjugated Systems by a Quantification of the Resonance Concept", *Angew. Chem. Int. Ed. Engl.* **24** (1985) 687 – 689.

21-4. M. G. Hutchings and J. Gasteiger, "Residual Electronegativity - An Empirical Quantification of Polar Influences and its Application to the Proton Affinity of Amines", *Tetrahedron Lett.* **24** (1983) 2541 – 2544.

21-5. J. Gasteiger and M. G. Hutchings, "Quantification of Effective Polarisability. Applications to Studies of X-Ray Photoelectron Spectroscopy and Alkylamine Protonation", *J. Chem. Soc. Perkin 2* (1984) 559 – 564.

21-6. For further information on PETRA see:
http://www2.ccc.uni-erlangen.de/software/petra/

21-7. H. Bauknecht, A. Zell, H. Bayer, P. Levi, M. Wagener, J. Sadowski and J. Gasteiger, "Locating Biologically Active Compounds in Medium-Sized Heterogeneous Data Sets by Topological Autocorrelation Vectors: Dopamine and Benzodiazepine Agonists", *J. Chem. Inf. Comput. Sci.* **36** (1996) 1205 – 1213.

21-8. J. Gasteiger, X. Li, C. Rudolph, J. Sadowski and J. Zupan, "Representation of Molecular Electrostatic Potentials by Topological Feature Maps", *J. Am. Chem. Soc.* **116** (1994) 4608 – 4620.

21-9. S. Anzali, J. Gasteiger, U. Holzgrabe, J. Polanski, J. Sadowski, A. Teckentrup and M. Wagener, "The Use of Self-Organizing Neural Networks in Drug Design", in *3D QSAR in Drug Design*, Vol. 2, Eds.: H. Kubinyi, G. Folkers and Y. C. Martin, Kluwer/ESCOM, Dordrecht, NL, 1998, pp. 273 – 299.

21-10. J. Sadowski and J. Gasteiger, "From Atoms and Bonds to Three-Dimensional Atomic Coordinates: Automatic Model Builders", *Chem. Reviews* **93** (1993) 2567 – 2581.

21-11. J. Sadowski, J. Gasteiger and G. Klebe, "Comparison of Automatic Three-Dimensional Model Builders Using 639 X-Ray Structures", *J. Chem. Inf. Comput. Sci.* **34** (1994) 1000 – 1008.

21-12. CORINA can be accessed on the internet: *http://www2.ccc.uni-erlangen.de/software/corina/*

21-13. J. H. Schuur, P. Selzer and J. Gasteiger, "The Coding of the Three-dimensional Structure of Molecules by Molecular Transforms and Its Application to Structure - Spectra Correlations and Studies of Biological Activity", *J. Chem. Inf. Comput. Sci.* **36** (1996) 334 – 344.

21-14. M. Hemmer, V. Steinhauer and J. Gasteiger, "Deriving the 3D Structure of Organic Molecules from Their Infrared Spectra", *Vibrat. Spectroscopy*, **19** (1999) 151-164.

21-15. As review of different coding systems see for example: J. E. Ash, W. A. Warr and P. Willett, *Chemical Structure Systems*, Ellis Horwood, New York, USA, 1991.

21-16. M. A. Johnson and G. M. Maggiora (Eds.), *Concepts of Molecular Similarity*, Wiley Interscience, New York, USA, 1990.

21-17. J. Zupan and M. Novic, "General Type of a Uniform and Reversible Representation of Chemical Structures", *Anal. Chim. Acta* **348** (1997) 409 – 418.

22 Prospects of Neural Networks for Chemical Applications

A review of neural network applications in chemistry appeared a few years ago in the journal *Analytica Chimica Acta* (**248** (1991), 1 – 30); it was titled "Neural Networks: A New Tool for Solving Chemical Problems, or Just a Passing Phase?". Today, the question in the title can be answered quite clearly. The very fact that we have written an introduction on it shows that we are convinced that there is a bright future for the application of neural networks in chemistry. In the last few years, neural networks have become an established tool in applied and theoretical chemistry. Their versatility and ease of use have inspired numerous applications, new ideas and new perspectives in the field itself, as well as in the handling of chemical data.

Now, with the publication of the second edition of this book, all this can only be emphasized. The very fact that a second edition is asked for underlines the important role neural networks have gained in chemistry and will continue to do so. We have passed through the initial hype of using neural networks in chemistry, have seen some disillusions and skepticism, and see now a steady and well-established use of these tools in analyzing chemical information.

One of the most valuable and interesting consequences of chemical applications of neural networks is that it forces us to reconsider the representation and interpretation of our data.

The representation of data is very important for the extraction of vital or interesting information from experimental data. However, the directness of the neural network approach has brought this fact to new prominence; whether in Kohonen learning or in the back-propagation of errors, the representation of data is crucial. Furthermore, the representation of outputs, or targets associated with outputs, is critical as well, because the target information needed to train, adapt, or label the networks plays a central role in changing the weights, the most important parameters in any neural network.

Transforming a variable into a new one, changing the scale, or normalizing a variable's values can completely change the properties of the network. The correction of weights, in spite of all the well-defined equations, has become an "art" with the introduction of arbitrary parameters like "learning rate" and "momentum". Variable learning rate constants are not the exclusive attribute of back-propagation learning procedures only: they are encountered in ABAM, Kohonen, and counter-propagation learning algorithm as well.

Controlling the learning rate means controlling the amount of information necessary to push the network towards a proper solution. This sensitive procedure is already rather well developed and will become more and more important in the future.

While neural networks have enormous potential for making multicategory decisions on one level, there is at present an increasing tendency to solve complex problems in parts, and then put the partial solutions together; it is as if the authors are afraid of using large multicategory decision networks. While this does minimize training time, pasting together a decision hierarchy from smaller partial solutions is a very serious distortion of the nature of the problem. Apparently, because global-solution methods are not yet fully developed or tested, some authors trust the partial solutions more than general ones. This might become a trend in neural network research: global vs. local models. In our opinion, the only real generalization of solutions is possible by cyclic or stepwise use of both unsupervised and supervised learning methods, and by exchanging the results and adapting the representations to actual local needs.

Another rapidly growing trend from the neural network research is the appearance of small special-application networks built into user-friendly shells (black boxes). Just as ordinary people are content to use the telephone without knowing how it works, end users of artificial intelligence systems are happy to get predictions, as long as they turn out to be beneficial.

There is a trend of patenting such dedicated solutions, an increasing number of which include the neural network applications. If this trend continues, soon less and less information will be coming from the industrial laboratories about what is happening in this field. Therefore, it is even more important for the universities, government and nonprofit institutions to support research in neural networks.

In the last few years it has been confirmed by many applications that artificial neural networks are an important tool for chemists to handle various types of problems. However, no matter how enthusiastic one might be about the new method, it should never be forgotten that neural networks are only mathematical tools which should carefully be designed, trained, and above all, verified. The reader should never skip or neglect the verification of the obtained artificial neural networks. Careful selection of the test set, various cross-validations either by 'leave-one' or 'leave-more-than-one-out', the comparison of results obtained by neural networks with results obtained by other methods, or any other standard validation technique should always be applied before the final judgment on the success or failure of neural networks is proclaimed.

One thing has to be mentioned at the end. As in all other data handling applications, neural networks are not exempt from sensitivity to badly chosen initial data. Never forget the golden rule: "garbage in – garbage out". If you extract only one thing from this book, let it be the importance of proper selection of data.

Appendices

In order to support the readers of this book in their endeavor to understand and utilize neural networks, we provide them with additional material in electronic form. Rather than including this material on diskette or CD-ROM in this book we have decided to install a website that contains this additional information. This allows us to constantly expand and update this material.

The URL for this website is

http://www2.ccc.uni-erlangen.de/ANN-book/

The additional information includes access to programs, data sets, presentation materials, tutorials, and publications.

A   Programs

A.1   A Hopfield Program, HOPF

A demo-program for the application of a Hopfield networks is provided on the website. The example detailed in Section 4.4 can be worked through explicitly. The program HOPF.EXE and a data file containing the patterns shown in Figures 4-2 and 4-5 can be downloaded and run on a personal computer. Furthermore, the user can input his/her own images in the format of 3x3, 4x4, 5x5, 6x6, 7x7, or 8x8 pixels. The program can handle up to 64 images.

A.2   An Adaptive Bidirectional Associative Memory Program, ABAM

A simulator program for an ABAM network is provided. The user can study the example explained in Section 5.5 by himself. The program ABAM.EXE and the two data files used to produce the results given in Figure 5-9 and Figure 5-10, respectively, can be downloaded from the website. The program can also be used to study data of one´s own. Up to 20 different pairs of input images / output targets can be handled.

A.3   A Kohonen Network Simulator

This is a Kohonen network program that has been augmented with various software tools for chemical applications:

– a browser for chemical structures *csbr*

– a tool for the visualization of Kohonen maps

It is therefore particularly helpful for the study of chemical data sets.

A.4   A Suite of Neural Network Programs, SNNS

A program package that includes a large variety of neural network models has been developed by the group of Prof. Andreas Zell, first at the University of Stuttgart and now at the University of Tübingen, Germany.

It can be accessed and downloaded from the website at:
http://www.informatik.uni-stuttgart.de/ipvr/bv/projekte/snns/snns.html

A.5   Telecooperation in Spectroscopy, TeleSpec

The methods presented in Sections 18.7 - 18.10 for the simulation of infrared spectra and the prediction of the 3D structure of a molecule from its infrared spectrum have been implemented on the internet on the TeleSpec site at *http://www2.ccc.uni-erlangen.de/research/ir/*.

For the simulation of an infrared spectrum, the user only has to input the constitution of a molecule. For this purpose, a molecule editor is provided. A 3D model of the molecule will automatically be generated by CORINA and transformed either into a 3D-MoRSE code or an RDF code. A counter-propagation network will be trained with similar structures and their corresponding infrared spectra, as retrieved from the SpecInfo© database. Input of the query structure into this counter-propagation network will produce a simulated infrared spectrum.

By the same token, the reverse mode of utilizing a counter-propagation network has been implemented. On input of an infrared spectrum, a 3D structure for the corresponding molecule is proposed. The various steps in these methods are briefly explained on the website; an interactive tutorial is provided.

In addition, access to a publicly available database of infrared spectra is provided. The database has been compiled by the users of the TeleSpec project and everybody is invited to send in her/his infrared spectra and associated structures to contribute to this service to the scientific community that is to the benefits of anybody interested in using infrared spectra for structure identification and elucidation.

A.6   The 3D Structure Generator CORINA

Molecules are three-dimensional objects and, therefore, many physical, chemical, and biological properties depend on the 3D structure. This has been demonstrated in various places throughout this book. Several methods for the encoding of the 3D structure of molecules have been presented.

In order to allow the investigation of the relationships between a property and the 3D structure, access to the Cartesian coordinates of the atoms of a molecule is necessary. Such an access is provided by the universal automatic 3D structure generator CORINA. Access to the 3D structure of up to 1,000 compounds is provided for free at *http://www2.ccc.uni-erlangen.de/software/corina/free_struct.html.* For the conversion of larger data set of molecules the company Molecular Networks GmbH should be contacted at *http://www.mol-net.de/.*

A.7   Calculation of Physicochemical Descriptors, the PETRA System

A variety of empirical methods for the calculation of all-important physicochemical effects, such as charge distribution, inductive, resonance, or polarizability effect has been developed since the mid-seventies in the research group of Prof. J. Gasteiger. These methods have been collected in the program package PETRA (Parameter Estimation for the Treatment of Reactivity Applications). The benefits of the results from these methods have been demonstrated at various places in this book. Details on these methods and access to the manual and the program is provided on the website at: *http://www2.ccc.uni-erlangen.de/software/petra/.*

B   Data Sets

B.1   Italian Olive Oils

The data on the 512 samples of Italian olive oils, studied in Chapter 10 are stored on the website and can be downloaded. The data comprise the analysis of the contents of eight fatty acids in olive oil samples and give the region of origin.

We thank Prof. M. Forina, University of Genova, Italy for providing us with this data set and giving us permission to further distribute these data.

B.2   Steroids Binding to the CBG Receptor

A data set of 31 steroids, binding to the corticosteroid binding globulin (CBG) receptor was studied in Sections 13.5 to 13.8 and 19.5.

This data set is widely used; c.f. E. A. Coats "The CoMFA Steroids as a Benchmark Data Set for Development of 3D QSAR Methods" in *3D QSAR in Drug Design,* Vol. 3, Eds.: H. Kubinyi, G. Folkers and Y. C. Martin, Kluwer/ESCOM, Dordrecht, NL, 1998, pp. 199 – 214.

However, the structural formulas in printed publications and in electronic data file all contained errors. After going through the original publications, all coding errors have been eliminated (M. Wagener, J. Sadowski and J. Gasteiger, *J. Am. Chem. Soc.* **117** (1995) 7769 – 7775). The data set is here made accessible with the structures coded as connection tables with stereochemical information in the Molfile format. Furthermore, the activity values and the activity classification for binding to the CBG receptor are given (cf. Table 13.2).

B.3   Combinatorial Library

In order to also provide a larger data set to the community, the combinatorial libraries studied in Sections 20.5 – 20.6 comprising 65,341 dimethylxanthene, 11,191 cubane, and 11,191 adamantane derivatives are made accessible. Because of the size of the data set we refrain from giving the connection tables as Molfile (ASCII) data sets.

Rather, we provide for each of the compounds the 12 autocorrelation coefficients obtained from the molecular electrostatic potential on the van der Waals surface.

C   Presentation Material

We have prepared two slide shows, one on the basic concepts of neural networks, and another one on various types of applications. These slide shows have been prepared quite some time ago with the program DrHalo. This program and the data sets of the slide shows can be downloaded and run on any PC under DOS or Windows.

D   Publications

A list of publications on the application of neural networks to chemical problems from our two research groups is provided. Furthermore, for some of these publications in printed form the copyright has been raised so that these papers can be reproduced in full form.

E   Tutorials

A tutorial on the use of counter-propagation networks is contained on the website.

Index